D0847328

DISCRETE EVENT SIMULATION

A Practical Approach

CRC Press
Computer Engineering Series

Series Editor
Udo W. Pooch
Texas A&M University

Published books:

Telecommunications and Networking
Udo W. Pooch, Texas A&M University
Denis P. Machuel, Telecommunications Consultant
John T. McCahn, Networking Consultant

Spicey Circuits: Elements of Computer-Aided Circuit Analysis
Rahul Chattergy, University of Hawaii

Discrete Event Simulation: A Practical Approach
Udo W. Pooch, Texas A&M University
James A. Wall, Simulation Consultant

Microprocessor-Based Parallel Architecture for Reliable Digital Signal Processing Systems
Alan D. George, Florida State University
Lois Wright Hawkes, Florida State University

Forthcoming books:

Algorithms for Computer Arithmetic
Alan Parker, Georgia Institute of Technology

Handbook of Software Engineering
Udo W. Pooch, Texas A&M University

DISCRETE EVENT SIMULATION

A Practical Approach

Udo W. Pooch
Texas A&M University

James A. Wall
Simulation Consultant

CRC Press
Boca Raton Ann Arbor London Tokyo

Library of Congress Cataloging-in-Publication Data

Catalog record is available from the Library of Congress.

Direct all inquiries to CRC Press, Inc., 2000 Corporate Blvd., N. W., Boca Raton, Florida, 33431.

International Standard Book Number 0-8493-7174-0

Printed in the United States 2 3 4 5 6 7 8 9 0
Printed on acid-free paper

PREFACE

The objective of this text is to provide the potential researcher, student, or professional in the field with the principles behind one of the most widely used tools available for solving problems using the digital computer. The simulation of the operation of a system, whether it be economics, management sciences, or one of the engineering disciplines, has rapidly become one of the most useful and common applications. There are many reasons for this. Among them, the fact that simulation allows the assessment of the potential performance before a newly designed system is operable. Moreover, simulation permits the comparisons of various operating strategies of existing systems without affecting the day-to-day operations and performance of the system. Finally, simulation allows time compression or expansion. That is, it is possible to simulate months or years of activity by a system in a few minutes of computer time.

Along with the widespread use of simulation has come a great deal of misuse. A simulator is only as good as the underlying techniques used in constructing it, and the validity of the results gained from a simulation experiment is influenced by such factors as the techniques used in the collection of data and the analysis techniques used in summarizing the data. Furthermore, prior to its use, a simulation model must be validated, or shown to actually model the system being studied. We have presented in this text a broad body of theory and foundations necessary for the proper development and operation of a system simulation model.

This book is suitable as a general reference on simulation for the practitioner or as a general text for a course on system simulation at the undergraduate or first-year graduate level. The slant of the examples, and applications, is toward the engineering disciplines and computer science. An effort has been made to moderate the mathematical formalism and theoretical details to enhance its usefulness in other disciplines such as economics, management sciences, and business administration. The text is self-contained in that the background necessary to understand and use the developed theory is minimal. The ability to program in some high-level language (preferably FORTRAN) is essential, since most of the examples and exercises involve writing programs. An introductory course in differential/integral calculus is necessary to provide the link between, for example, the probability density function and the cumulative distribution function for some random variable. A standard introductory course in probability and statistics would prove helpful, although the pertinent subjects in that area are covered in some detail in this text.

The text evolved from lecture notes used in a one-semester graduate level course at Texas A&M University. A particular effort has been made to present a balanced treatment of the basic aspects of simulation. The emphasis here is on simulation. These aspects include modeling fundamentals, probability and statistical concepts, random number generation, sampling, test procedures, design of experiments, validation/verification, and output analysis of a completed model. The chapters on validation, experimental designs, output analysis as well as some discussion of the pitfalls of data collection should provide the practitioner and student alike, with information seldom mentioned or emphasized in typical simulation texts.

Chapter 1 of the text is an introduction to the simulation process. The authors intended this book to be on simulation, not on languages or on queueing theory. Thus we have delineated the simulation process in Chapter 1, in very great detail. No specific tools or mathematics were emphasized. In fact, in our approach simulation languages and queueing theory do not appear until very late in the book, and then only to satisfy completeness. It is our belief that simulation in its own right can be used as a tool to solve problems. We do not believe in simulation languages per se, because too often applications have been "shoe horned" into language constructs, without knowing or understanding the underlying statistical basis or implications. Queueing theory too often has been used as a crutch to explain simulation, simply because queueing models have answers. The authors believe that simulation is a very good and useful tool, without these ancillary concerns. While simulation languages and queueing theory are excellent devices, the authors believe that they bias the discussions on simulation. Thus, we have deferred their discourse to the later chapters of the book.

Chapters 2 through 6 present the fundamental concepts of probability and statistics, at the level normally presented in an introductory probability/statistics course. The concepts of distributions, random sampling, statistical tests, tests of hypothesis as well as a discussion of some of the more common distributions are discussed. Estimation and hypothesis analysis are presented in Chapter 6. Those readers with a sound background in these concepts may skim over these chapters.

Chapter 7 concerns the generation of random numbers. It includes a survey of some of the more commonly used generation techniques as well as a discussion of some common tests employed to assess the randomness of a generated sequence.

Chapters 8 through 12 are concerned directly with the construction of simulation models, and as such comprise the main thrust of the text. Discrete event simulation is discussed in Chapter 8, model validation in Chapter 9, design of experiments and techniques for determining optimal parameters in Chapters 10 and 11. Chapter 12 presents a description of output analysis, in terms of the verification of the simulation model.

Chapter 13 compares and contrasts the general-purpose languages such as FORTRAN with the simulation languages GPSS, SIMSCRIPT, and GASP IV. Specifically discussed are the motivation behind the development of the simulation languages and the features each provides to support the development of simulation models and various world views.

Distributed simulation, a relatively new topic, is discussed in Chapter 14. This methodology does not add to the general principles previously covered in the book. Rather it addresses the issue of how to speed up or accelerate complex simulation models.

Finally, Chapter 15 comprises an introductory look at queueing theory. This chapter is included for completeness sake. Queueing systems are among the most common types being simulated. A brief treatment of the analytic modeling of such systems is presented to aid in developing and validating simulation models of these systems.

In preparing this text, we have attempted to establish a methodology useful in developing and applying simulation. This methodology includes probability and distribution theory, statistical estimation and inference, the generation of random variates, systems theory, time management methods, verification and validation techniques, experimental design and programming language considerations. Without a doubt, there are simulators which were developed using few of the techniques discussed, and which yield "good" results. There are also "bad" simulators which were developed using the techniques discussed. However, we feel that these instances are isolated exceptions rather than the rule.

We take great pleasure in closing this Preface by acknowledging those individuals that make this book possible. Many individuals have contributed to this book, and we wish space were available to acknowledge and thank all of them. However special thanks must be made to Norman Ma, Bob Jarvis, David Hess, Willis Marti, Kwok-Yam Tam, Mike Vidlak, John Desoi, Alok Pancholi, Jose Salinas, Jeremy Green, and Douglas Schales.

The Department of Computer Science, Texas A&M University is to be acknowledged for its implicit support and for a stimulating environment in which to write this book. This book evolved from the simulation classes in the CS Department; we thank the many students who have passed through these classes for their stimulation, curiosity, good humor and enthusiasm. Our final acknowledgments go to the many researchers, practitioners, and colleagues who have contributed to the development of the field of simulation. As always we thank all contributors, named or otherwise, while acknowledging our responsibility for accuracy and/or the lack of.

Udo W. Pooch
James A. Wall
College Station, Texas
July 15, 1992

To Marie:

Thanks for bearing up with the gorilla again, but 15 in 20 is not bad.

U. W. P.

To Dianne:

Always and forever...

J. A. W.

TABLE OF CONTENTS

1

BASIC CONCEPTS AND TERMINOLOGY

The concept of modeling, on which simulation is based, has been in use for many years. Some examples of historical models are Newton's second law, $F = \overline{m}a$, relating the force exerted on a body to its acceleration; Kepler's model of the universe in which he postulated that the sun rather than the earth was the center of the universe; and Einstein's theory of relativity. Since the concept of modeling is so old, what has happened to heighten interest in this problem–solving method in recent years? The answer is that the electronic computer was invented. With the advent of the computer, simulation has been applied to nearly every field of human endeavor. Problems in fields as diverse as business, politics, law enforcement, and nuclear engineering have been successfully solved with the use of simulation.

Simulation has been defined by Shannon (1.4) as "the process of designing a computerized model of a system (or process) and conducting experiments with this model for the purpose either of understanding the behavior of the system or of evaluating various strategies for the operation of the system." Though the literature gives many definitions for simulation, this definition seems to encompass the more important aspects of this problem–solving process. Of particular importance is the linking of simulation with the traditional model–building approach to problem solving. More will be said about this later. Suffice it to say at this point that the main impact of computers on the model–building approach is that the form of the model has changed.

1.1 Concept of a System

Central to any simulation study is the idea of a system. To model a system, one must first understand what a system is. The term **system** is defined in *Funk* *and Wagnall's Standard Dictionary* as "an orderly collection of logically related principles, facts or objects." When used in the context of a simulation study, the term system generally refers to a collection of objects with a well–defined set of interactions among them. An example is the solar system. The planets and the sun form the collection of objects (elements) of the system. This definition of a system is general enough to allow its application to many problems.

Systems can be defined more broadly than simply as a collection of objects and interactions. For example, one could include in the consideration all external factors capable of causing a change in the system. These external factors form the **system environment**. The **state of a system** is the minimal collection of information with which its future behavior can be uniquely predicted in the absence of chance events. Since the inclusion of time in the consideration of a system implies that the state of a system changes, there must be some process or event that prompts this change. Such a process or event is called an **activity**. The system state may change in response to activities internal to the system or to activities external to the system. Activities external to the system are referred to as **exogenous**, while activities internal to the system are referred to as **endogenous**. Although it is convenient to distinguish between exogenous and endogenous activities, it is not always possible to do so. When one is defining a given system, it is not always apparent which factors are internal to the system and which are external. Furthermore, with a given system definition an exogenous activity may prompt a series of endogenous activities. The resulting system state may in turn trigger another exogenous activity. Thus, in many cases very little distinction can be made between endogenous and exogenous activities; instead it is the change in the system state induced by an activity that is of primary interest.

There are a number of ways of classifying systems. An obvious classification distinguishes between systems that are natural and those that are man–made. The solar system is a natural system, while the automobile is a man–made system. Other classifications that can be used include continuous versus discrete, deterministic versus stochastic, and open versus closed.

1.1.1 Continuous versus Discrete Systems

The terms continuous and discrete applied to a system refer to the nature or behavior of changes with respect to time in the system state. A system whose changes in state occur continuously over time are **continuous systems**; systems whose changes occur in finite quanta, or jumps, are **discrete**. Some systems possess the properties of both continuous and discrete systems. Some of the system state variables may vary continuously in response to events while others may vary discretely. Such systems can validly be labeled **hybrid** systems.

Example 1.1 Consider a system comprising the automobiles on a given five–mile stretch of a busy freeway. One state variable of interest might be the number of cars. This variable varies discretely with time. If this is the only variable of interest, the system can be considered a discrete system. On the other hand, suppose that the variable of interest is the average distance between automobiles. Since this quantity varies continuously over time (and one assumes that it is possible to measure the distance with infinite accuracy), this state variable is continuous. If it is the only state variable of interest, the system can be classified as continuous. Of course, since both variables are contained within a state description of this system, the system can also be called a hybrid system.

1.1.2 Stochastic versus Deterministic Systems

A **deterministic** system is a system in which the new state of the system is completely determined by the previous state and by the activity. Considered in another way, a given system evolves in a completely deterministic manner from one state to another in response to a given activity. This type of system is depicted in Figure 1.1, where S_0 refers to the system state before activity A and S_n refers to the system state after the occurrence of the activity.

A **stochastic** system contains a certain amount of randomness in its transitions from one state to another. In some cases it might not be possible to assign a probability to the state that the system will assume after a given state and activity. In other cases these probabilities are known or can be determined. A stochastic System is depicted in Figure 1.2, where $S_{n'}$ and $S_{n''}$ are two possible states that the system can enter after the state S_0 in response to activity A. Thus a stochastic system is nondeterministic in the sense that the next state cannot be unequivocally predicted if the present state and the stimulus (activity) is known.

Example 1.2 Consider a system made up of a single–bay service station in which customers are serviced immediately if the server is idle or join a waiting line (queue) if the server is busy. If the time required to serve a customer varies randomly, the system is stochastic. The state variable of interest, an arriving customer's waiting time, cannot be deterministically calculated from the information that there are X customers in the system before the activity "customer Y arrives."

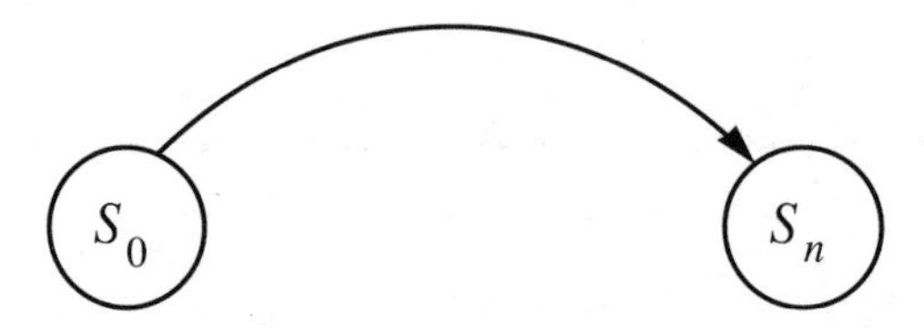

Figure 1.1. A deterministic system.

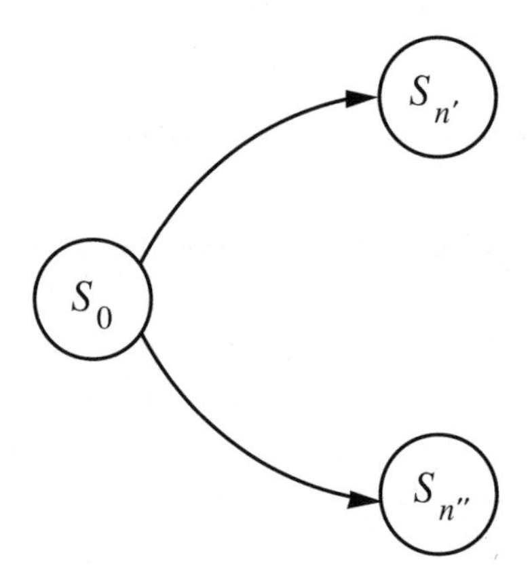

Figure 1.2. A stochastic system.

1.1.3 Open versus Closed Systems

A **closed** system is a system in which all state changes are prompted by endogenous activities. In contrast, **open** systems are systems whose states change in response to both exogenous and endogenous activities. It is sometimes difficult to distinguish between endogenous and exogenous activities, and even if the distinction can be made, they are handled in the same way in most simulation studies. It is also difficult to distinguish between open and closed systems. This distinction is mentioned only for completeness, and it is doubtful that we will need to refer to it again.

1.2 System Methodology

Simulation is based on a problem–solving method that has been is use for many years, sometimes referred to as the **model–building** method or more commonly the **scientific method**. Thus, when system simulation is used to solve a problem, the following time–tested steps, or stages, are applied.

1. Observation of the system.
2. Formulation of hypotheses or theories that account for the observed behavior.
3. Prediction of the future behavior of the system based on the assumption that the hypotheses are correct.
4. Comparison of the predicted behavior with the actual behavior.

The system being studied may impose constraints on certain steps of this scientific method. Consider the simulation of a system that does not yet exist. Obviously the observation of such a system is not possible, but the simulation of such a system may still be possible if the analysis is carefully conducted and if the ultimate system requirements are known. The scientific method's requirement for prior observation of the system has resulted in the development of a slightly different approach to problem solving, called **system methodology**. This methodology illustrated in Figure 1.3, consists of four phases: planning, modeling, validation, and application. The correspondence between the various phases of the system methodology and the appropriate steps of the scientific method should be obvious.

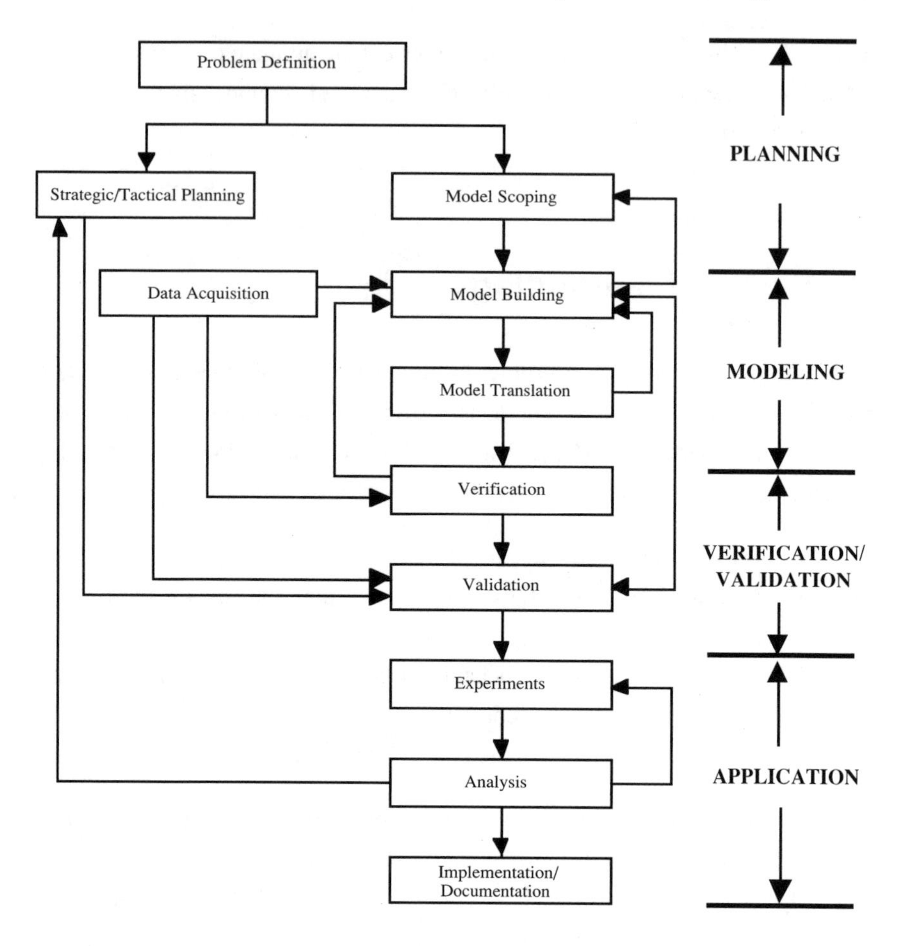

Figure 1.3. Simulation methodology with feedback.

1.2.1 Planning

The initial phase of this problem–solving process is planning (see Table 1.1). The **planning**, or **pre–modeling phase** includes the initial encounter with the system, the problem to be solved, and the factors pertaining to the system and its environment that are likely to affect the solution of the problem.

The planning phase mainly consists of problem definition, analysis of system and data requirements, and the availability of resources. This may involve the development of rough time estimates and process analysis, consideration with regard to the nature of the type of system (stochastic, dynamic, etc.), and a determination of the important aspects of the system. A careful consideration of model and system objectives is important.

a. Problem Formulation

The first step in any analysis effort is to clearly define the problem and the scope. This includes an assessment of the available time frame; what information is desired and how it will be used; a definition of the boundaries of the system, as well as an analysis of the assumptions required to adhere to the boundary definitions; and a preliminary plan (conceptual plan) on how to accomplish this analysis.

Table 1.1 Simulation Methodology

Planning	Problem Formulation Problem definition including a statement of problem–solving objective. Resource Estimation Time, cost, personnel management. System and Data Analysis	• Time schedule • Performance measures • System boundaries • Preliminary project plan.
Modeling	Model Building Abstraction of the system into mathematical (symbolic) relationships with the problem formulation. Data Acquisition	

Table 1.1 Simulation Methodology (continued)

	The identification, specification, and collection of data. Model Translation Preparation and debugging of the model for computer processing.	
Verification/ Validation	Verification The process of establishing that the program executes as intended. Validation Establishing a desired level of accuracy between the model and the real system	
Application	Experimentation The execution of the simulation model to obtain output values (design of experiments). Analysis Analyzing the results of the experimentation and to draw inferences and make recommendations for problem resolution. Implementation/Documentation Process of implementing decisions resulting from the simulation and documenting the model and its use.	

A clear **definition** will facilitate communication, clarify expectations and provide a ground work for the development of the preliminary or conceptual plan. A **feasible plan** will allow for cost estimation, scheduling, personnel assignments, and for checkpoint definitions. The feasible plan must also include **time frame restrictions** based on economic cycles, resource justifications, contractual agreements, management edicts, and most of all the realization that with the development of the simulation model there will be iterative refinements, modifications and enhancements of various degrees of granularity established at checkpoints. Deadlines for preliminary and final presentations should be established. These should, of course, allow for some flexibility resulting from later knowledge.

Another aspect of problem formulation has to do with determining how the end user is expecting to utilize the simulation model results. It is important to note that simulation models can provide not only average of performance indicators, but also variances, extremes, and even time series. This is important since system design and operation is frequently concerned with extremes and variations, as much as or more than average. Examples of typical performance measures are given in Table 1.2.

Table 1.2 Example Performance Measures

Performance Measure	*Example*
Throughput	• Number, volume, weight of entities produced in a period of time (system or component) – by entity type. • Number, volume, weight of entities produced or processed per unit time (production rate) – by entity type. • Time between entity or batched entity departures from system or component.

Table 1.2 Example Performance Measures (continued)

Ability to meet deadlines	• Time to produce or process a specified number of entities. • Time in the system for entities (flowtime). • Entity lateness (time between completion and due date of completion). • Entity tardiness (time between completion and due date of completion if completion occurred after the due date; otherwise no tardiness is observed).
Operation work cycles	• The number of entities that were blocked. • The time that entities spent blocked. • The time between entity arrivals. • The cycle time per entity processed • The number of times a resource was preempted from an operation.
Resource utilization	• Fraction of time a particular resource is busy, idle, inoperative, blocked, or waiting. • The number or proportion of resources which are busy, idle, inoperative, blocked, or waiting.
Storage utilization	• The number of entities waiting for a particular resource. • The number of entities waiting for any resource. • The number of entities which balk (spillover) from a given storage area.
Costs	• Costs to operate a unit for an amount of time. • Inventory holding costs. • Scrap costs. • Costs per amount of material produced.

Table 1.2 Example Performance Measures (continued)

Yield	• Amount of scrap generated during a period of time. • Number of rejects generated during a period of time. • Percent shippable entities. • Percent shippable weight.

While simulation models are capable of producing virtually any type of output or detail of information that is desired, there is often a considerable cost difference. Thus the simulationist should define information that is necessary to support any required decision process, prioritize any remaining possibilities, and maintain this shopping list during the model scoping effort.

Another task necessary during problem formulation, is that of establishing **system boundaries**. System boundaries deal with establishing system extensions and assumptions required to adhere to system boundary definitions. The initial extent of a simulation should be limited, independent of any given time frame, resulting in quicker feedback, testing the acceptability of results and formats, and providing preliminary results.

Finally, a **preliminary project plan** must be established. This is necessary because the simulation modeling process is of an interdisciplinary nature, with multiple reporting levels, and many decision points used to assess progress. While this plan is preliminary and will necessarily be reviewed and reused as the project proceeds it must include length of activities, required and available resources, and the availability of required personnel.

b. Resource Estimation

Once the problem has been clearly defined, an **estimate of the resources** required to collect data and analyze the problem (system) can be made. Resources such as time, money, personnel, and special equipment should be considered. If crucial resources are not available, solution of the problem can be judged infeasible before a significant amount of time or money is spent. The alternative to discontinuing the project is to modify the definition of the problem. The problem can be restated so that it can be successfully solved with available resources. In some cases this constraint may mean solving a less ambitious problem. The importance of this initial resource estimate cannot be overemphasized; it is clearly more desirable to modify objectives at an early stage than to fall short because crucial resources are unavailable.

c. System and Data Analysis

The next major task in the planning phase is to **analyze** the system. In this phase the analyst attempts to become familiar with all relevant aspects of the problem being studied. A thorough literature search to discover previous approaches to similar problems may prove valuable in choosing possible courses of action. Consultation with other qualified persons may provide insight into aspects of the problem that the primary researcher has overlooked or misunderstood. The importance of this initial problem analysis should not be underestimated. The analyst must be willing to spend the time and the effort necessary to fully understand the system or problem under study. Many projects have failed or incurred overruns in cost or time because of an inadequate understanding of the project.

At this stage any objective function(s) to compute performance measures for evaluating alternative simulation model solutions, and all preliminary parameters and variables will have been identified. Furthermore, parameters and variables will have to be identified with respect to initial conditions and availability of sources. Data will be needed to estimate values of constants and parameters, stating values for all variables, and for simulation output validation.

Data on the behavior of the exogenous, uncontrollable variables and parameters may be obtained from historical data, theoretical methods, and subjective estimates of distribution. Historical data to be used for endogenous, uncontrollable variables as well as validation must be broken up into two distinct independent groups (or as much independent as is possible). A summary of the various types of variables is given in Table 1.3, while their logical interrelationship is illustrated in Table 1.4.

Table 1.3 Types of Variables

Relevant Variables	Characteristic or attribute of system that takes on different values and that affects the measure of performance.
Parameter	Characteristic or attribute with only one value. It may change as different alternatives are studied.
Exogenous Factor	Parameter or variable having a value that affects, but is itself unaffected by the system. Such as a factor in the environment of the system.
Endogenous Factor	Parameter or variable having a value determined by others in the system.
Controllable Variables	Set by decision maker during analysis. These would be set according to some experimental design or search procedure.
Uncontrollable Variables	Input to the model to represent the relevant parts assumed to be external to the system. Specific value of each is required as starting condition.
Estimate of Dynamic Exogenous	• Time series • Distribution • Theoretical considerations • Statistical analysis of historical data
Estimate of Dynamic Endogenous	Variables to be predicted by the model. Estimates of their values used to start the simulation.

Before model building should commence, the project team must consider the level of detail of the analysis, of the results, and the limits of the analysis. At this stage, the plan is preliminary with emphasis on strategic considerations. As the project proceeds more information will allow iterative refinement of the plan. Just as the scope of the system needed to be limited and understood, the scope of the analysis needs to be limited. Conclusions beyond model boundaries are not appropriate; sensitivity analysis are infinite and must be limited; rigorous statistical evaluations may or may not be feasible or even desirable; and finally relative comparisons may be appropriate rather than assuming that the outputs are absolute indicators.

Table 1.4 Interrelationship of Simulation Variables

<table>
<tr><td></td><td colspan="2">Controllable</td><td colspan="2">Uncontrollable</td></tr>
<tr><td></td><td>Static</td><td>Dynamic</td><td>Static</td><td>Dynamic</td></tr>
<tr><td rowspan="2">Exogenous</td><td>Set by search procedure</td><td>Policies change with time</td><td colspan="2" rowspan="2">Must estimate to determine starting values and changes over time</td></tr>
<tr><td colspan="2">Historical data for validation</td></tr>
<tr><td rowspan="2">Endogenous</td><td colspan="2">Do not exist</td><td>Parameter</td><td>Includes performance measure</td></tr>
<tr><td colspan="2">Historical data for validation</td><td colspan="2">Must estimate starting values</td></tr>
</table>

1.2.2 Modeling

The second phase of the problem–solving process is the **modeling phase** (see Table 1.5). In this phase the analyst constructs a **system model**, which is a representation of the real system. A model may be described as the body of information about a system gathered for the purpose of studying the system. That is, it is not only a collection of information, but also an orderly representation or orderly structuring of the information.

Table 1.5 Simulation Methodology

Planning	Problem Formulation Problem definition including a statement of problem–solving objective. Resource Estimation Time, cost, personnel management. System and Data Analysis	• Time schedule • Performance measures • System boundaries • Preliminary project plan.
Modeling	Model Building Abstraction of the system into mathematical (symbolic) relationships with the problem formulation. Data Acquisition The identification, specification, and collection of data. Model Translation Preparation and debugging of the model for computer processing.	• Model scoping • Levels of detail • Submodels • Variable/parameter estimation • Programming • Debugging
Verification/ Validation	Verification The process of establishing that the program executes as intended. Validation	

Table 1.5 Simulation Methodology (continued)

	Establishing a desired level of accuracy between the model and the real system	
Application	Experimentation The execution of the simulation model to obtain output values (design of experiments). Analysis Analyzing the results of the experimentation and to draw inferences and make recommendations for problem resolution. Implementation/Documentation Process of implementing decisions resulting from the simulation and documenting the model and its use.	

The characteristics of this model should be representative of the characteristics of the real system. Note the use of the term **representative** rather than identical. Many real systems are so complex that a model with characteristics identical to those of the real system would be equally complex and thus unmanageable or prohibitively expensive. One of the goals of this phase, then, is to select some minimal set of the system's characteristics so that the model approximates the real system yet remains cost–effective and manageable. The model includes these characteristics and a well–defined set of relations among these characteristics.

Models can be used as an explanatory device to define a system or problems, as a communications mechanism to determine system elements, as a design assessor to synthesize and evaluate proposed solutions to problems, as a predictor to forecast and aid in planning future developments, as a documentation medium, as a training device, and as a control mechanism. The model building procedure is given in Table 1.6.

Table 1.6 Model Building

1. Preliminary simulation model diagrams.
2. Construct and develop system flow diagrams.
3. Review model diagrams with appropriate project team personnel.
4. Initiate data collection.
5. Modify the top–down design, test and validate for the required degree of granularity.
6. Complete data collection.
7. Iterate through steps 5 and 6 until final granularity has been reached.
8. Final system diagrams, translation, and verification

There are many types of models, including descriptive models, physical models, mathematical models, flowcharts, schematics, and computer programs. **Descriptive models** are simply verbalizations of the system's composition and its response to a given stimulus. **Physical** or **iconic models** are scaled facsimiles of the system being analyzed; examples are models used to evaluate aircraft designs in wind tunnel experiments and mock–ups of landscaping and buildings in architectural design studies. **Mathematical models** are abstract expressions of the relationships among system variables; examples of these models are Newton's laws of motion and the description of the stress factors of a dam as it is filled with water. The advantage of mathematical models is that they can be manipulated with paper and pencil. With these models it is not necessary to construct an expensive mock–up of the system to study its behavior. Their disadvantage is that the accurate description of a given system may require an expression so complex that it would make even the most facile mathematician shudder. **Flowcharts** and **schematics** display the basic logical interactions between system components and may be as detailed as desired. When used as a prelude to computer program models, flowchart models are valuable aids to programming and program documentation.

a. Model Scoping

Model scoping is the process of determining what process, operation, equipment, etc., within the system should be included in the simulation model, and to what level of detail. Each component should be independently considered for inclusion with the following guidelines in mind: Will the inclusion of the component into the model significantly affect the results of the analysis? Will the inclusion of the component into the model affect the model's salability? As components are considered for inclusion, the following must be estimated:

1. The accuracy of the overall model.
2. The accuracy required for effective analysis.
3. The operational effects of including this component.
4. The availability and accuracy of data concerning the component.
5. The effect of removing the component from the model.
6. The cost, both in terms of time and effort, to include the component.

b. Level of Detail

The **level of detail** with which to include model components must be determined by the component's effect on the results of the analysis, and the component's effect on the salability of the analysis. As the level of detail is considered, the previous considerations for model scoping must be reconsidered and reexamined. The use of standard submodels to represent fixed parts of a system may save time, but may lead to mixed levels of detail. Consequently the appropriate level of model detail will vary based upon the model's objective. Because two models, both representing a like component, may implement the component in a different way, each component must be evaluated from the standpoint of inclusion/omission and level of detail.

After scoping, the preliminary project plan should again be carefully reevaluated and refined as required.

c. Subsystem Modeling

If the system being studied is so complex that no representative model can be used, then the well-known problem reduction technique, or **subsystem modeling**, is an alternate approach. In this approach the system is divided into a collection of less complex subsystems. Each of these subsystems is modeled, and an overall system model is constructed by linking the subsystem models appropriately. This approach of course requires that subsystems may be readily modeled, identified, and isolated.

Three general approaches have been used in defining or identifying subsystems. The first method, called the **flow approach**, has been used to analyze systems characterized by the flow of physical or information items through the system. If the system to be modeled is an automobile assembly plant, the flow approach would appear to be a feasible way of breaking this system into subsystems. In the flow approach subsystems are identified by grouping aspects of the system that produce a particular physical or information change in the flow entity.

A second technique used in identifying subsystems is the **functional approach**. This technique is useful when there are no directly observable "flowing" entities in the system. Instead, a logical sequence of functions being performed must be identified. System characteristics that perform a given function are grouped to form a subsystem. This approach can be used in manufacturing processes that do not use assembly lines.

The third method for identifying or isolating subsystems is called the **state–change approach**. This procedure is useful in systems that are characterized by a large number of interdependent relationships and which must be examined at regular intervals to detect state changes. System characteristics that respond to the same stimulus or set of stimuli are then grouped to form a subsystem.

Once the subsystems have been identified and isolated, they must be **modeled**. Of course, the subsystems could in turn be subdivided into sub–subsystems, and so on, as far as necessary. The goal is to subdivide the total system in such a way that each subsystem is easily comprehensible and readily identified. If subsystems cannot be identified and isolated, the analyst has no choice but to model the entire system with a single model and hope that this large model will be manageable. If the subsystem approach has been used, the submodels must be linked into an overall system model. The logical interactions of the subsystems must be identified and implemented in a form compatible with the overall model.

d. Variable and Parameter Estimation

Another task in the modeling (programming) phase of a simulation study is the **estimation of the system variables and parameters**. At this point real–world data are summarized into a manageable statistical description of the system's characteristics. This is commonly done by collecting data over some period of time and then computing a frequency distribution for the desired variables. A number of pitfalls await the unsuspecting researcher in the collection and interpretation of data. The cost in terms of time, money, and personnel can become prohibitive; the data may be incomplete or inaccurate; the data may contain unsuspected interdependencies, periodicities, or complexities; the method of collection may introduce an inadvertent bias into the data. For example, values may be represented as time series as other complex interactions. The data could be one that resulted from the interaction of several variables that are exogenous to the model but are related in the external world (i.e. **multicollinearity**) or where there is a relationship between the variances in the error terms (**hederosedarity**). Failure to recognize and identify such complications can cause the simulation output to mispredict the behavior of the system. Other than the inherent cost of collecting data, most of these pitfalls can be avoided through the use of experimental design procedures.

Data acquisition is the identification, specification, and collection of data to support the model. This task is performed to support the defined model. Much other data had been previously reviewed, but to define boundaries, model scope, and the level of detail.

Data collection and verification is often the most time consuming and expensive task in simulation. Data is generally the limiting factor in model scope, level of detail and practicality. Some important guidelines are given in Table 1.7.

If the model is a computer program, as is the case in a simulation model, the modeling phase also involves the selection of the language in which the program is to be written. A number of considerations determine the choice of a programming language, including the difficulty of translating the model and interrelationships into the language, the presence or absence of facilities in the language to support such routine activities as queue management, generation of random numbers, and output formatting, and the analyst's familiarity with the language. A crucial step in using a computer program to model a system is the construction of detailed system flowcharts. Although flowcharting has fallen into disfavor in recent years (some are even written after the program), flowcharts provide a visual representation of the program's logic and serve as program documentation after the logic has been translated into code. As a general rule, the better the formulation of the model when the flowchart is made, the easier it is to create a complete computer program model of the system.

Table 1.7 Data Collection Guidelines

1. Data collection is often the bottleneck of the simulation modeling process.
2. Not all data is available simultaneously.
3. Simulation models can be exercised with incomplete data (especially during the verification and validation of components).
4. Simulation models can be used to evaluate sensitivity with respect to data *a priori*.
5. Great care must be taken in the distribution of model results when incomplete data is used.

Submodels should be chosen in such a way that they are as nearly independent as possible from other submodels. Each can then be independently written and tested, and the output can be compared to historical data on a component basis.

1.2.3 Validation / Verification

A **model** is validated by proving that the model is a correct representation of the real system. **Validation** should not be confused with **verification** (see Table 1.8). When a computer program is verified, for example, the program is checked to ensure that the logic does what it was intended to do. A verified computer program can in fact represent an invalid model. The program may do exactly what the programmer intended, yet it may not represent the operation of the real system.

Table 1.8 Simulation Methodology

Planning	Problem Formulation Problem definition including a statement of problem–solving objective. Resource Estimation Time, cost, personnel management. System and Data Analysis	• Time schedule • Performance measures • System boundaries • Preliminary project plan.
Modeling	Model Building Abstraction of the system into mathematical (symbolic) relationships with the problem formulation. Data Acquisition The identification, specification, and collection of data. Model Translation Preparation and debugging of the model for computer processing.	• Model scoping • Levels of detail • Submodels • Variable/parameter estimation • Programming • Debugging

Table 1.8 Simulation Methodology (continued)

Verification/ Validation	Verification The process of establishing that the program executes as intended. Validation Establishing a desired level of accuracy between the model and the real system	
Application	Experimentation The execution of the simulation model to obtain output values (design of experiments). Analysis Analyzing the results of the experimentation and to draw inferences and make recommendations for problem resolution. Implementation/Documentation Process of implementing decisions resulting from the simulation and documenting the model and its use.	

Validation of computer simulation models is a difficult task. There is considerable disagreement about what constitutes validation. Despite this disagreement, there are techniques that have proven useful in this phase of the simulation process. One technique is to compare the results of the simulation model with the results historically produced by the real system operating under the same conditions. A second technique is to use the simulator to predict results. The predictions are then compared with the results produced by the real system during some future period of time.

Whichever validation technique is used, the method of comparing the outputs of the real system and those of the simulated model is basically the same. Various statistical procedures can be used to determine whether the data produced by the simulator could have been produced by the real systems and vice versa. Two such methods are the chi–square goodness–of–fit test and the Kolmogorov–Smirnov test.

Validation of a simulation model is not easy, but it must be performed before the simulation model can be used. The lack of precision in the validation process is evidenced in the literature by statements such as "90% of the predicted values lie within 5% of the observed values." Whether this statement implies that the simulator is a good one or a bad one is left to the interpretation of the analyst. Such a statement does convey considerably more information than a statement such as "the results obtained were quite good."

Just as a good experimental design can aid in the data collection in the modeling phase so can validation aid in correctness of the simulation model. Most standard experimental designs require that observations be taken on the system variables that can be controlled. The simulation model must operate under identical conditions. Only then can valid information be drawn about the relationship between the resulting output (observations) of the real system and the outputs of the simulation model. In most designs several factors can be varied simultaneously, and thus some information on interactive effects of those factors can be obtained. Several effects may be estimated from the same data. Careful design of the simulation experiment will prove beneficial in the long run.

1.2.4 Application

Once the model has been properly validated, it can be applied to solving the problem at hand. However, the development of the simulation model may still not be complete (see Table 1.9).

Table 1.9 Simulation Methodology

Planning	Problem Formulation Problem definition including a statement of problem–solving objective. Resource Estimation Time, cost, personnel management. System and Data Analysis	• Time schedule • Performance measures • System boundaries • Preliminary project plan.
Modeling	Model Building Abstraction of the system into mathematical (symbolic) relationships with the problem formulation. Data Acquisition The identification, specification, and collection of data. Model Translation Preparation and debugging of the model for computer processing.	• Model scoping • Levels of detail • Submodels • Variable/parameter estimation • Programming • Debugging
Verification/ Validation	Verification The process of establishing that the program executes as intended. Validation	

Table 1.9 Simulation Methodology (continued)

	Establishing a desired level of accuracy between the model and the real system	
Application	Experimentation The execution of the simulation model to obtain output values (design of experiments). Analysis Analyzing the results of the experimentation and to draw inferences and make recommendations for problem resolution. Implementation/Documentation Process of implementing decisions resulting from the simulation and documenting the model and its use.	

a. Design of Experiments

An **experimental design** requires observation of the system under specific combinations of those variables capable of being manipulated. Simulation runs are made under these control conditions and inferences are drawn about the relationship between the controllable variables and the measures of performance. Prior to experimental design runs the strategic plan should be reviewed and finalized. Similarly, the tactical plan must be reviewed. It may require some initial experimentation to determine the validity of the tactical plan and to finalize it.

The extent of the experimentation will depend on the cost to estimate the performance measures, the sensitivity of the performance measures to specific variables, and the extent of the interdependencies between the control variables. In any case, experimentation may be serial or phased according to the strategic/tactical plan.

b. Analysis

Analysis is the process of analyzing the results of the simulation model's executions, from which to draw inferences and make recommendations for problem resolution. The project team should determine what cases to evaluate and what necessary outputs are required. Unfortunately situations can easily arise in simulation modeling that result in misinterpreting the data and the corresponding analysis. Factors for consideration must include the following:

1. Poorly chosen pseudorandom number generator.
2. Inappropriate random variate generation technique.
3. Incorrect input parameter specification.
4. Model programming errors.
5. Specification errors for the model.
6. Length of simulation.
7. Sensitivity to key parameters.
8. Data collection errors in simulation.
9. Optimization parameter errors.
10. Problematic optimization procedures.
11. Incorrect experimental design.
12. Poor choice of descriptors (parameters) to estimate.
13. Peculiarities of the estimation method.
14. Influence of the initial conditions on data.
15. Probability distribution and parameter estimates.
16. Influences of final conditions on data and on estimation method.
17. Misuse of estimates.

As large as this list might seem to be, the chapters in the remainder of the book will provide enough guidance to weather these potential influences.

c. Implementation

Implementation is the process of putting into practice decisions resulting from the simulation and documenting the simulation model and its use. The likelihood of implementation is directly associated to the effectiveness of the analysis and the reporting of the results. Model simulation and analysis procedures provide little, if any, automatic documentation. This task is the responsibility of the project team, and should consist of a full record of the total project activity, not just a user's guide.

Observation of the output from the model may reveal programming errors, the desirability of a change in the way the simulator has been implemented, or even the need to reformulate or modify the original problem statement. It has been said that no computer program is completely free of errors. This is especially true of simulation models. A simulator may work error–free for a long period of time until it encounters a new and perhaps unique combination of program parameters that generate the next error.

1.3 Advantages and Disadvantages of Simulation

Simulation has been applied to nearly every field of human activity in the past few years. One might ask, if simulation is so good, why is any other type of modeling used? The answer is that simulation is not applicable in many cases, and even in some cases in which it does apply, there may be easier and cheaper ways of solving the problem. Phillips, Ravindran, and Solberg (1.2) state that simulation is one of the easiest tools of management science to use but probably one of the hardest to apply properly and perhaps the most difficult one from which to draw accurate conclusions. Whether one agrees with this statement or not, one should realize that there are distinct advantages and disadvantages to simulation.

Adkins and Pooch (1.1) list five advantages of simulation modeling.

1. It permits controlled experimentation. A simulation experiment can be run a number of times with varying input parameters to test the behavior of the system under a variety of situations and conditions.

2. It permits time compression. Operation of the system over extended periods of time can be simulated in only minutes with ultrafast computers.
3. It permits sensitivity analysis by manipulation of input variables.
4. It does not disturb the real system. This is a great advantage, since most managers would be reluctant to try experimental strategies on an on–line system.
5. It is an effective training tool.

They also list four disadvantages to using the simulation approach to problem solving.

1. A simulation model may become expensive in terms of manpower and computer time. This is not surprising if the magnitude of the problems being attempted is considered. For example, consider the simulation of messages through a large–scale (1000–node) communication network. Just the bookkeeping requirements for a problem of this magnitude are staggering. The cost of a simulation experiment can be minimized through in–depth understanding of the system being simulated *before* the model is developed and through careful design of the simulation experiment.
2. Extensive development time may be encountered. Most simulation models are quite large and, like any large programming project, take time. Strategies such as the chief programmer team, top–down design, and modular programming, which have been applied to other large programming projects, are likely to be useful in the development of system simulators and could reduce the development time.
3. Hidden critical assumptions may cause the model to diverge from reality. Ideally this phenomenon should be discovered in the validation phase of the simulation process, but it might go undetected, depending on the severity of the problem and the diligence with which the model is validated.
4. Model parameters may be difficult to initialize. These may require extensive time in collection, analysis, and interpretation.

Thus, although simulation has proved an effective approach to problem solving, it has its drawbacks. The researcher should be aware of these drawbacks *before* becoming committed to this approach.

It should be noted that simulation only supports the entire decision–making process. To effectively use simulation, the user must understand the decision–making and simulation modeling process, where simulation can be applied within the decision process, where and when these tools are most applicable, and how to apply simulation.

1.4 Simulation Terminology

Since the use of simulation modeling has become so widespread, a terminology unique to the field has evolved. This terminology is by no means standard, but enough authors have used it that it has gained some acceptance. Some of the terms have been introduced in the course of this chapter; others will be used as the need arises.

1. About the model

   a. A real–world object is called an **entity**.
   b. Characteristics or properties of entities are called **attributes**.
   c. Any process that causes changes in a system is called an **activity**.
   d. A description of all the entities, attributes, and activities, as they exist at some point in time is called the **state of the system**.

2. About the environment

   a. The objects and processes (entities and activities) surrounding the system are called the system environment.
   b. Activities that occur within the system are called endogenous activities.
   c. Activities in the environment that affect the system are called exogenous activities.

d. The classification of all activities as either endogenous or exogenous establishes the system boundary.

e. A system with no exogenous activities is called a closed system; otherwise the system is open.

3. About the system

a. **Continuous** systems include variables that can assume any real value in a prescribed set of intervals; these systems are characterized by discontinuous changes in the system state.
b. **Discrete** systems include variables that can assume only particular values from among a finite set of alternatives; these systems are characterized by discontinuous changes in the system state.
c. A system whose response is completely determined by its initial state and input is said to be **deterministic**.
d. A system whose response may take a range of values given the initial system state and input is said to be **stochastic**.

4. About the simulation

a. **Validation** refers to the proof that the model is a correct representation of the real system.
b. **Verification** refers to the proof that the simulation program is a faithful representation of the system model.
c. **Experimental** design refers to a sequence of simulation runs in which parameters are varied, with both economy and sound statistical methodology considered in achieving some specified goal.

1.5 Summary

The use of the simulation approach to problem solving has become widespread since the development of the electronic computer. This chapter has surveyed the systems approach to problem solving, on which simulation is founded. The usual sequence of steps that occur before an attempt is made to solve the problem is to identify the problem, perform a systems analysis, and restate the problem within the context of the system definition. Once this has been done, according to Pritsker (1.3), there are four basic tasks that should be performed in a simulation project.

1. Determine that the problem requires simulation. The crucial factors are the cost, the feasibility of conducting real–world experiments, and the possibility of mathematical analysis.
2. Build a model to solve the problem.
3. Write a computer program that converts the model into an operating simulation program. This would include components or activities such as problem description, file maintenance, time management, modeling the probabilistic elements, statistical collection, and output reporting.
4. Use the computer simulation program as an experimental device to resolve the problem.

The next few chapters survey some of the fundamental methods used in simulation. Readers familiar with these subjects may want to merely skim those chapters before moving on to later chapters.

1.6 References

1.1 Adkins, Gerald, and Pooch, Udo W. "Computer Simulation: A Tutorial", *Computer*, 10, 4, April 1977.

1.2 Phillips, D. T.; Ravindran, A.; and Solberg, J.J., *Operations Research Principles and Practice*, New York: John Wiley and Sons, 1976.

1.3 Pritsker, A. A. B., *The GASP IV Simulation Language*, New York: John Wiley and Sons, 1974.

1.4 Shannon, Robert E. "Simulation: A Survey with Research Suggestions", *AIIE Transactions*, 7, 3, September 1975.

2

PROBABILITY CONCEPTS IN SIMULATION

One of the problems commonly encountered while modeling real–world systems is that few systems exhibit constant, predictable behavior. Observations or measurements in a physical system often vary depending on, among other things, the time of the observation. These variations often appear to be random or chance variations. Despite this inherent unpredictability, some model or mathematical structure is required to describe the system. The system is generally described by a probability model that in most cases is developed through experimentation. Experiments are performed on the system, the outcomes of those experiments are noted, and a model is then postulated on the basis of these outcomes. Inferences based on this model are then made, and additional experiments are conducted in an attempt to validate the model. In general, the more experiments performed on the system, the more likely the model is to accurately reflect the true nature of the system.

Since the development of a probability model is so important in the simulation of any system, this chapter is devoted to probability and probability models.

2.1 Probability

To understand the development of a probability model and the importance of this model in describing physical systems, one must understand the concepts of probability. These concepts, many of which have their origins in games, are simple and intuitive in most cases and can usually be grasped easily through an example.

Probability is a measure of uncertainty. This uncertainty is conveyed through phrases like "probably", "in all likelihood", "there is a good chance that". This definition, however, is not precise enough for our purposes, and a more precise definition will be given after an example.

Example 2.1 Consider the simple experiment of tossing a coin. If we assume that the coin is thin enough that the possibility of its landing on its edge can be ignored, then there are two possible outcomes to this experiment. Either it will land heads up, or it will land tails up. Can you predict with certainty which side it will land on if it is tossed? The answer, most probably, is no. The side on which it lands is unpredictable, or uncertain, because the outcome is influenced by the construction of the coin, the manner of tossing, the regularity of the surface on which it lands, and so on (i.e., it is memoryless). The outcome of the tossed coin is left to chance. With respect to its outcome we do know that it will land on one of two sides (for certain), but either side has an equal chance of occurring.

An experiment of counting observations or measurements such as that in Example 2.1, in which the outcome is uncertain or unpredictable, is called a **random experiment**. The outcomes of this random experiment are called **simple random events** (i.e., memoryless). Thus one simple random event for the experiment of Example 2.1 is the event "the coin lands heads up." The other is the event "the coin lands tails up." Each of these two events will have some measure of likelihood or certainty attached to it – the **probability** of its occurrence. This measure of likelihood (with confidence level) is determined through experimentation. The experiment is conducted a number of times, and its outcomes are noted. The relative frequency of occurrence of a given outcome should stabilize, or tend to a limiting value. This limiting value or tendency is defined as the **probability** of the event.

Once all possible outcomes (events) of a random experiment are determined and a probability has been attached to each, a **probability (stochastic) model** is said to have been determined.

Thus, probability is the "art" of making determinable statements about non–determinable events. For example, it can be based on observing many occurrences of real world events; it can be based on mathematical or theoretical events; it is always expressed between 0 and 1; and the sum of the probability of all events is 1.0 (certainty). Probability is the "degree of certainty" that an individual has about the outcome of an uncertain event.

Example 2.2 Consider the example of tossing a die. The outcomes of this experiment are { ⚀, ⚁, ⚂, ⚃, ⚄, ⚅ }. For convenience, represent these outcomes as 1, 2, 3, 4, 5, and 6 respectively. Suppose that experimentation has shown that the likelihood that a given face will appear is proportional to the number of spots on the face. Then the probability model for this experiment is

Event	1	2	3	4	5	6
Probability	1 / 21	2 / 21	3 / 21	4 / 21	5 / 21	6 / 21

In Example 2.2 the probability of each of the events is different. This is generally the case; however, there are experiments in which the likelihood of each of the events is the same. In this case the events are said to be **equally likely** or **equiprobable**.

Example 2.3 Consider the well–known wheel of fortune, which is a circular disk with pegs placed around its perimeter, and a pointer (flapper) protruding into the slots between the pegs (Figure 2.1). The disk is pinned through its center to some fixed surface, and each slot is numbered. The experiment consists of spinning the wheel and noting which number is under the pointer when it stops. The outcomes (events) are thus {"1 is under the pointer", "2 is under the pointer", "3 is under the pointer", and "4 is under the pointer"}. Again, for convenience, denote these events as 1, 2, 3, and 4 respectively. Suppose that experimentation revealed these events to be equally likely. Then the probability model for this experiment is

Event	1	2	3	4
Probability	1 / 4	1 / 4	1 / 4	1 / 4

In the examples given thus far the outcomes (events) were nonnumerical, although a numerical correspondence was evident. It is nearly always advantageous to envision the outcomes of a random experiment as numerical. For this reason, the idea of a random variable is introduced. Simply stated, a **random variable** is a quantity whose value is determined by the outcome of a random experiment.

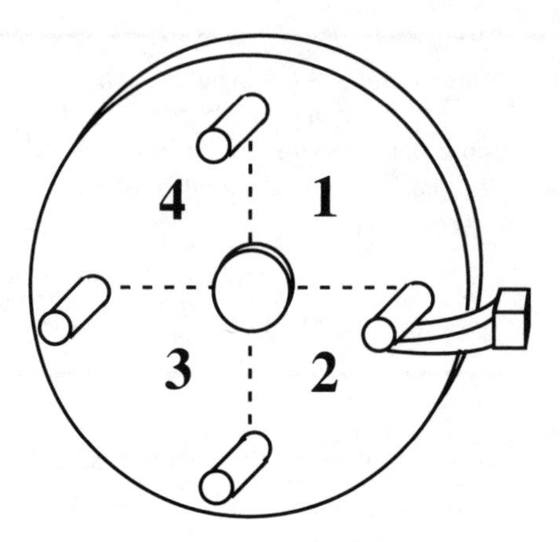

Figure 2.1. The wheel of fortune.

More precisely, the random variable **X** is a function whose domain is the event space (set of all possible outcomes of a random experiment) and whose range is some subset of the real numbers. This definition is summarized by **X**: $E \rightarrow R$.

Example 2.4 Consider the event space (list of simple outcomes) of the experiment described in Example 2.2.

$$E = \{ ⚀, ⚁, ⚂, ⚃, ⚄, ⚅ \}$$

A random variable **X**: $E \rightarrow R$ can be defined by rules of the form

$$\begin{array}{ll} A \rightarrow 1 & D \rightarrow 4 \\ B \rightarrow 2 & E \rightarrow 5 \\ C \rightarrow 3 & F \rightarrow 6 \end{array}$$

In fact, this is precisely what was used in the example.

Example 2.4 illustrates that there is nothing deep, dark, or mysterious about random variables. Nor is there anything highly significant about them. In most cases they are defined for convenience.

2.2 Set Theory, Compound Events

As defined in the previous section, a **simple random event** is a single outcome of a given random experiment. In some cases we are less interested in individual events than in combinations of these events. This section reviews some of the fundamentals of set theory as they relate to combining simple random events into composite events (hereafter simply referred to as events).

DEFINITION 2.1

*An **event** is some subset of the event space of a random experiment.*

The **event space** for a random experiment, defined as the collection of all possible outcomes of the experiment, will be denoted by Ω.

Example 2.5 Consider the experiment of tossing two dice and noting the sum of the faces showing. The event space is $\Omega = \{2, 3, 4, 5, 6, 7, 8, 9, 10, 11, 12\}$ and some events are

$E_1 = \{2\}$ ("snake eyes" appear)
$E_2 = \{3, 4, 5\}$ (the sum is 3, 4, or 5)
$E_3 = \{2, 4, 6, 8, 10, 12\}$ (an even number appears)

This definition of an event, then, includes all possible subsets of the event space – the simple events, compound events, and even the event space itself.

It is sometimes of interest to know the probability that an event E will not occur rather than that it will. This leads to the following definition.

DEFINITION 2.2

*The **complement** of an event E, denoted $\overline{E}$, is the set of elements that are in Ω but not in E.*

Example 2.6 The complements of the events defined in Example 2.5 are

$$\overline{E}_1 = \{3, 4, 5, 6, 7, 8, 9, 10, 11, 12\}$$

$$\overline{E}_2 = \{2, 6, 7, 8, 9, 10, 11, 12\}$$
$$\overline{E}_3 = \{3, 5, 7, 9, 11\}$$

There are two basic ways of combining two events to form a third event: union and intersection

DEFINITION 2.3
The ***intersection*** *of two events E_1 and E_2, denoted $E_1 \bullet E_2$, is defined as the outcomes that the events have in common. Two events that have no outcome are said to be* ***mutually exclusive***, *and $E_1 \bullet E_2 = \varnothing$, where the symbol, $\varnothing$, is the null or empty set.*

DEFINITION 2.4
The ***union*** *of two events E_1 and E_2, denoted , $E_1 + E_2$ is defined as the outcomes in either E_1 or E_2 or both.*

Example 2.7 The union and intersection of the events defined in Example 2.5 are

$$E_1 + E_2 = \{2, 3, 4, 5\} \qquad E_1 \bullet E_2 = \varnothing$$
$$E_1 + E_3 = \{2, 4, 6, 8, 10, 12\} \qquad E_1 \bullet E_3 = \{2\}$$
$$E_2 + E_2 = \{2, 3, 4, 5, 6, 8, 10, 12\} \qquad E_2 \bullet E_3 = \{4\}$$

Thus one can use these rules for combining events to construct a number of related events from any given initial set. For any event E note that

$$E + \overline{E} = \Omega$$

and

$$E \bullet \overline{E} = \varnothing \text{ (null set)}$$

It should be clear that the probability that some event in the event space will occur is 1 (certainty) whereas the probability that *no* event in the even space will occur is 0 (impossibility).

PROPERTY 2.1

For any random experiment, it must be true that

$$P(\Omega) = 1 \text{ and } P(\varnothing) = 0.$$

If the outcomes of a random experiment are **equally likely**, calculation of the probability of some event composed of a number of different outcomes is straightforward.

DEFINITION 2.5

Suppose that a random experiment has N equally likely outcomes. Further, suppose that some event E is composed of n outcomes. Then the probability of the event E, denoted P(E), is given by

$$P(E) = n/N.$$

The probability of the complement of E, $P(\overline{E})$, is given by

$$P(\overline{E}) = 1 - P(E) = (N - n)/N.$$

Example 2.8 Consider the events defined in Example 2.5. If the outcomes of the random experiment are equally likely (the probability of each is 1/11), the probabilities of E_1, E_2, and E_3 are

$$P(E_1) = 1/11 \qquad P(\overline{E}_1) = 10/11$$
$$P(E_2) = 3/11 \qquad P(\overline{E}_2) = 8/11$$
$$P(E_3) = 6/11 \qquad P(\overline{E}_3) = 5/11$$

A fundamental result relating the probability of the union and intersection of two events E_1 and E_2 has been derived.

PROPERTY 2.2

If E_1 and E_2 are two events from a random experiment, then

$$P(E_1 + E_2) = P(E_1) + P(E_2) - P(E_1 \bullet E_2).$$

Example 2.9 Suppose the outcomes of the experiment discussed in Example 2.5 are equally likely. Consider the events defined in Example 2.7. Then

$$\begin{aligned} P(E_1+E_2) &= P(E_1)+P(E_2)-P(E_1 \bullet E_2) \\ &= 1/11+3/11-0 = 4/11 \\ P(E_2+E_3) &= P(E_2)+P(E_3)-P(E_2 \bullet E_3) \\ &= 3/11+6/11-1/11 = 8/11 \\ P(E_1+\overline{E}_1) &= P(E_1)+P(\overline{E}_1)-P(E_1 \bullet \overline{E}_1) \\ &= 1/11+10/11-0 = 1 \end{aligned}$$

Note that in this special case (equally likely outcomes), these probabilities could be calculated without recourse to the formula of Property 2.2 simply by enumerating the outcomes contained in each event and applying Definition 2.5 directly. Property 2.2 is most useful for random experiments that do not have equally likely outcomes.

2.3 Conditional Probability, Independent Events

At times one has certain prior knowledge concerning the outcome of a random experiment. This prior knowledge may make a given event either more or less likely. Of course, complete knowledge of the outcome is not possible, otherwise the probability of any event containing that outcome is 1, while the probability of any event not containing that outcome is 0. We wish to consider the case where there is partial knowledge beforehand. An example may help clarify this notion.

Example 2.10 Suppose that the two dice are thrown. Before you learn the exact outcome, someone tells you that the sum is even. This knowledge obviously affects the probability that the exact sum is 2 or that the exact sum is 3. In fact, if the dice are fair (all outcomes are equally likely), the probability that the sum is 2 is 1/6 with the knowledge that the sum is even, 1 / 11 without this knowledge. The probability that the sum is 3 is 0 with the knowledge that the sum is even, 1 / 11 without this knowledge.

When advance knowledge affects the probability that a given event has occurred, this knowledge is said to have conditioned the probability of that event.

DEFINITION 2.6
If E_1 and E_2 are two events from a random experiment, then the probability that event E_1 has occurred given that event E_2 has occurred is called the ***conditional probability*** *of E_1 given E_2. This probability is denoted $P(E_1 | E_2)$, and may be calculated as*

$$P(E_1 \mid E_2) = P(E_1 \cdot E_2)/P(E_2).$$

This definition is quite intuitive when E_2 is viewed as having conditioned (reduced) the event space. That is, once it is known that event E_2 has occurred, only a reduced subset of the original event space is possible. The probability of E_1 must now be calculated from the reduced event space.

Example 2.11 Consider the random experiment of drawing a card from an ordinary, well–shuffled deck of 52 cards. Suppose that after the draw you are told that the card is red. What is the possibility that it is the two of hearts?

Define event E_1 as the event that the card is the two of hearts, and event E_2 as the event that the card is red. Then $E_1 \bullet E_2$ is the event that the card is both red and the two of hearts, the event that the card is the two of hearts. Then

$$P(E_1 | E_2) = \frac{P(E_1 \bullet E_2)}{P(E_2)} = \frac{1/52}{1/2}$$
$$= 4/52 = 1/26$$

In this example the knowledge that the card was red conditioned the event space so that it contained only 26 instead of 52 equally likely outcomes. The event that the two of hearts was drawn was then just one of these 26 equally likely simple random events.

Sometimes the knowledge that one event has occurred does not affect the probability that another event has occurred. In this case the events are said to be independent.

DEFINITION 2.7

Two events E_1 and E_2 are said to be ***independent*** *if*

$$P(E_1 \mid E_2) = P(E_1)$$

Combining this definition with Property 2.2 gives the following property.

PROPERTY 2.3

If E_1 and E_2 are independent events, then

$$P(E_1 \bullet E_2) = P(E_1) \bullet P(E_2)$$

Example 2.12 Suppose that a box contains eight black balls and two white balls. A random experiment is conducted in which two balls are drawn from the box. Define E_1 as the event that a black ball is drawn on the first draw and E_2 as the event that a black ball is drawn on the second draw.

If the first ball drawn is replaced before the second one is drawn, the two events are independent. The probability of drawing two consecutive black balls is then

$$P(E_1 \bullet E_2) = P(E_1) \bullet P(E_2) = (8/10)(8/10) = 64/100$$

If the first ball is not replaced before the second is drawn, the events are not independent. Hence the probability of drawing two consecutive black balls is

$$P(E_1 \bullet E_2) = P(E_2 | E_1) \bullet P(E_1) = (7/9)(8/10) = 56/90$$

For the convenience of the reader, we have summarized some of the more common properties about probabilities in Table 2.1.

Table 2.1 Common Probability Features

1	Probability of an event E_1 cannot be negative	$P(E_1) \geq 0$
2	Probability of an event E_1 cannot be greater than one	$P(E_1) \leq 1$
3	The probability of all outcomes is one, and the probability of the null set is zero	$P(\Omega) = 1$ $P(\varnothing) = 0$
4	The probability of the union of two events	$P(E_1 + E_2) = P(E_1) + P(E_2) - P(E_1 \bullet E_2)$
5	The conditional probability (of event E_1, given that E_2 has occurred)	$P(E_1 \mid E_2) = P(E_1 \bullet E_2) / P(E_2)$
6	Events E_1 and E_2 are independent if	$P(E_1 \mid E_2) = P(E_1)$ $P(E_2 \mid E_1) = P(E_2)$
7	If events E_1 and E_2 are independent, then	$P(E_1 \bullet E_2) = P(E_1)P(E_2)$
8	Addition Theorem When E_1 and E_2 are mutually exclusive	$P(E_1 + E_2) = P(E_1) + P(E_2) - P(E_1 \bullet E_2)$ $P(E_1 + E_2) = P(E_1) + P(E_2)$
9	Multiplication Theorem If E_1 and E_2 are independent	$P(E_1 \bullet E_2) = P(E_2) + P(E_1 \mid E_2)$ $P(E_1 \bullet E_2) = P(E_1) \bullet P(E_2)$
10	Addition of Probabilities Theorem If "p" of one of "n" mutually exclusive events to occur is equal to the sum of their separate probabilities	$p(x_1 + x_2 + \ldots + x_n) =$ $p(x_1) + p(x_2) + \ldots + p(x_n)$ $= \sum_{i=1}^{n} p(x_i)$
11	Multiplication of Probabilities Theorem "p" for stochastically independent events to occur together is equal to the product of "p" of each occurrence	$p(x_1 \bullet x_2 \bullet \ldots \bullet x_n) =$ $p(x_1) \bullet p(x_2) \bullet \ldots \bullet p(x_n)$ $= \prod_{i=1}^{n} p(x_i)$

Table 2.1 Common Probability Features (continued)

12	Combinations – the number of comb–inations of "*n*" things taken "*k*" at a time r! = r(r–1)(r–2) . . . (2)(1) 0! = 1	$C_{n,k} = \binom{n}{k} = \frac{n!}{k!(n-k)!}$
13	Permutations – the number of perm–utations of "*n*" things taken "*k*" at a time	$P_{n,k} = \frac{n!}{(n-k)!}$

2.4 Discrete Distributions

As defined earlier, a **random variable** is a function that associates a numerical value with each simple random event (outcome) of a random experiment. The random variable is either discrete or continuous depending on the type of value assigned to the outcomes. If a random variable assumes a discrete number (finite or countably infinite) of values, it is called a **discrete random variable.** Otherwise it is called a **continuous random variable.** Examples of discrete random variables include the number of parts processed by a machine in a day, the number of defects discovered by an inspection process, the number of items in inventory at a given point in time, and the number of absentees on a given shift. Examples of continuous random variables include the time required to perform a task, the time an item spends in inventory, the dimension of a finished part, and the average daily throughput on a computer system.

Example 2.13 The value of the sum of two thrown dice varies discretely between 2 and 12. The random variable that assigns this sum to each outcome is a **discrete random variable.**

Example 2.14 The time t between failures of a particular electronic component varies continuously, $0 < t < \infty$. The random variable assigning this time to each failure is a **continuous random variable.**

When a frequency or probability function is associated with a discrete random variable, one is said to have defined a **discrete probability distribution**. This distribution tells how the outcomes of a random experiment (reflected in the values assumed by the random variable) are distributed. In a simulation of a discrete stochastic system, many of the inputs and the outputs will be represented by distributions of values rather than by single numbers.

Example 2.15 Consider the random experiment of tossing a single die. Define **X** as the random variable that "counts" the spots of the side facing up. Then **X** can assume the values 1, 2, 3, 4, 5, or 6. Assume that the die is loaded so that the probability that a given face lands up is proportional to the number of spots on the face. The discrete probability distribution for this random experiment is

x	1	2	3	4	5	6
$P(\mathbf{X} = x)$	1 / 21	2 / 21	3 / 21	4 / 21	5 / 21	6 / 21

This distribution can be depicted graphically as in Figure 2.2.

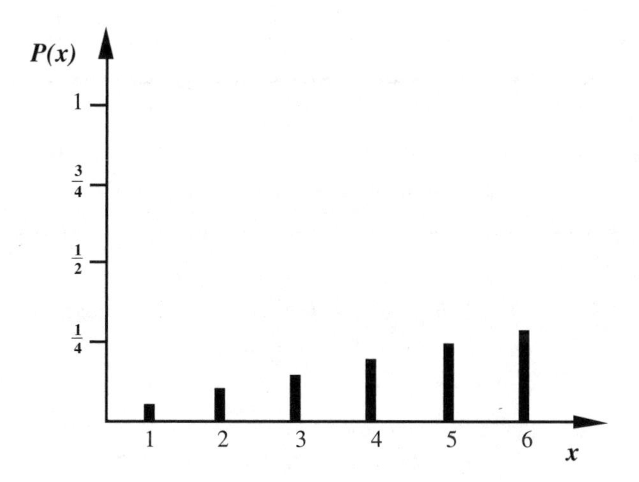

Figure 2.2. Probability function for the loaded die problem.

The above example illustrates some properties that every discrete probability function must have.

PROPERTY 2.4

*Let **X** be a discrete random variable that takes the values $x_1, x_2, \ldots, x_n$, and let P be the probability function, with $P(x_i) = P(\mathbf{X} = x_i)$. Then*

1. $P(x_i) \geq 0, \text{for all } x_i, i = 1, 2, \ldots, n$
2. $\sum_{i=1}^{n} P(x_i) = 1$

Another useful function is the **cumulative distribution function (cdf)**. This function, normally denoted $F(x)$, measures the probability that the random variable **X** assumes a value less than or equal to x. That is, $F(x) = P(\mathbf{X} \leq x)$.

Example 2.16 Consider the experiment described in Example 2.15. The cumulative distribution is

x	1	2	3	4	5	6
$F(x)$	1 / 21	3 / 21	6 / 21	10 / 21	15 / 21	21 / 21

Depicted graphically, this function appears as in Figure 2.3. This sketch points out some properties that any cumulative distribution function must have.

PROPERTY 2.5

*Let **X** be a discrete random variable, and let F be the associated **cumulative distribution function**, with $F(\infty) = P(\mathbf{X} \leq x)$. Then*

1. $0 \leq F(x) \leq 1, \ -\infty < x < \infty$
2. If $x_1 \leq x_2$, then $F(x_1) \leq F(x_2)$. That is, F is nondecreasing.
3. $\lim_{x \to \infty} F(x) = F(\infty) = 1, \quad \lim_{x \to -\infty} F(x) = F(-\infty) = 0$

Note that the probability function P is defined only for the values that **X** can assume, while F is defined for all values of x. The jumps in the step function F occur at values in the range of **X** that have nonzero probabilities.

One way of classifying a discrete probability distribution is by its **expectation**, or long–run value.

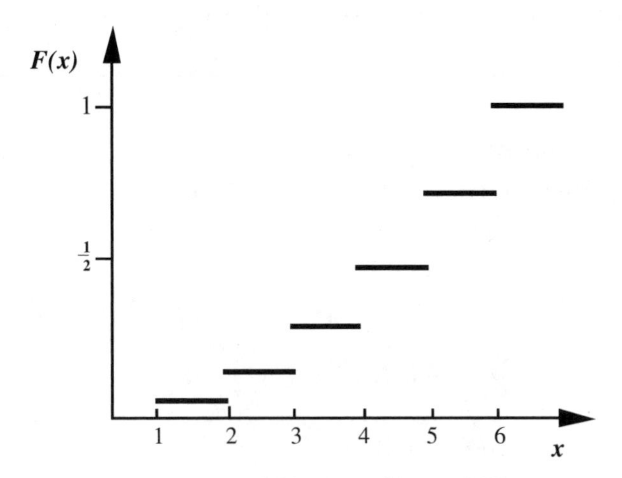

Figure 2.3. Distribution function for the loaded die problem.

DEFINITION 2.8

Let **X** *denote a discrete random variable that takes the values* $x_1, x_2, \ldots, x_n$, *and let P denote the associated probability function. The* ***expectation****, or* ***expected value****, of the random variable, denoted E(***X***) is given by*

$$E(\mathbf{X}) = x_1 P(x_1) + x_2 P(x_2) + \ldots + x_n P(x_n)$$
$$= \sum_{i=1}^{n} x_i P(x_i)$$

Example 2.17 Consider the experiment discusses in Example 2.15. The expectation of the random variable **X** is

$$E(\mathbf{X}) = (1)(1/21) + (2)(2/21) + (3)(3/21) + (4)(4/21) + (5)(5/21) + (6)(6/21)$$
$$= 1/21 + 4/21 + 9/21 + 16/21 + 25/21 + 36/21$$
$$= 91/21 \doteq 4.33$$

One would expect the long–run average to be 4.33. Obviously the die would never show 4.33, but if the experiment is conducted a large number of times and the outcomes averaged, this computed average should approach the expectation.

Another way of classifying a discrete random variable is through specification of its variance. This quantity is a **measure of dispersion**; it indicates how the random variable is dispersed about its expected value.

DEFINITION 2.9

*Let **X** denote a discrete random variable takes the values $x_1, x_2, \ldots, x_n$, P the associated probability function, and E the expectation operator. The **variance** of **X**, denoted V(**X**) is given by*

$$V(\mathbf{X}) = (x_1 - E(\mathbf{X}))^2 P(x_1) + \ \ldots \ + (x_n - E(\mathbf{X}))^2 P(x_n)$$
$$= \sum_{i=1}^{n} (x_i - E(\mathbf{X}))^2 P(x_i)$$
$$= E[(\mathbf{X} - E(\mathbf{X}))^2]$$

Example 2.18 Consider the experiment discussed in Example 2.15. The expectation of the random variable **X** was calculated in Example 2.17 as $E(\mathbf{x}) = 91/21$. The variance of **X** is

$$\begin{aligned} V(\mathbf{X}) &= (1-91/21)^2(1/21)+(2-91/21)^2(2/21)+(3-91/21)^2(3/21) \\ &\quad +(4-91/21)^2(4/21)+(5-91/21)^2(5/21)+(6-91/21)^2(6/21) \\ &= (-70/21)^2(1/21)+(-49/21)^2(2/21)+(-28/21)^2(3/21)+(-7/21)^2(4/21) \\ &\quad +(14/21)^2(5/21)+(35/21)^2(6/21) \\ &\doteq 0.5291+0.5185+0.2540+0.0212+0.1058+0.7936 \\ &\doteq 2.2222 \end{aligned}$$

The **variance**, simply stated, measures the average squared deviation from the mean (expected value). Along with the expectation, it provides a convenient way of summarizing the important aspects of a probability distribution.

As is apparent from the definition, the variance is in terms of squared units. Because this proves unwieldy in some instances, the standard deviation has been developed.

DEFINITION 2.10

Let $\mathbf{X}$ *denote a discrete random variable and V(X) its variance. The* ***standard deviation*** *of* $\mathbf{X}$, *denoted S(*$\mathbf{X}$*) is given by*

$$S(\mathbf{X}) = [V(\mathbf{X})]^{1/2}.$$

The computation of $V(\mathbf{X})$ and hence of $S(\mathbf{X})$ is made somewhat simpler by using the linearity of the expectation operator E.

$$\begin{aligned} V(\mathbf{X}) &= E[(\mathbf{X} - E(\mathbf{X}))^2] \\ &= E[\mathbf{X}^2 - 2\mathbf{X}E(\mathbf{X}) + (E(\mathbf{X}))^2] \\ &= E[\mathbf{X}^2] - 2E(\mathbf{X})E(\mathbf{X}) + (E(\mathbf{X}))^2 \\ &= E[\mathbf{X}^2] - (E(\mathbf{X}))^2 \end{aligned}$$

Example 2.19 The variance in Example 2.18 can be alternately calculated as follows

$$V(\mathbf{X}) = E(\mathbf{X}^2) - [E(\mathbf{X})]^2$$

$$= \sum_{i=1}^{6} x_i^2 P(x_i) - [E(\mathbf{X})]^2$$

$$= [(1)(1/21) + (4)(2/21) + (9)(3/21) + (16)(4/21) + (25)(5/21) + (36)(6/21)] - (91/21)^2]$$

$$= \left[\frac{1+8+27+64+125+216}{21}\right] - \frac{8281}{441}$$

$$= \frac{441}{21} - \frac{8281}{441} = 21 - 18.7777$$

$$= 2.2222$$

The standard deviation of $\mathbf{X}$ is then $S(\mathbf{X}) = [V(\mathbf{X})]^{1/2} = [2.2222]^{1/2} = 1.4907$.

The expected value $E(\mathbf{X})$ is normally denoted by the Greek letter μ, while the standard deviation is normally denoted by the Greek letter σ. The importance of these summary measures in characterizing a distribution will become apparent later.

2.5 Continuous Distributions

A **continuous random variable** is a random variable that assumes a continuum of values.

Example 2.20 A manufacturer of a given electronic component is trying to determine the length of the appropriate warranty to offer customers. The variable of interest is the time before failure of the component. Since this variable can theoretically assume any value greater than 0, it is a continuous random variable.

Since a continuous random variable can assume an uncountably infinite number of values, calculation of the probabilities by enumeration, as was done in the discrete case, is not possible. In fact, when working with continuous random variables, one encounters the seeming contradiction that the probability that the random variable assumes at any particular value in its range is 0, while the probability that it assumes some value in its range is 1. Thus to gain any measure of probability, one must resort to the continuous analog of the cumulative distribution function, given by

$$F(x) = P(\mathbf{X} \leq x).$$

PROPERTY 2.6

The ***cumulative distribution function (cdf)*** *of a continuous random variable* $\mathbf{X}$ *has the following properties.*

1. *$F(x)$ is continuous.*
2. *$F'(x)$ exists except at most a finite number of points.*
3. *$F'(x)$ is continuous, at least piecewise.*

Without getting too deeply into the mathematical aspects of these properties, one can observe that they imply that the range of $\mathbf{X}$ consists of one or more intervals, that the cdf is a smooth curve over each interval, and that the derivative can be integrated over all intervals of interest.

DEFINITION 2.11

The ***probability density function****, denoted f(x), is the derivative of the cumulative distribution function.*

$$f(x) = \frac{d}{dx} F(x)$$

The term **density** refers to the analog of computing the mass in a physical sense. That is, the probability density function gives the distribution of the probability of **X** over its range. Note, however, that the probability density function f evaluated at a point does not give the probability that the random variable **X** will assume the value.

PROPERTY 2.7

The ***probability density function*** *f of a random variable* ***X*** *has the following properties.*

1. $f(x) = 0$ if x is not in the range of **X**.
2. $f(x) \geq 0$.
3. $\int_{-\infty}^{\infty} f(x) dx = 1$.
4. $F(X_1) = \int_{-\infty}^{X_1} f(x) dx = 1$.

Example 2.21 Suppose that the lifetime in months of the electronic component referred to in Example 2.20 was found to have the probability density function

$$\begin{aligned} f(x) &= e^{-x}, && 0 \leq x < \infty \\ &= 0, && \text{otherwise} \end{aligned}$$

This function may be plotted graphically as in Figure 2.4. The corresponding cumulative distribution function appears graphically as in Figure 2.5.

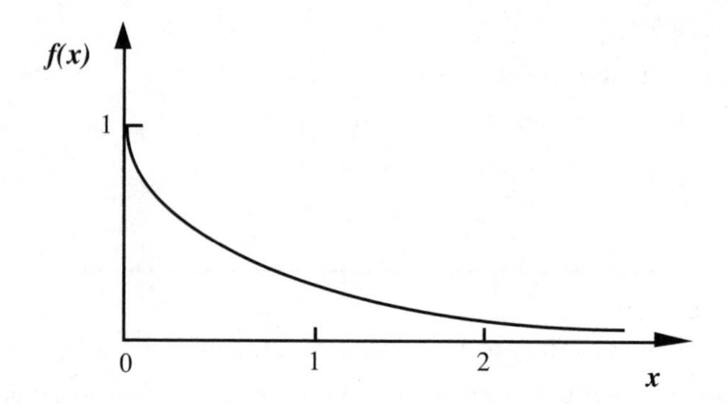

Figure 2.4. Density function for the component lifetime problem.

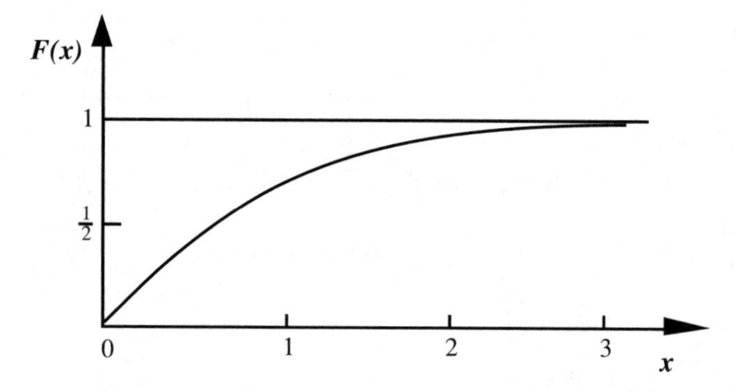

Figure 2.5. Distribution function for the component lifetime problem.

One can compute the probability that the component will fail within ten months as follows.

$$P(\mathbf{X} \leq 10) = F(10) = \int_{-\infty}^{10} f(x)dx$$

$$= \int_{-\infty}^{0} 0dx + \int_{-\infty}^{10} e^{-x}dx$$

$$= -[e^{-x}]_0^{10} = -e^{-10} + 1$$
$$= 1 - 0.000045$$
$$= 0.99995$$

This corresponds to finding the area under the curve of *f(x)* between 0 and 10.

Just as with discrete distributions, a continuous distribution can be summarized using the expectation (mean) and variance.

DEFINITION 2.12

*Let **X** be a continuous random variable, and let f(x) denote its probability density function. The **expectation** of **X**, denoted E(**X**), is given by*

$$E(\mathbf{X}) = \int_{-\infty}^{\infty} x f(x) dx$$

Example 2.22 Consider the experiment described in Examples 2.20 and 2.21. The expected life of the component is

$$\begin{aligned} E(\mathbf{X}) &= \int_{-\infty}^{\infty} x f(x) dx \\ &= \int_{-\infty}^{0} 0 dx + \int_{0}^{\infty} x e^{-x} dx \\ &= -x e^{-x} \Big|_{0}^{\infty} - \int_{0}^{\infty} (-e^{-x}) dx \\ &= e^{-x} \Big|_{0}^{\infty} = 1 \end{aligned}$$

DEFINITION 2.13

*Let **X** be a continuous random variable, and let f(x) denote its probability density function. The **variance** of **X**, denoted V(**X**), is given by*

$$\begin{aligned} V(\mathbf{X}) &= \int_{-\infty}^{\infty} (x - E(\mathbf{X}))^2 f(x) dx \\ &= \int_{-\infty}^{\infty} x^2 f(x) dx - E(\mathbf{X})^2 \end{aligned}$$

*The **standard deviation** S(**X**) is given by*

$$S(\mathbf{X}) = [V(\mathbf{X})]^{1/2}$$

Example 2.23 Consider the experiment described in Examples 2.20–2.22. The variance in lifetime for the electronic component can be computed by

$$\begin{aligned} V(\mathbf{X}) &= \int_{-\infty}^{\infty} (x - E(\mathbf{X}))^2 f(x)dx \\ &= \int_{-\infty}^{\infty} x^2 f(x)dx - E(\mathbf{X})^2 \\ &= \int_{-\infty}^{\infty} x^2 e^{-x} dx - 1 \\ &= [-x^2 e^{-x} - 2xe^{-x} - 2e^{-x}]_0^{\infty} - 1 \\ &= 2 - 1 = 1 \end{aligned}$$

The standard deviation $S(\mathbf{X})$ is given by $S(\mathbf{X}) = [V(\mathbf{X})]^{1/2} = 1^{1/2} = 1$.

Figure 2.6 provides a simple graphical illustration comparing the continuous and discrete probability distributions and cumulative distribution functions.

It is not possible to achieve a continuous distribution in a real–world experiment, since all known measuring devices yield discrete values. However, when a discrete random variable is measured at enough points that it begins to take on the appearance of a continuous random variable, it is convenient to approximate the discrete distribution by a continuous one. In that way one can use the tools of calculus to calculate probabilities, means, and variances rather than tediously sum up a large number of fractional numbers.

CONTINUOUS

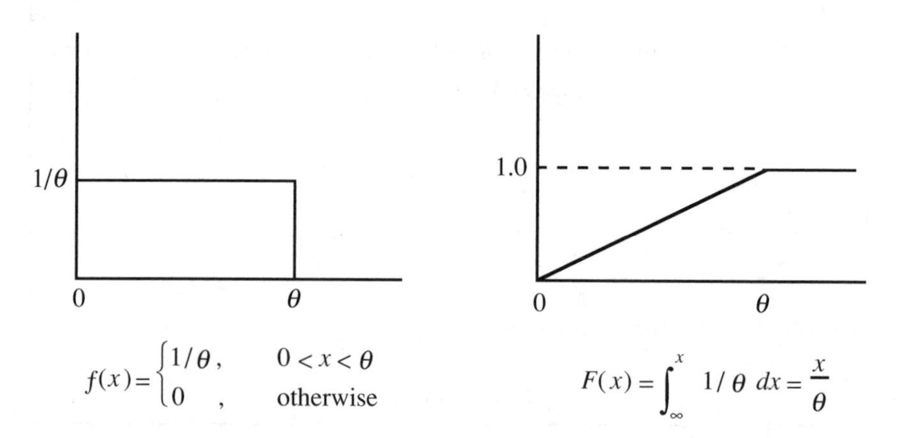

Figure 2.6. Graphical illustration of probability distributions.

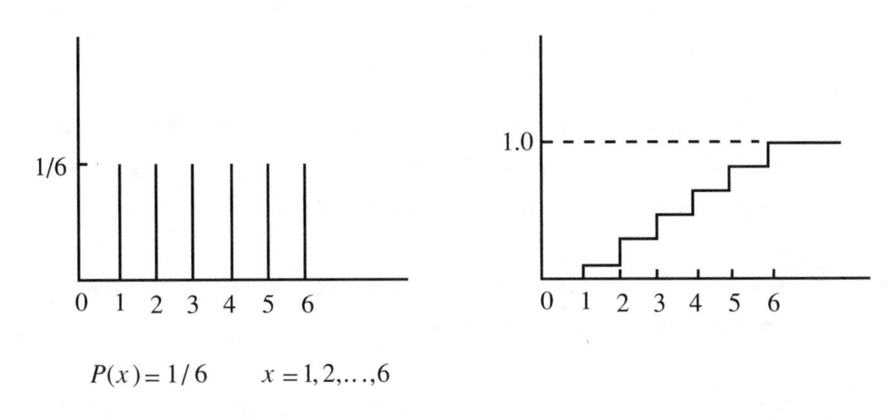

Figure 2.6. Graphical illustration of probability distributions (continued).

2.6 Summary

In this chapter we have surveyed some of the fundamental aspects of probability and probability models. This material, along with statistical inference, random number generation, and queueing theory is fundamental to the simulation of systems. More detail on these concepts can be found in any of the many excellent references on probability.

2.7 Exercises

2.1 Three fair dice are tossed. Define the random variable **X** to denote the sum of the spots showing on all three dice. Determine the probability distribution, the expected value, and the variance of **X**.

2.2 A box contains six balls – three red, two black, and one white.

a. Determine the probability that the white ball is drawn on a single draw.

b. Determine the probability that the second ball drawn is black given that the first ball was white and that it was not replaced.
c. Rework part assuming that the white ball was replaced.

2.3 Let the random variable $\mathbf{X}$ count the number of boys in a family of six children. Assuming equal probabilities for boys and girls, determine the distribution, expected value, and variance of $\mathbf{X}$.

2.4 Determine the probability that a student guesses seven of ten questions correctly on a true–false examination.

2.5 If the probability of a head on a single toss of a fair coin is 0.3, determine the expected number of tosses until the first tail appears.

2.6 The average number of calls to a switchboard is 30 per hour. Determine the probability that 10 calls will come in during the next 15 minutes.

2.7 A random variable $\mathbf{X}$ is distributed normally with mean 10 variance 1. Determine the probability that an observation will

a. One standard deviation of the mean.
b. Two standard deviation of the mean
c. Three standard deviations of the mean

2.8 A random variable $\mathbf{X}$ has mean 2 and variance 3. Find the mean and variance of

a. $\mathbf{Y} = 2\mathbf{X} + 3$
b. $\mathbf{Z} = \mathbf{X} - 3$

2.9 Assume that the height of men in the United States is distributed normally with a mean of 70 inches and a standard deviation of 2 inches. Determine the probability that a man selected at random is

a. Exactly 72 inches tall
b. Less than 68 inches tall
c. Between 68 and 72 inches tall

2.10 A random variable **X** has probability density function $f(x) = 2^{-2x}$, $x > 0$. Determine the expected value and standard deviation of **X**.

3

RANDOM VARIABLES AND PROBABILITY DISTRIBUTIONS

3.1 Functions of a Random Variable

In many applications, once a random experiment is conducted and the distributions of the random variable determined, some related quantity must be examined. In such cases it is sometimes convenient to define a new random variable as a function of the original random variable.

Example 3.1 Suppose that a random experiment is conducted in which the quantity of interest is the temperature of a liquid. Let **X** denote the random variable that measures this temperature in degrees centigrade (Celsius), and assume that the distribution of **X** is known. Suppose now that we want to investigate the distribution of the temperature in degrees Fahrenheit. Rather than conduct the experiment again, we can simply modify the original observations using the relationship $\mathbf{Y} = 9/5\ \mathbf{X} + 32$. The new random variable **Y**, then, is a function of the original random variable **X**.

This example illustrates one of the simpler functions of random variables, a linear function.

DEFINITION 3.1

*Let **X** denote a random variable corresponding to the outcomes of some random experiment. Suppose the random variable **Y** is related to **X** by the relation*

$$Y = aX + b,$$

where a and b are arbitrary constants. Then ***Y*** *is said to be a* ***linear function of X***.

If the distributions of **X** is known, it is quite straightforward to determine the distribution of **Y**. Each observation of **X** is multiplied by the constant a and then added to the constant b. The corresponding probabilities remain the same.

In Chapter 2 we saw that the expected value of a discrete variate **X** was defined as the average value of the variate over "all" possible distinct values x_i, $i = 1, 2, 3, \ldots, N \leq \infty$.

If we let $f(x_i)$ represent the relative frequency that each value of x_i will occur, then

$$E(\mathbf{X}) = \sum_{i=1}^{N} x_i f(x_i)$$

for the discrete case, and

$$E(\mathbf{X}) = \int_{-\infty}^{\infty} x f(x) dx$$

for the continuous case.

It is also quite easy to calculate the **expected value and variance for simple linear functions of random variables**. For example, suppose that **X** is a continuous random variable corresponding to the outcomes of a random experiment and that $\mathbf{Y} = a\mathbf{X} + b$. Then

$$
\begin{aligned}
E(\mathbf{Y}) &= E(a\mathbf{X}+b) \\
&= \int_{-\infty}^{\infty}(ax+b)f(x)dx \\
&= a\int_{-\infty}^{\infty}xf(x)dx + b\int_{-\infty}^{\infty}f(x)dx
\end{aligned}
$$

$$
\boxed{E(\mathbf{Y}) = aE(\mathbf{X})+b}
$$

Similarly

$$
\begin{aligned}
V(\mathbf{Y}) &= V(a\mathbf{X}+b) \\
&= \int_{-\infty}^{\infty}(ax+b)^2 f(x)dx - (E(a\mathbf{X}+b))^2 \\
&= \int_{-\infty}^{\infty}a^2x^2 f(x)dx + 2ab\int_{-\infty}^{\infty}xf(x)dx + b^2\int_{-\infty}^{\infty}f(x)dx - (E(a\mathbf{X}+b))^2 \\
&= \int_{-\infty}^{\infty}a^2x^2 f(x)dx + 2abE(\mathbf{X}) + b^2 - (aE(\mathbf{X})+b)^2 \\
&= a^2\left[\int_{-\infty}^{\infty}x^2 f(x)dx - (E(\mathbf{X}))^2\right]
\end{aligned}
$$

$$
\boxed{V(\mathbf{Y}) = a^2V(\mathbf{X})}
$$

These results may seem surprising, but they are really quite intuitive. For example, the additive constant b has no effect on the variance. This makes sense intuitively; the addition of a constant shifts all observations uniformly and thus does not affect their relative positions with respect to the mean. These results also hold for discrete random variables, with substitution of the summation sign Σ for the integral sign $\int$, and the probability function P for the density function f.

Example 3.2 Suppose that for the random experiment described in Example 3.1, the random variable **X** was distributed with mean (expected value) equal to 30 degrees and variance equal to 64 degrees squared. Then

$$
\begin{aligned}
E(\mathbf{Y}) &= (9/5)E(\mathbf{X})+32 \\
&= (9/5)(30)+32 \\
&= 86 \text{ degrees}
\end{aligned}
$$

$$
\begin{aligned}
V(\mathbf{Y}) &= (9/5)^2 V(\mathbf{X}) \\
&= (3.24)(64) = 207.36 \text{ degrees squared}
\end{aligned}
$$

The standard deviation of **Y** is then

$$S(\mathbf{Y}) = [207.36]^{1/2} = 14.4 \text{ degrees}$$

Using the technique illustrated for a simple linear function, one can calculate the **expected value of any arbitrary function** of a random variable. If $\mathbf{Y} = g(\mathbf{X})$, then

$$E(\mathbf{Y}) = E(g(\mathbf{X})) = \sum_i g(x_i)P(x_i) \quad \text{in the discrete case}$$

or

$$E(\mathbf{Y}) = E(g(\mathbf{X})) = \int_{-\infty}^{\infty} g(x)f(x)dx \quad \text{in the continuous case}$$

and

$$V(\mathbf{Y}) = V(g(\mathbf{X})) = \sum_i (g(x_i) - E(g(x_i)))^2 P(x_i) \quad \text{in the discrete case}$$

or

$$V(\mathbf{Y}) = V(g(\mathbf{X})) = \int_{-\infty}^{\infty} (g(x) - E(g(x\ ))^2 f(x)dx \quad \text{in the continuous case}$$

For the convenience of the reader we have summarized some common properties of expectations in Table 3.1.

Table 3.1 Common Properties of Expectations

1	The expected value of a constant	$E(c) = c$
2	The expected value of a constant times an arbitrary function of a random variable	$E[c \cdot g(\mathbf{X})] = c \cdot E[g(\mathbf{X})]$
3	The expected value of the sum of two arbitrary functions of a random variable	$E[g_1(\mathbf{X}) + g_2(\mathbf{X})] =$ $E[g_1(\mathbf{X})] + E[g_2(\mathbf{X})]$
4	Ordering of expected variables	$E[g_1(\mathbf{X})] \leq E[g_2(\mathbf{X})]$ if $g_1(\mathbf{X}) \leq g_2(\mathbf{X})$ for all values of $\mathbf{X}$
5	The expected value of the product of two functions of a random variable	$E[g_1(\mathbf{X}) \cdot g_2(\mathbf{Y})] =$ $E[g_1(\mathbf{X})] \cdot E[g_2(\mathbf{Y})]$ only if $\mathbf{X}$ and $\mathbf{Y}$ are independent

3.2 Moments

The previous section introduced the notion of a function of the random variable **X**. One simple function whose expectation is of interest is the **power function** $g(\mathbf{X}) = \mathbf{X}^r$, for $r = 1, 2, 3, \ldots$.

DEFINITION 3.2
The expectation of the power function $g(\mathbf{X}) = \mathbf{X}^r$, $r = 1,2,3,\ldots$, is called the ***rth moment of the random variable X*** *taken about zero and is denoted μ_r.*

Applying the definition of expectation, one can calculate the rth moment of a random variable **X** taken about zero as

$$E[g(x)] = E(x^r) = \mu_r = \sum_{i} x_i^r P(x_i) \quad \text{if } \mathbf{X} \text{ is discrete}$$

or

$$E[g(x)] = E(x^r) = \mu_r = \int_{-\infty}^{\infty} x^r f(x)dx \quad \text{if } \mathbf{X} \text{ is continuous}$$

Moments can also be calculated **about a point other than zero**, say some constant c. The computational formula about a constant c is

$$E[g(x)] = E(x^r) = \mu_r' = \sum_i (x_i - c)^r P(x_i) \quad \text{if } \mathbf{X} \text{ is discrete}$$

or

$$E[g(x)] = E(x^r) = \mu_r' = \int_{-\infty}^{\infty} (x - c)^r f(x)dx \quad \text{if } \mathbf{X} \text{ is continuous}$$

Furthermore, by letting $g(x) = (x - \mu_1')^r$, the **moments of** x **about the mean**, become

$$E[g(x)] = E[(x - \mu_1')^r] = \mu_1' = \sum_i (x_i - \mu_1')^r P(x_i) \quad \text{if } \mathbf{X} \text{ is discrete}$$

or

$$E[g(x)] = E[(x - \mu_1')^r] = \mu_1' = \int_{-\infty}^{\infty} (x - \mu_1')^r f(x)dx \quad \text{if } \mathbf{X} \text{ is continuous}$$

These quantities are called moments because of their obvious analogy to moments of inertia in physics. Some of these moments have significance in summarizing various characteristics of the distribution.

First moment. The first moment about zero simply gives the expected value or mean of the distribution. It is denoted by the Greek letter μ and serves as a

measure of the central tendency or **location of the distribution**. It is computed as follows.

$$E(x) \equiv \mu_1' = \mu = \sum_i x_i P(x_i) \quad \text{if } \mathbf{X} \text{ is discrete}$$

or

$$E(x) \equiv \mu_1' = \mu = \int_{-\infty}^{\infty} x f(x) dx \quad \text{if } \mathbf{X} \text{ is continuous}$$

Second moment. The second moment about the mean gives the variance of the distribution. It is denoted by σ^2 and gives a **measure of dispersion** or the **spread of the distribution**. The square root of the variance is called the **standard deviation (root mean square)**. The variance (**moment of dispersion**) is calculated as follows.

$$E[(x-\mu_1')^2] \equiv \mu_2 = \sigma^2 = \sum_i (x_i - \mu)^2 P(x_i) \quad \text{if } \mathbf{X} \text{ is discrete}$$

or

$$E[(x-\mu_1')^2] \equiv \mu_2 = \sigma^2 = \int_{-\infty}^{\infty} (x-\mu)^2 f(x) dx \quad \text{if } \mathbf{X} \text{ is continuous}$$

$$\mu_2 = \int_{-\infty}^{\infty} x^2 f(x) dx - (\mu_1')^2 = \mu_2' - (\mu_1')^2 = E(\mu_2')^2 - (\mu_1')^2$$

Associated with the variance is the **standard deviation** σ, defined by

$$\sigma = \sqrt{\mu_2}$$

Third moment. The third moment about the mean is a **measure of the symmetry** of the distribution. If this is positive, the distribution is asymmetrical and positively skewed (the peak of the distribution is to the left of the mean). If it is zero, the distribution is symmetric. If it is negative, the distribution is asymmetric and negatively skewed. These three cases are depicted in Figure 3.1. The third moment about the mean is calculated by

$$E\left[(x-\mu_1')^3\right] \equiv \mu_3' = \sum_i (x_i-\mu)^3 P(x_i) \quad \text{if } \mathbf{X} \text{ is discrete}$$

or

$$E\left[(x-\mu_1')^3\right] \equiv \mu_3' = \int_{-\infty}^{\infty} (x-\mu)^3 f(x)dx \quad \text{if } \mathbf{X} \text{ is continuous}$$

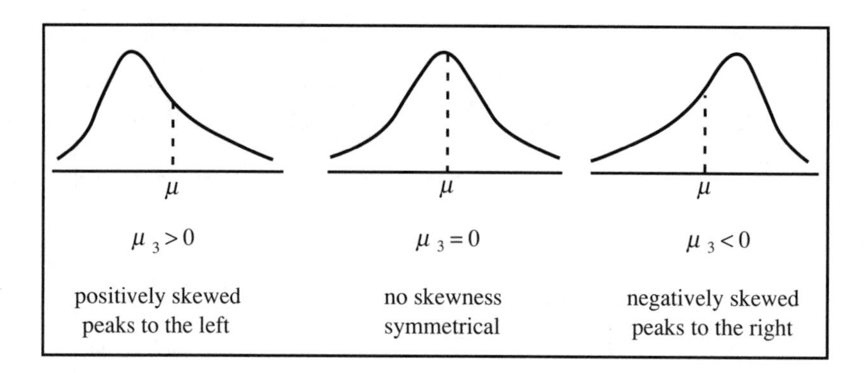

Figure 3.1. Illustration of the third moment about the mean.

From the definition of skewness, it can be seen that scaling the random variable **X** to be β **X** changes the skewness by the factor β^3. To avoid this problem, the **coefficient of skewness**

$$\gamma_1 = \frac{\mu_3}{\sigma^3} = \frac{E[(x-\mu')^3]}{(\mu)^{3/2}}$$

is often used instead of μ_3. It is the skewness of $(x - \mu_1') / \sigma$ and thus a location and scale invariant measure.

Fourth moment. The fourth moment about the mean is a **measure of kurtosis** or the **flatness** or **peakedness of the distribution**. It is calculated by

$$E\left[(x - \mu_1')^4\right] \equiv \mu_4' = \sum_i (x_i - \mu)^4 P(x_i) \quad \text{if } \mathbf{X} \text{ is discrete}$$

or

$$E\left[(x - \mu_1')^4\right] \equiv \mu_4' = \int_{-\infty}^{\infty} (x - \mu)^4 f(x)dx \quad \text{if } \mathbf{X} \text{ is continuous}$$

If $\sigma = 1$ and the fourth moment is greater than 3, the frequency curve is tall and thin (called ***leptokurtic***). When the fourth moment is equal to 3 the frequency curve is medium or ***mesokurtic***. When the fourth moment is less than 3 the frequency curve is flat or ***platykurtic***. Example shapes of these curves are illustrated in Figure 3.2.

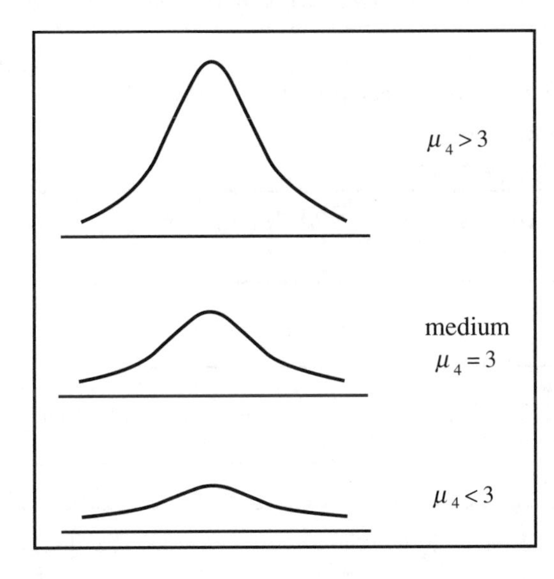

Figure 3.2. Example frequency curves for the fourth moment.

As with skewness, we define the **coefficient of kurtosis** to be a location and scale invariant measure. Thus we define

$$\gamma_2 = \frac{\mu_4}{\sigma^4} - 3$$

Dividing by σ^4 makes the quantity unitless, and subtracting 3 results in γ_2 being equal to 0 for a normal random variable.

3.3 Generating Functions

Another function of the random variable **X** which is of interest is the function $g(\mathbf{X}) = e^{\mathbf{X}\theta}$, where θ is some variable.

DEFINITION 3.3
The function $M(\theta) = E[e^{X\theta}]$, where E is the expectation operator, is termed (if it exists) the ***moment–generating function*** *of the random variable* ***X***.

Using the concept of expectation developed earlier, one can calculate the moment–generating function by

$$M(\theta) = E[e^{\mathbf{X}\theta}] = \sum_i e^{x_i\theta} P(x_i) \quad \text{if } \mathbf{X} \text{ is discrete}$$

or

$$M(\theta) = E[e^{\mathbf{X}\theta}] = \int_{-\infty}^{\infty} e^{\mathbf{X}\theta} f(x)dx \quad \text{if } \mathbf{X} \text{ is continuous}$$

Example 3.3 Suppose **X** is a continuous random variable that has the probability density function $f(x) = e^{-x}$, $0 < x < \infty$ and 0 elsewhere (see Example 2.21). Then the moment–generating function is

$$M(\theta) = E[e^{\mathbf{X}\theta}] = \int_{-\infty}^{\infty} e^{x\theta} \cdot e^{-x} dx$$

$$= \int_{0}^{\infty} e^{-(1-\theta)x} = \frac{1}{1-\theta}, \quad \theta \neq 1$$

The function M is called the **moment–generating function** because all moments of **X** can be readily calculated from it.

In particular, $e^{\mathbf{X}\theta}$ can be expanded in an infinite series as

$$e^{\mathbf{X}\theta} = 1 + \mathbf{X}\theta + \frac{\mathbf{X}^2\theta^2}{2!} + ... + \frac{\mathbf{X}^n\theta^n}{n!} + ...$$

Then applying the expectation operator gives

$$M(\theta) = E[e^{\mathbf{X}\theta}] = 1 + E(\mathbf{X})\theta + \frac{E(\mathbf{X}^2)\theta^2}{2!} + ... + \frac{E(\mathbf{X}^n)\theta^n}{n!} + ...$$

$$= 1 + \mu_1\theta + \frac{\mu_2\theta^2}{2!} + ... + \frac{\mu_n\theta^n}{n!} + ...$$

Thus the moments of **X** can be calculated by expanding $M(\theta)$ in an infinite series and noting the coefficients of the powers of θ.

Example 3.4 Consider the distribution discussed in Example 3.3.

$$M(\theta) = \frac{1}{1-\theta} = 1 + \theta + \theta^2 + \theta^3 + ... + \theta^n + ...$$

$$= 1 + (1)(\theta) + (2!)\frac{\theta^2}{2!} + ... + (n!)\frac{\theta^n}{n!} + ...$$

Then

$$\mu = \mu_1 = 1, \quad \mu_2 = 2, \quad ... \quad \mu_n = n!$$

Suppose that **X** is an integer–valued discrete random variable with probability function P. Then the moment–generating function can be simplified using the transformation $Z = e^{\theta}$. The resulting function, called the **probability generating function** or the **Z–transform** of **X**, is given by

$$\Psi(z) = E[\mathbf{Z}^x] = \sum_i z^i P(x_i)$$

This function is used to calculate the moments of a distribution and will be useful in developing the steady–state equations for a simple queueing system in Chapter 15.

3.4 Multivariate Distributions

Everything we have done thus far in this chapter has assumed that a simple random variable **X** was defined on the event space of some random experiment. In the simulation of real–world systems it is sometimes convenient or even necessary to define more than one random variable on the event space. We must then consider the **joint distribution** of the random variables. In this section we consider the special case of two or more random variables. Results are easily extended to the case in which more than two or more random variables are defined on the same even space.

If **X** and **Y** are two random variables defined on the same event space, then outcomes of the random experiment consist of ordered pairs (x, y) in two–dimensional space. Events then are formed by combining these two–dimensional outcomes.

Example 3.5 Consider the random variables **X** and **Y** defined on the same event space. If **X** can assume the values 0, 1, 2, 3, 4 and **Y** can assume the values 0 and 1, then the random variables can jointly assume the values

$$\Omega = \{(0,0),(0,1),(1,0),(1,1),(2,0),(2,1),(3,0),(3,1),(4,0),(4,1)\}$$

In the case of discrete random variables a probability function P_{XY} assigns the probability to some event E.

DEFINITION 3.4

For any two discrete random variables **X** *and* **Y** *the* ***joint probability function*** *$P_{XY}(x, y)$ is a function such that*

$$F_{\mathbf{XY}}(x, y) = P(\mathbf{X} = y \text{ and } \mathbf{Y} = y)$$

That is, $P_{\mathbf{XY}}$ gives the probability that the random variable **X** assumes the value x at the same time that **Y** assumes the value y.

The **joint cumulative distribution function** for a pair of random variables can be analogously defined.

DEFINITION 3.5

For any two random variables **X** *and* **Y** *defined on the same event space, the* ***joint distribution function*** *$F_{XY}(x, y)$ is a function defined such that*

$$F_{\mathbf{XY}}(x, y) = P(\mathbf{X} \le y \text{ and } \mathbf{Y} \le y)$$

Note that this function is defined for all values of x and y. Furthermore, this definition holds for continuous as well as discrete random variables.

The probability that a continuous random variable assumes any particular value in its range is 0. This observation carries over the **multivariate distributions**. If **X** and **Y** are continuous random variables defined over the same event space, then $P(\mathbf{X}) = x$ and $\mathbf{Y} = y) = 0$ for any x, y. This leads to development of a **joint density function**.

DEFINITION 3.6

Let **X** *and* **Y** *be two continuous random variables defined on the same event space. The* ***joint density function*** *$f_{XY}(x, y)$ is a function defined by*

$$f_{\mathbf{XY}}(x, y) = \frac{\partial^2 F_{\mathbf{XY}}(x, y)}{\partial x \partial y}$$

whenever this derivative exists. This definition gives a means of calculating the ***cumulative distribution function*** *at a particular point (x, y).*

$$F_{\mathbf{XY}}(x, y) = \int_{-\infty}^{x} \int_{-\infty}^{y} f(u, v)\,du\,dv$$

where u and v are dummy variables.

Although in most cases we are interested in the joint distribution of **X** and **Y**, it is also possible to consider the distribution of either random variable alone. This type of distribution is called the **marginal distribution**.

DEFINITION 3.7

Let ***X*** *and* ***Y*** *be continuous random variables defined on the same event space, with joint density function* f_{XY}*. The* ***marginal density functions*** *for* ***X*** *and* ***Y*** *are given by*

$$f_{\mathbf{X}}(x) = \int_{-\infty}^{\infty} f(x, y)\,dy, \qquad f_{\mathbf{Y}}(y) = \int_{-\infty}^{\infty} f(x, y)\,dx$$

Similar formulas apply if ***X*** *and* ***Y*** *are discrete random variables.*

This definition of the **marginal distributions** allows us to consider what it means for two jointly distributed random variables to be independent.

DEFINITION 3.8

Let ***X*** *and* ***Y*** *be two random variables defined on the same event space. The random variables are* ***said to be independent*** *if*

$$F_{\mathbf{XY}}(x, y) = F_{\mathbf{X}}(x) F_{\mathbf{Y}}(y)$$

The marginal density functions for continuous random variables (probability functions for discrete random variables) allow calculation of the means and variances of the random variables.

DEFINITION 3.9

Let **X** *and* **Y** *be continuous random variables defined on the same event space. Then the mean and variance are respectively*

$$E(\mathbf{X}) = \mu_\mathbf{X} = \int_{-\infty}^{\infty} x f_\mathbf{X}(x)dx, \qquad V(\mathbf{X}) = \int_{-\infty}^{\infty} (x-\mu_\mathbf{X})^2 f_\mathbf{X}(x)dx$$

Similar formulas hold for **Y**, *and when* **X** *and* **Y** *are discrete.*

The variance of a random variable **X** is a measure of dispersion. When we are dealing with joint random variables **X** and **Y**, another quantity of interest is the **covariance**.

DEFINITION 3.10

Let **X** *and* **Y** *be jointly continuous random variables with* ***joint density function*** $f_{XY}(x, y)$. *The* **covariance** *of* **X** *and* **Y**, *denoted* σ_{XY}, *is given by*

$$\begin{aligned}\sigma_\mathbf{XY}(x,y) &= E[(\mathbf{X}-\mu_\mathbf{X})(\mathbf{Y}-\mu_\mathbf{Y})] \\ &= \int_{-\infty}^{\infty}\int_{-\infty}^{\infty}(x-\mu_\mathbf{X})(y-\mu_\mathbf{Y})f_\mathbf{XY}(x,y)dx\,dy\end{aligned}$$

An analogous definition exists for discrete random variables.

The covariance is a measure of the linear association of the random variables **X** and **Y**. In particular, a quantity known as the **correlation** may be defined as

$$\rho_\mathbf{XY} = \sigma_\mathbf{XY} / \sigma_\mathbf{X}\sigma_\mathbf{Y}$$

A value of $\sigma_\mathbf{XY} = 0$ implies no correlation between **X** and **Y**, whereas a correlation of +1 or –1 implies perfect linear correlation.

Of particular importance in statistical analysis is the expectation and the variance of the sums of jointly distributed random variables.

DEFINITION 3.11

Let $\mathbf{X}$ *and* $\mathbf{Y}$ *be jointly distributed random variables, and define* $\mathbf{U} = a_1\mathbf{X} + b_1\mathbf{Y}$. *Then the mean and variance is respectively*

$$E(\mathbf{U}) = a_1\mu_{\mathbf{X}} + b_1\mu_{\mathbf{Y}}$$
$$V(\mathbf{U}) = a_1^2\sigma_{\mathbf{X}}^2 + b_1^2\sigma_{\mathbf{Y}}^2 + 2a_1b_1\sigma_{\mathbf{XY}}$$

If $\sigma_{\mathbf{XY}} = 0$ *(*$\mathbf{X}$ *and* $\mathbf{Y}$ *are uncorrelated), the latter expression simplifies somewhat.*

Note that independent random variables are uncorrelated. The converse of this is not true in general.

In this section we have briefly examined jointly distributed random variables. The treatment was necessarily brief. The importance of these concepts should become clear as we proceed.

3.5 Exercises

3.1 If the joint density function for two continuous random variables is given by

$$f_{\mathbf{XY}}(x,y) = e^{-(x+y)}, \quad x,y > 0$$

find

a. $F_{\mathbf{XY}}(x,y)$
b. $F_{\mathbf{X}}(x)$
c. $F_{\mathbf{Y}}(y)$

3.2 If $\mathbf{X}$ is a chi–square random variable, find

a. $P(\mathbf{X} \geq 26.119)$ if $\upsilon = 14$

b. $P(\mathbf{X} \le 3.841)$ if $\upsilon = 1$

c. $P(\mathbf{X} < 27.587)$ if $\upsilon = 17$

3.3 If **X** is a *t* random variable, find

a. $P(\mathbf{X} > 1.372)$ if $\upsilon = 10$

b. $P(1.333 < \mathbf{X} < 1.740)$ if $\upsilon = 17$

3.4 If X is an *F* random variable, find

a. $P(\mathbf{X} \le 19.5)$ if $\upsilon_1 = 40,\ \upsilon_2 = 2$

b. $P(\mathbf{X} > 200)$ if $\upsilon_1 = 2,\ \upsilon_2 = 1$

3.5 Specifications for the diameter of a shaft are 1 ± 0.1 inches. If the manufacturing process produces shafts normally distributed with a mean of 1 inch and a variance of 0.0025, what is the probability that a shaft chosen at random will meet the specifications?

4

DETAILED ANALYSIS OF COMMON PROBABILITY DISTRIBUTIONS

A number of common distributions have emerged from the study of real–world systems. Although in some cases the outcomes of experiments do not precisely follow these distributions, they are often used as first approximations to the real distribution. In this chapter we will explore in some detail these common distributions and attempt to indicate their utility in the study of real–world systems.

4.1 Bernoulli Distribution

The Bernoulli distribution, a discrete distribution, is one of the simplest and is useful in experiments having only two outcomes. Assume that a random variable **X** has been defined and that it takes the value 0 with probability p and the value 1 with the probability $q = 1 - p$. Then the distribution of **X** is

x	$P(x)$
0	p
1	q

The cumulative distribution function for this random variable is

$$F(x) = \begin{cases} 0, & x < 0 \\ p, & 0 \le x < 1 \\ 1, & x \ge 1 \end{cases}$$

The expected value, variance, and standard deviation are

$$\mu = E(\mathbf{X}) = 0 \cdot p + 1 \cdot q = q$$
$$\sigma^2 = V(\mathbf{X}) = E(\mathbf{X}^2) - [E(\mathbf{X})]^2 = 0 \cdot p + 1 \cdot q - q^2 = q(1-q)$$
$$\sigma = S(\mathbf{X}) = [q(1-q)]^{1/2}$$

Figure 4.1 shows the Bernoulli distribution.

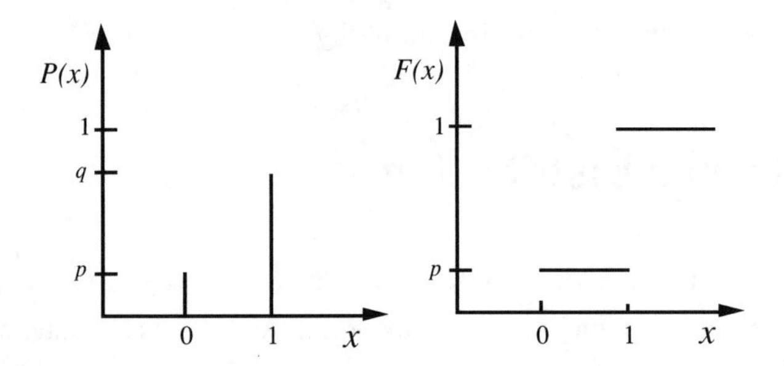

Figure 4.1. The Bernoulli distribution.

Example 4.1 Consider the experiment of tossing a fair coin. Let **X** assume the value 0 if a head appears, 1 if a tail appears. Then $p = q = 1/2$, and

$$\mu = 1/2,\ \sigma^2 = 1/4,\ \sigma = 1/2$$

4.2 Binomial Distribution

The binomial distribution is a discrete distribution that results from a sequence of independent Bernoulli trials. Assume that an experiment that has two outcomes is conducted n times, $n > 0$. Assume also that the probability of one outcome on any trial (call the outcome 0) is p and that the probability of the other outcome (call it 1) is $q = 1 - p$. Then the outcome 0 can occur any number of integral times from 0 to n, as can the outcome 1. The probability that the outcome 0 occurs precisely k times (1 occurs $n - k$ times) is

$$P(\mathbf{X}=k)=\binom{n}{i}p^i(1-p)^{n-i}=\binom{n}{i}p^i(q)^{n-i}, \qquad i=0,1,\ldots,n$$

$$=\frac{n!}{k!(n-k)!}\,p^k q^{n-k}, \qquad k=0,1,\ldots,n$$

The total events is equal to $(p + q)^n$.

The cumulative distribution function is given by

$$F(k)=\sum_{s=0}^{k}\frac{n!}{s!(n-s)!}\,p^s q^{n-s}, \qquad k=0,1,\ldots,n$$

The binomial distribution is illustrated in Figure 4.2 for $n = 5, p = 0.5$.

The expected value and variance are easily calculated by treating the binomial as the sum of n Bernoulli random variables $\mathbf{X}_1, \mathbf{X}_2, \ldots, \mathbf{X}_n$. Then

$$\begin{aligned} m &= \mathrm{E}(\mathbf{X}) = \mathrm{E}[\mathbf{X}_1+\mathbf{X}_2+\ldots+\mathbf{X}_n] \\ &= q+q+\ldots+q = nq \\ \sigma^2 &= V(\mathbf{X}) = V[\mathbf{X}_1+\mathbf{X}_2+\ldots+\mathbf{X}_n] \\ &= pq+pq+\ldots+pq \\ &= npq \end{aligned}$$

The third and fourth moments are

$$\begin{aligned} \mu_3 &= npq(q-p) = np(1-p)(1-2p) \\ \mu_4 &= npq[1+3pq(n-2)] \\ &= 3n^2p^2(1-p)^2 + np(1-p)[1-6p(1-p)] \end{aligned}$$

where the probability of failure, q, is equal to $1 - p$, and therefore the **measure of skewness** is

$$\gamma_1 = \frac{\mu_3}{(\mu_2)^{3/2}} = \frac{1-2p}{\sqrt{np(1-p)}} ,$$

and the **measure of kurtosis** is

$$\gamma_2 = \frac{\mu_4}{(\mu_2)^2} = \frac{[1-6p(1-p)]}{np(1-p)} .$$

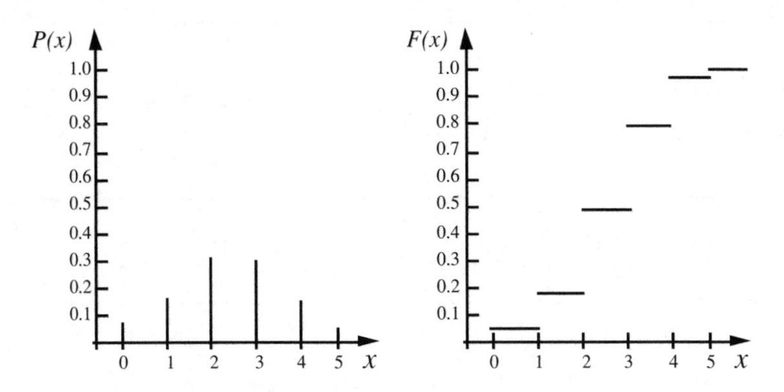

Figure 4.2. Binomial distribution ($n = 5, p = 0.5$).

The extension of the binomial distribution to that of the multinomial distribution is a discrete generalization with k–variables. Using the multiplication formula we have the following results

$$f(x_1, x_2, \ldots, x_k) = n! \prod_{i=1}^{k} \left\{ \frac{p_i^{x_i}}{x_i!} \right\}$$

and

$$F(m_1, m_2, \ldots, m_k) = \sum_{x_i=0}^{m_i} n! \prod_{i=1}^{k} \left\{ \frac{p_i^{x_i}}{x_i!} \right\}$$

Example 4.2 Suppose five fair dice are tossed, and one wants to compute the probability of 0, 1, 2,..., 5 sixes. If the outcomes 0 signifies "not a six" and 1 signifies "a six", then $p = 5/6$ and $q = 1/6$ and the mean and variance are given by

$$\mu = E(\mathbf{X}) = (5)(1/6) = 5/6$$
$$\sigma^2 = V(\mathbf{X}) = (5)(5/6)(1/6) = 25/36$$

4.3 Geometric Distribution

The geometric distribution is also a discrete distribution that occurs as a result of n independent Bernoulli trials, just as the binomial distribution does. The geometric distribution measures the number of trials before the first success or failure. If p is taken as the probability of the 1 outcome (success) on a given trial and q is the probability of the 0 outcome (failure), then the probability function for the geometric distribution is

$$P(k) = P(\mathbf{Y} = k) = pq^{k-1}$$
$$\text{or} \quad P(k) = p(1-p)^{k-1}$$

where $\mathbf{Y}$ measures the number of trials until the first success, $k = 1, 2, ...$ The cumulative distribution function is clearly

$$F(k) = \sum_{i=0}^{x} p(1-p)^i = 1-(1-p)^{x+1} = 1-q^{x+1}$$

Figure 4.3 shows the geometric distribution for $p = 0.7$.

The mean and variance for the geometric distributions are

$$\mu = \sum_{n=1}^{\infty} np(1-p)^{n-1} = \frac{1}{p}$$

$$\sigma^2 = \frac{1-p}{p^2} = \frac{q}{p^2}$$

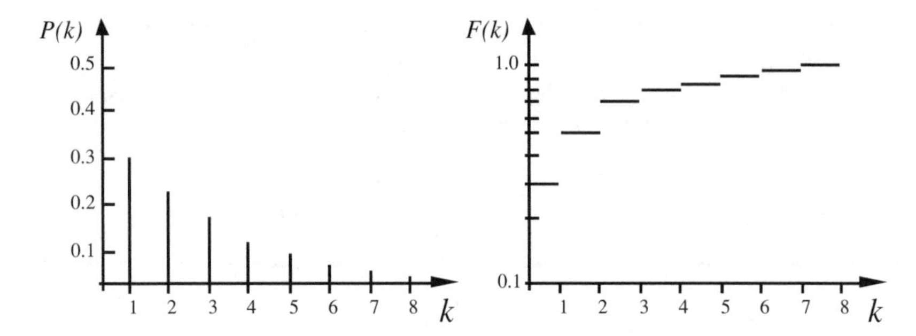

Figure 4.3. Geometric distribution (p = 0.7).

Example 4.3 Suppose a random experiment is conducted in which a biased coin is tossed until the first head appears. Further suppose the probability of a head on any toss is 1/5, while the probability of a tail is 4/5. Then

$$\mu = 1/(1/5) = 5$$
$$\sigma^2 = (4/5)/(1/25) = 20$$

4.4 Poisson Distribution

The Poisson distribution, another discrete distribution, has been widely used to model arrival distributions and other seemingly random events. It was originally developed to model telephone calls to a switchboard. The probability function is given by

$$P(x) = \frac{\lambda^x e^{-\lambda}}{x!}, \qquad x = 0,1,2,\ldots$$

where $\lambda > 0$ is a constant called the **parameter** of the distribution. The cumulative distribution function is given by

$$F(x) = \sum_{i=0}^{x} \frac{\lambda^i e^{-\lambda}}{i!}$$

while its mean, variance, third, and fourth moments are given by

$$\mu = \lambda, \quad \sigma^2 = \lambda$$
$$\mu_3 = \lambda$$
$$\mu_4 = 3\lambda(\lambda + 1/3)$$

Figure 4.4 illustrates the Poisson distribution for $\lambda = 4.5$.

The extension of the Poisson distribution to that of the multivariate case is a discrete generalization with k–variables. Using the multiplication formula we have the following results

$$f(x_1, x_2, \ldots, x_k) = e^{-(\lambda_1 + \lambda_2 + \ldots)} \prod_{i=1}^{k} \left\{ \frac{\lambda_i^{x_i}}{x_i!} \right\}$$

and the distribution function is

$$F(m_1, m_2, \ldots, m_k) = \prod_{i=1}^{k} \left[e^{\lambda_i} \sum_{x_i=0}^{m_i} \frac{\lambda_i^{x_i}}{x_i!} \right]$$

The Poisson distribution is applied in "stochastic" modeling whenever the probability of events occurring over a period of time are Poisson distributed.

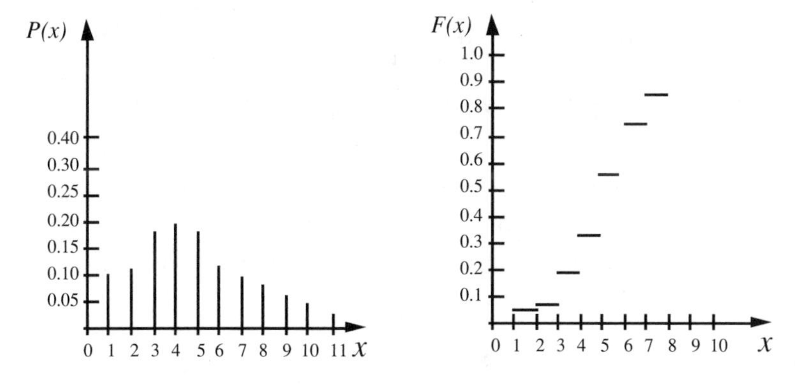

Figure 4.4. Poisson distribution ($\lambda = 4.5$).

Example 4.4 Calls to a switchboard are thought to be Poisson distributed with an average of five calls per minute. The probability of ten calls in the next minute is

$$P(\mathbf{X}=10)=P(10)=\frac{\lambda^{10}e^{-5}}{10!}$$

$$=\frac{5^{10}e^{-5}}{10!}=0.018$$

4.5 Uniform Distribution

A continuous distribution, the uniform distribution is one in which the density function is a constant.

$$f(x)=\begin{cases}1/(b-a), & a \le x \le b\\ 0, & \text{otherwise}\end{cases}$$

This distribution is sometimes called the **rectangular distribution**. The interval [a, b] is called the **range of the distribution**. The cumulative distribution function is given by

$$F(x)=\int_{-\infty}^{x} f(t)dt=\begin{cases}0, & x<a\\ \dfrac{x-a}{b-a} & a\le x\le b\\ 1, & x>b\end{cases}$$

The mean , variance, third, and fourth moments for this distribution are

$$\mu = \frac{a+b}{2}, \qquad \sigma^2 = \frac{(b-a)^2}{12}$$

$$\mu_3 = \int_0^1 (x-0.5)^3\, dx = 0$$

$$\mu_4 = \int_0^1 (x-0.5)^4\, dx = 0.0125$$

Figure 4.5 shows the uniform distribution.

This distribution is used to model truly random events. If a sequence of values is chosen at random on the interval $a \le x \le b$, it has the uniform distribution.

A particular uniform distribution in which $a = 0$, $b = 1$, called the **standard uniform distribution**, is of interest in generating random numbers. The standard uniform distribution is a flat distribution with equal probability, having

$$f(x) = 1 \qquad 0 \le x \le 1$$

and

$$F(x) = \int_0^x dx$$

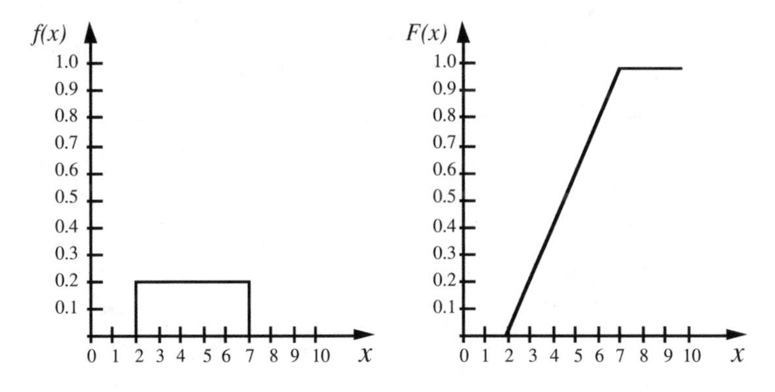

Figure 4.5. Uniform distribution (a = 2, b = 7).

There is an equal likelihood of any measurement occurring at any point along the abscissa of the frequency function.

$$\mu = \int_0^1 x dx = 0.5$$

$$\sigma^2 = \int_0^1 (x - 0.5)^2 dx = 0.0833$$

and

$$\sigma = 0.2887 \ .$$

Thus, 57.74% of all measurements fall within $\pm 1\sigma$. The frequency function can be extended beyond the usual (0–1) range, by appropriate transformations.

The extension to the bivariate case follows readily with

$$f(x,y) = f(x)f(y) = 1 \qquad \begin{cases} 0 \le x \le 1 \\ 0 \le x \le 1 \end{cases}$$

and

$$F(\mathbf{X},\mathbf{Y}) = \int_0^{\mathbf{X}} \int_0^{\mathbf{Y}} dx\, dy = 1.$$

There is an equal likelihood of any point (x, y) within the square bounded by (0, 0), (1, 0), (1, 1), and (0, 1).

$$\mu = \int_0^1 \int_0^1 xydxdy \qquad \begin{cases} \mu_x = 0.5 \\ \mu_y = 0.5 \end{cases}$$

and

$$\sigma^2 = \int_0^1 \int_0^1 (x-0.5)^2 (y-0.5)^2 \, dxdy \qquad \begin{cases} \sigma_x^2 = 0.0833 \\ \sigma_y^2 = 0.0833 \end{cases}$$

This is illustrated in Figure 4.6.

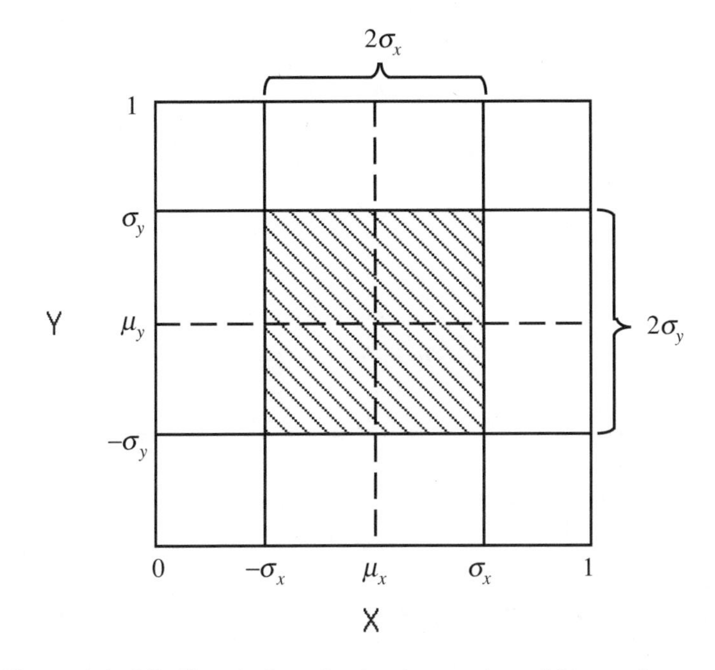

Figure 4.6. Likelihood of a point (x, y) occurring within a unit square.

Thus, with $\sigma_x = \sigma_y = 0.2887$, one–third of all measurements fall within $\pm 1\sigma_x$, $\pm 1\sigma_y$.

Example 4.5 Consider a finite set of "n" samples of measurements x_i. The mean ($\bar{x}$), the variance (S^2), and the unbiased standard deviation (S) may be obtained from the sample using

$$\bar{x} = \frac{1}{n} \sum_{i=1}^{n} x_i$$

$$S^2 = \frac{1}{n} \sum_{i=1}^{n} (x_i - \bar{x})^2 = \frac{1}{n} \sum_{i=1}^{n} x_i^2 - \bar{x}^2$$

$$S = \left[\frac{1}{n-1} \sum_{i=1}^{n} (x_i - \bar{x})^2 \right]^{1/2}$$
$$= \left[\frac{1}{n-1} \left(\sum_{i=1}^{n} x_i^2 - \bar{x}^2 \right) \right]^{1/2}$$

The approximate spread for this sample will show that $\pm 1\sigma$ includes 68.3% of all cases, $\pm 2\sigma$ includes 95.5% of all the cases, and $\pm 3\sigma$ includes 99.7% of all the cases. The **probable error** (*PE*) which includes 50% of all cases is defined as

$$PE = 0.6745\sigma$$

while the **mean absolute error** (MAE) is

$$\text{MAE} = 0.7979\sigma$$

Thus,

$$\boxed{\sigma = 1.4826\, PE = 1.2533\, MAE}$$

More will be said about the standard uniform distribution in Chapter 7.

4.6 Normal Distribution

Probably the most common continuous distribution is the normal distribution. It has been found useful in modeling most measurement phenomena, such as scores on a test, heights and weights, and errors made in manufacturing processes. The distribution is symmetrical, with a strong central tendency but still has some variety in shape according to the parameter σ. The probability density function is

$$\boxed{f(x) = \frac{1}{\sigma\sqrt{2\pi}} \exp\left\{ -\frac{(x-\mu)^2}{2\sigma^2} \right\}, \qquad -\infty < x < \infty}$$

The parameter μ only shifts the center as is show in Figure 4.7.

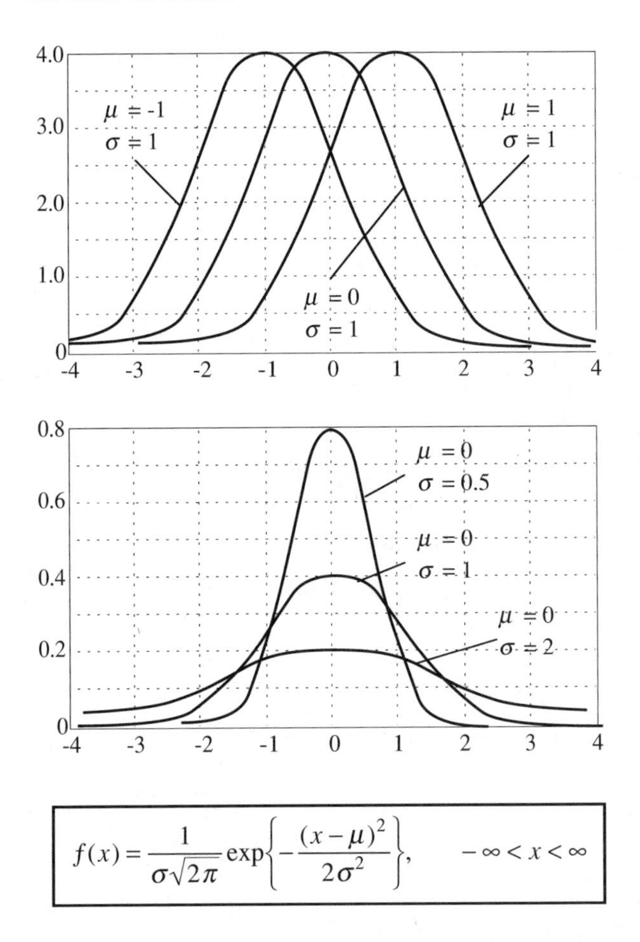

$$f(x) = \frac{1}{\sigma\sqrt{2\pi}} \exp\left\{-\frac{(x-\mu)^2}{2\sigma^2}\right\}, \qquad -\infty < x < \infty$$

Figure 4.7. Normal density function.

This distribution is unusual in that its cumulative distribution function cannot be computed exactly. Its cumulative distribution function is given by

$$\begin{aligned} F(x) &= \int_{-\infty}^{x} f(x)dx \\ &= \int_{-\infty}^{x} \frac{1}{\sigma\sqrt{2\pi}} \exp\left\{-\frac{(t-\mu)^2}{2\sigma^2}\right\}dt \end{aligned}$$

but this integral does not exist in closed form. As a result numerical methods must be employed, and since it is impractical for us to employ them repeatedly, tables have been devised. Since it would be impractical have tables for all values of μ and σ^2, a "**standard normal table**" was formed for a standard normal distribution with $\mu = 0$ and $\sigma^2 = 1$; that is

$$f(z) = \frac{1}{\sqrt{2\pi}} e^{-z^2/2}$$

This function can be integrated numerically to obtain approximations for cumulative probabilities.

Cumulative probabilities for nonstandard normal distributions may be calculated from the standard tables of values. If **X** is normally distributed with mean μ and variance σ^2, then the random variable

$$\mathbf{Z} = (\mathbf{X} - \mu) / \sigma$$

is normally distributed with mean 0 and variance 1. Figure 4.8 shows the normal distribution with $\mu = 5.2$, $\sigma = 2.2$.

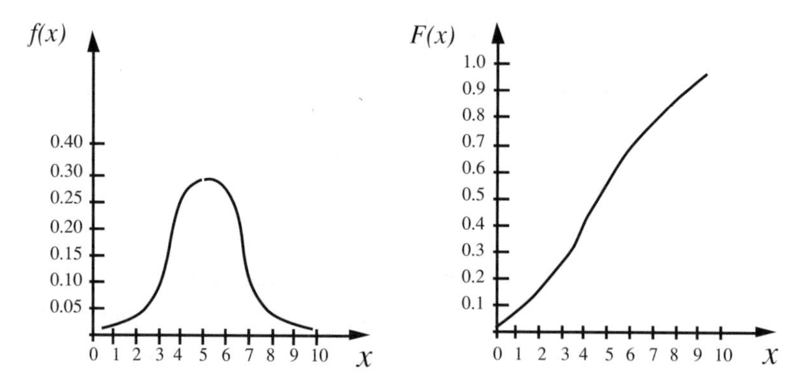

Figure 4.8. Normal distribution ($\mu = 5.2$, $\sigma = 2.2$).

Example 4.6 The grades on the final examination in a mathematics course were determined to be approximately normally distributed with mean 75 and variance 25.

Then the standardized grades computed using $\mathbf{Z} = (\mathbf{X} - 75)/5$ are approximately normal with mean 0 and variance 1.

Example 4.7 Example calculations using normal distributions.

(a) Find $Pr\{x \leq 150\}$ for $\mu = 120$, $\sigma^2 = 100$

Using $z = \dfrac{x-\mu}{\sigma} = \dfrac{150-120}{10} = 3.0$,

from the tables of normal distributions,

$$Pr\{z \leq 3.0\} = 0.99865$$

(b) Find $Pr\{x \leq 90\}$ for $\mu = 105$, $\sigma^2 = 100$

Using $z = \dfrac{x-\mu}{\sigma} = \dfrac{90-105}{10} = -1.5$,

by symmetry,

$$\begin{aligned}\Pr\{z \leq -1.5\} &= 1 - \Pr\{z \leq 1.5\} \\ &= 1 - 0.93319 = 0.07781\end{aligned}$$

(c) Find $Pr\{100 \leq x \leq 120\}$ for $\mu = 110$, $\sigma^2 = 64$

$$z_L = \frac{100-110}{8} = -1.25$$
$$z_R = \frac{120-110}{8} = 1.25$$

Thus,

$$\begin{aligned}Pr\{-1.25 \leq z \leq 1.25\} &= z(1.25) - z(-1.25) \\ &= z(1.25) - [1 - z(1.25)] \\ &= 2z(1.25) - 1 \\ &= 2(0.89435) - 1 \\ &= 0.78870\end{aligned}$$

The extension to the bivariate normal or normal probability density function $f(x, y)$ describing stochastically independent phenomena in two dimensions can be obtained from the following

$$f(x,y) = f(x)f(y) = \frac{1}{2\pi\sigma_x\sigma_y}\exp\left[-\frac{(x-\mu_x)^2}{2\sigma_x^2} - \frac{(y-\mu_y)^2}{2\sigma_y^2}\right]$$

and

$$F(\mathbf{X},\mathbf{Y}) = \int_{-\infty}^{\mathbf{X}}\int_{-\infty}^{\mathbf{Y}} f(x,y)dxdy$$
$$= \frac{1}{2\pi\sigma_x\sigma_y}\int_{-\infty}^{\mathbf{X}}\int_{-\infty}^{\mathbf{Y}} \exp\left[-\frac{(x-\mu_x)^2}{2\sigma_x^2} - \frac{(y-\mu_y)^2}{2\sigma_y^2}\right]dxdy$$

where $\frac{1}{2\pi} = 0.1592$

This general case, an **elliptical normal distribution**, may be specialized to a **circular normal distribution** when $\sigma_x^2 = \sigma_y^2$ and the correlation coefficient $\rho = 0$. Thus, we obtain

$$f(x,y) = \frac{1}{2\pi\sigma^2}\exp\left[-\frac{\left[(x-\mu_x)^2 + (y-\mu_y)^2\right]}{2\sigma^2}\right]$$

and

$$F(\mathbf{X},\mathbf{Y}) = \int_{-\infty}^{\mathbf{X}}\int_{-\infty}^{\mathbf{Y}} f(x,y)dxdy$$
$$= \frac{1}{2\pi\sigma^2}\int_{-\infty}^{\mathbf{X}}\int_{-\infty}^{\mathbf{Y}} \exp\left[-\frac{\left[(x-\mu_x)^2 + (y-\mu_y)^2\right]}{2\sigma^2}\right] dxdy$$

The circular normal distribution $f(x, y)$ and $F(\mathbf{X}, \mathbf{Y})$ may be transformed using

$$r = \left[(x-\mu_x)^2 + (y-\mu_y)^2\right]^{1/2}$$

from rectangular to polar coordinates, where

$$f(r) = \frac{r}{\sigma^2} e^{-r^2/2\sigma^2}$$

and

$$F(R) = \int_0^R f(r)dr = \frac{1}{\sigma^2}\int_0^R e^{-r^2/2\sigma^2} r dr$$

This distribution, in its new form, may be recognized as the **Rayleigh distribution**, frequently occurring in network and data communication simulations. The spread of this distribution about the origin or center (mean, zero in x and y) in σ units is illustrated in Figure 4.9. As can readily be seen, 1σ, 2σ, and 3σ include 39.5%, 86.5%, and 98.5% of all cases, respectively. Furthermore, the **circular probable error** (CPE) includes 50% of all cases and is given by

$$\text{CPE} = 1.177\sigma$$

while the **mean radial error** (MRE) is given to be $1.253\sigma = 1.065$ CPE.
Thus,

$$\sigma = 0.8493 \text{ CPE} = 0.7979 \text{ MRE}$$

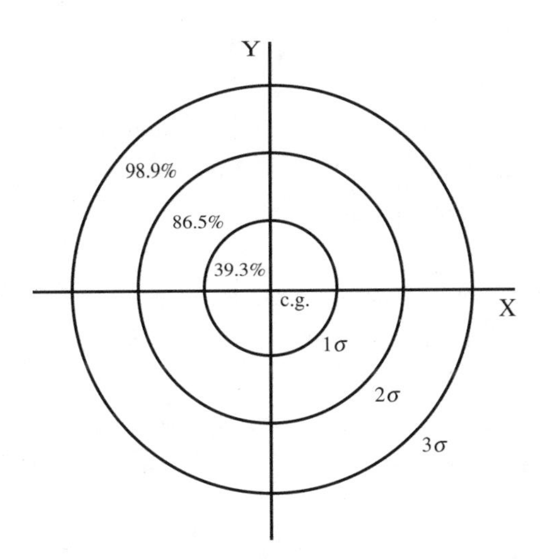

Figure 4.9. The Rayleigh distribution.

The normal distribution is probably the most useful distribution in modeling, not only because it accurately models many phenomena, but also because of the following reasons.

1. The sum of n independently and identically distributed random variables tends to be normally distributed as n tends to infinity. This result is known as the **central limit theorem**. The central limit theorem (CLT) actually comes in two flavors: strong and weak.

   (a) **Strong CLT:** Let $S_n = x_1 + x_2 + \ldots + x_n$ be the sum of N independent, identically distributed random variables, with mean μ and variance σ^2.

   Then as $N \rightarrow \infty$, the distribution of S_n tends to become normal with mean $N\mu$ and variance $N\sigma^2$, regardless of the distribution of the x_i's.

   (b) **Weak CLT:** The CLT holds in essence when the variables are "not" independent, identically distributed. However, the variance is not easily calculated, and the convergence to be far slower.

   The importance of the CLT is that it will often justify the assumption that data or simulation outputs are normally distributed and will simplify some model building and analysis tasks.

2. If **X** is distributed binomially, then as n tends toward infinity, **X** tends to be normally distributed with $\mu = nq$ and $\sigma^2 = npq$.

3. If **X** is distributed according to the Poisson distribution, then as n gets large, **X** tends to be normally distributed with $\mu = \lambda$.

Before ending this section we would like to provide the reader with some pragmatic considerations on the CLT and the normal distribution:

1. The CLT comes into play very rapidly ($N = 4$) if the distributions of the x_i's are well–behaved.

2. The CLT comes into play moderately ($N = 12$) when the distributions are fairly well–behaved (i.e., not discontinuous).

3. The CLT comes into play slowly ($N \geq 100$) for ill–behaved distributions of x_i.

4. The effective range (99+% of the area) of the distribution is $\pm 3\sigma$.

5. When process times are considered, which are composed of several "independent" subtasks, that time will usually be well approximated with a normal distribution.

6. Error processes will generally follow normal distributions.

7. The normal distribution is a good approximation to the binomial distribution when $np < 5$ and $p \leq 1/2$ or $n(1 - p) > 5$ and $p > 1/2$.

Thus the normal distribution is useful not only in its own right but also as an approximation to many other distributions.

4.7 Exponential Distribution

Another continuous distribution that has wide utility is the (negative) exponential distribution. This distribution has been used to model "sudden and catastrophic" failures due to manufacturing defects and light bulbs burning out. It has also been used to characterize service times and interarrival times in queueing systems. The probability density function for the exponential distribution is

$$
\begin{aligned}
f(x) &= \alpha e^{-\alpha x}, & & 0 \leq x < \infty \\
f(x) &= 0, & & \text{elsewhere}
\end{aligned}
$$

where α is a positive parameter. The cumulative distribution function is

$$F(x) = \alpha\int_0^x e^{-\alpha t}dt = 1 - e^{-\alpha x}, \qquad x \geq 0$$
$$F(x) = 0, \qquad x < 0$$

The mean and variance are

$$\mu = E(\mathbf{X}) = \alpha\int_0^\infty xe^{-\alpha x}dx = 1/\alpha$$

$$\sigma^2 = V(\mathbf{X}) = \alpha\int_0^\infty (x - 1/2)^2 e^{-\alpha x}dx = 1/\alpha^2$$

Thus the exponential distribution's mean and standard deviation are the same. The exponential distribution is illustrated in Figure 4.10 for $\alpha = 1$.

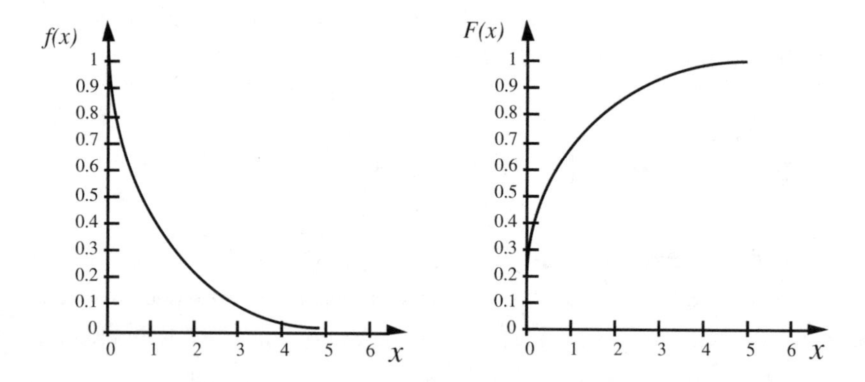

Figure 4.10. Exponential distribution ($\alpha = 1$).

Probably the most important aspect of the exponential distribution in modeling is that it has the **"forgetfulness" property**. This is summarized as follows

PROPERTY 4.1

*Assume that the random variable **X** has an exponential distribution. Then for any* $x_1, x_2 > 0$,

$$P(\mathbf{X} \leq x_2 | \mathbf{X} \geq x_1) = P(0 \leq \mathbf{X} \leq x_2 - x_1).$$

The converse is also true.

It is also interesting to note the relationship between the Poisson and exponential distributions.

PROPERTY 4.2

Assume that the arrival process of customers to a ***queueing system*** *follows the Poisson distribution, with arrival rate* λ. *Then the interarrival distribution (time between arrivals) will be exponential with parameter* $1/\lambda$.

This relationship will be explored more fully in Chapter 15.

Example 4.8 The variable discussed in Examples 2.20 and 2.21 has an exponential distribution with mean and variance equal to 1.

4.8 Chi–square Distribution

A distribution that will be used extensively in goodness–of–fit testing is the chi–square distribution. The probability density function for this distribution is

$$f(\chi^2) = \frac{\chi^{2(\upsilon/2)-1} e^{-\upsilon/2}}{2^{\upsilon/2}\Gamma\left(\frac{\upsilon}{2}\right)}, \qquad \chi^2 > 0$$

In this formula, Γ is the standard Gamma function. The single parameter of this distribution is υ known as the **degrees of freedom**. This distribution arises often when the squares of standard normal distributions are combined. If $\mathbf{Z}_i$, $i = 1,2,\ldots,R$, are independent standard normal random variables, then $\mathbf{Z}_1^2 + \mathbf{Z}_2^2 + \ldots + \mathbf{Z}_R^2$ is a chi–square distribution with R degrees of freedom.

4.9 Student's *t*- distribution

The t–distribution has been shown to be useful in hypothesis testing. Let **Z** and **U** be independent random variables, where **Z** follows a standard normal distribution and **U** a chi–square distribution with υ **degrees of freedom**. Then $t = \mathbf{Z}/(\mathbf{U}/\upsilon)$ follows a t–distribution. It, too, has only one parameter υ, which again denotes degrees of freedom. The t–distribution arises quite often when normal distributions are sampled. If **X** denotes the sample mean, S the sample standard deviation, and N the size of the sample, the random variable

$$t = N^{1/2}(\mathbf{X} - \mu) / S$$

follows a t–distribution with $n - 1$ degrees of freedom.

4.10 *F*- distribution

The F–distribution is also useful in hypothesis testing. Let **U** and **V** be two independent chi–square random variables with υ_1 and υ_2 degrees of freedom respectively. Then $(\mathbf{U}/\upsilon_1)/(\mathbf{V}/\upsilon_2)$ is distributed as an F–distribution with υ_1 and υ_2 degrees of freedom. This distribution arises when one is sampling from two normal populations. Let sample 1 consist of N_1 points from a normal population with mean μ_1 and variance σ_1^2, and let S_1^2 be the sample variance. Let sample 2 consist of N_2 points from a normal population with mean μ_2 and variance σ_2^2, and let S_2^2 be the sample variance. Then

$$\boxed{(S_1^2 / \sigma_1^2) / (S_2^2 / \sigma_2^2)}$$

is distributed according to the F–distribution with $N_1 - 1$ and $N_2 - 1$ degrees of freedom.

5

STATISTICS AND RANDOM SAMPLES

At times a random variable **X** that is being used to represent some aspect of a simulation model is known to follow a particular distribution. If this is the case, the researcher's task is greatly simplified. More often than not, however, all that is known about the distribution of a random variable is what can be gleaned from the study of a set of sample values that has been collected through observations. Some technique is then needed to characterize the behavior of the random variable. Two general approaches have been taken in attempting to solve this problem. The first is to construct an **empirical distribution** using **least squares** or some other suitable curve–fitting technique described in this chapter. This approach should be used when the random variable does not appear to follow any of the common distributions. The second approach, to be described in the next chapter, is to hypothesize that the random variable follows a particular distribution and to use statistical methodology to **test the validity of this hypothesis**. This approach is the more common of the two and, if successful, yields a distribution function that may be expressed analytically and whose behavior in most cases is well known. The normal distribution has been widely used and studied, and its characteristics are quite well known. The fortunate analyst who finds that the random variable of interest follows a normal distribution can draw on a wealth of knowledge available on that distribution.

Once a random variable has been found to follow a given distribution, the analyst's next task is to describe the distribution in easily understood terms. This task normally involves the **estimation of the parameters** (summary measures) of the distribution. When working with the normal distribution, for example, one is interested in estimating its **mean** and **variance**. The t–distribution, on the other hand, is categorized by the single parameter **degrees of freedom**.

The purpose of this chapter is to review the procedures used in constructing empirical distributions and the basic concepts of **statistical inference** and **estimation**. Four of the more common statistical tests are included. These tests will be used extensively in the subsequent chapters.

5.1 Descriptive Statistics and Frequency Diagrams

In simulation and statistics we presume that we do not know the underlying distribution of some random process, or that if we are able to assume a distribution we still do not know its parameter values. Statistics is the science of using the tools of "**replication**" of an experiment and "**observation**" of the outcomes to draw **inference** about the random process. Statistics relies heavily on probability and general mathematics to gain as much insight as possible from a **set of observations** (data). In most cases we will insist that the data (**set of observations**) is a random sample. A collection of data will be a random sample if the conditions during collection were essentially constant, and successive **replications** were independent.

Assume that the data below represents the processing time (in msec) of a job on a particular computer system.

20.0	24.0	28.1	23.5	27.0
21.5	25.0	29.0	22.0	20.2
19.5	26.1	22.3	24.1	21.3
20.4	23.3	24.1	27.4	26.3
22.6	19.8	22.0	22.9	30.0
24.3	20.3	21.6	21.4	21.6
28.1	21.7	20.4	23.1	23.0
29.0	28.9	21.7	27.2	25.4
22.2	29.4	25.2	22.8	26.1
19.9	21.3	20.8	22.4	23.6
24.3	22.0	19.6	21.4	32.4
26.5	24.1	20.4	27.1	27.5
28.1	23.2	25.3	19.4	30.1

30.0	25.0	26.1	32.8	27.6
21.6	28.1	19.5	23.4	20.2
22.0	30.1	25.0	21.9	30.4
31.2	31.2	26.1	21.7	26.1
29.4	28.4	29.4	22.8	25.1
20.0	31.7	20.4	24.3	29.6
21.1	21.8	19.8	24.9	23.9

This data can then be organized into a **frequency distribution**, as follows:

Time in Machine		Frequency	Relative
Minimum	Maximum		Frequency
19	20	7	0.07
20	21	10	0.10
21	22	14	0.14
22	23	11	0.11
23	24	8	0.08
24	25	8	0.08
25	26	7	0.07
26	27	7	0.07
27	28	6	0.06
28	29	6	0.06
29	30	6	0.06
30	31	5	0.05
31	32	3	0.03
32	33	2	0.02

Finally, the data may be graphed and we can compare the **frequency diagram** and in particular the **relative frequency** diagram and relate the results to the concept of a **probability density function (pdf)**. In fact, the relative frequency diagram is an empirical approximation to a pdf, and will often be used in both model building and analysis.

5.1.1 Measure of Central Tendency

We previously discussed the mean for a pdf and saw that it was the first moment of the pdf, and one of many expectations we could calculate. Similarly we can calculate the mean, $\bar{x}$, or an average of a set of data.

$$\bar{x} = \frac{1}{n}(x_1 + x_2 + \ldots + x_n) = \frac{1}{n}\sum_{i=1}^{n} x_i$$

If the data has been placed in k cells (from the frequency diagram) where $\hat{x}_i$ is the center of the cell, and f_i is the number of observations for that cell, then $\bar{x}$ can be approximated by

$$\bar{x} \doteq \frac{\hat{x}_1 f_1 + \hat{x}_2 f_2 + \ldots + \hat{x}_k f_k}{f_1 + f_2 + \ldots + f_k} = \frac{1}{n}\sum_{i=1}^{k} \hat{x}_i f_i$$

where $f_1 + f_2 + \ldots + f_k = n$

In addition to the mean, $\bar{x}$, we often use two other **measures of central tendency,** the **median** and the **mode.**

DEFINITION 5.1
The **median** *is the observation that bisects the data into two equal parts; half the observations are greater than the median, half are less.*

DEFINITION 5.2
The **mode** *is the most likely value, i.e., the observation which has the highest frequency of occurrence.*

We all have a natural tendency to make comparisons among random processes by merely comparing or thinking of averages. This is because knowingly or unknowingly we have a tendency to assume that the underlying distribution is "symmetric" and in that case the "mean, median, and mode" are all "equal".

If we consider two cities which have the same average annual temperature, are their climates the same?

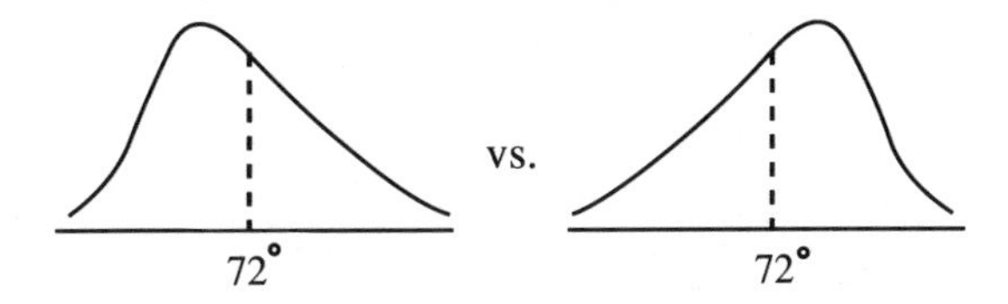

The reader can readily see the fault in this logic. The mean alone does not convey all information, particularly, if the distribution is not symmetric. Furthermore, it can be seen that **variability** or **dispersion** matters as well, and indeed is our next concern.

5.1.2 Measure of Dispersion

Similar to the mean or the first moment, the variance, σ^2, was also an expectation of the pdf, namely the second moment about the mean. We can also calculate an estimation of the variance S^2, called the **sample variance**, and the more often used square root, S, called the **standard deviation**.

The popularity of the standard deviation is largely due to the popularity of the normal distribution and the recognition that the range of the normal is effectively ±3 standard deviations. Thus, we physically relate more easily to the standard deviation than we do to the variance.

Mathematically the **sample variance** is given as

$$S^2 = \sum_{i=1}^{n} \frac{(x_i - \bar{x})^2}{n-1}$$

but is more easily computed by

$$S^2 = \frac{\sum_{i=1}^{n} x_i^2 - \frac{\left(\sum_{i=1}^{n} x_i\right)^2}{n}}{n-1}$$

For cellularized data (as from the frequency diagram), S^2 can be approximated by

$$S^2 \doteq \frac{\sum_{i=1}^{k} f_i \hat{x}_i^2 - \frac{1}{n}\left(\sum_{i=1}^{k} f_i \hat{x}_i\right)^2}{n-1}$$

where the parameters were previously defined.

Other simple measures of dispersion include the **maximum observation** (x_{max}), the **minimum observation** (x_{min}), and the **range**, $R = (x_{max} - x_{min})$. However, none of these measures provide the degree of information as does S.

The standard deviation is sometimes scaled so that two processes with significantly different means can be compared. This scaled measure, called the **coefficient of variation**, is defined as

$$CV = \left(\frac{1}{\bar{x}}\right) S$$

5.2 Statistics and Sampling Distributions

We have proceeded thus far without formally defining what a "**statistic**" is, although we have created some. Before proceeding any further, a formal definition is in order.

> DEFINITION 5.3
> *A **statistic** is any function of the observations of a random variable which does not depend on unknown parameters.*

Note that by that definition, $\bar{x}$, S^2, S, R, x_{min}, x_{max}, CV, mode, and median are all statistics. Note however that $\sum_{i=1}^{k}(x_i - \mu)^2$ is not a statistic. Why? This is because the true mean of the distribution is not known. On the other hand, $\sum_{i=1}^{k}(x_i - \bar{x})^2$ is a statistic because it can be computed. Since the observations in a random sample are themselves random variables, and a statistic is a function of the observations, it follows that a "statistic is a random variable". Further, it follows that a statistic has an associated distribution. Distributions of statistics are called "**sampling distributions**".

Let us consider the normal and the chi–square distributions in more detail. Why are these common sampling distributions? The central limit theorem provides the answer along with the concept of linear combinations of random variables.

Under modest assumptions, $\bar{x}$ will be normally distributed with

$$\mu_{\bar{x}} = \mu$$

and

$$\sigma_{\bar{x}} = \frac{\sigma}{\sqrt{n}}$$

Furthermore, many other sampling characteristics can be in terms of normal random variables.

Another important sampling distribution, the chi–square distribution is defined as follows: Let z_i be independent normally distributed with $\mu = 0$ and $\sigma^2 = 1$ (i.e., $N(0, 1)$) for $i = 1, 2, \ldots, k$. Then

$$\chi^2 \equiv z_1^2 + z_2^2 + \ldots + z_k^2$$

follows the chi–square distribution given by

$$f(\chi^2) = \frac{1}{2^{k/2}\Gamma\left(\frac{k}{2}\right)} (\chi^2)^{(k/2-1)} \; e^{\chi^2/2} \,, \qquad \chi^2 > 0$$

where

$$\Gamma\left(\frac{k}{2}\right) = \int_0^\infty \chi^{(k/2-1)} e^{-x} \, dx$$

and

$$\Gamma(k) = \left(\frac{k}{2} - 1\right)! \qquad \text{for } \frac{k}{2} \text{ an integer}$$

Note that the mean and variance are

$$\mu = k$$
$$\sigma^2 = 2k$$

The terrible form of this distribution need not disturb us, for there exists chi–square distribution tables similar to those for the normal distribution. The parameter k is called the **degrees of freedom** parameter. An important property of the chi–square distribution is

PROPERTY 5.1
The sum of the χ^2 variables is another χ^2.

Example 5.1 In some simulation studies it may not be possible or even necessary to establish that a random variable follows a particular known distribution. In this case, the data samples gained from observations can be used to construct the cumulative distribution function directly.

To see how this is done, assume the N observations of a random variable **X** have been noted, and let the observations be ordered in a monotonic nondecreasing sequence as follows.

$$x_1 \leq x_2 \leq .... \leq x_N$$

Next form the sequence of numbers

$$y_n = n / N, \qquad n = 1,2,...,N$$

and plot the points $(x_1, y_1),...,(x_N, y_N)$. By connecting these plotted points with some suitable function $y = F(x)$, where $F(x) = 0$ for $x < x_1$ and $F(x) = 1$ for $x > x_N$, one can obtain a curve satisfying all the properties of a cumulative distribution function. Such a function is known as an **empirical cumulative distribution function**, and the process described in Table 5.1 may be recognized as the **Lagrange interpolation process** for fitting a straight line through these two points y_k and y_{k+1}.

The Lagrange interpolation is frequently given in the following form for two points,

$$y = \frac{(x - x_1)}{(x_0 - x_1)} y_0 - \frac{(x - x_0)}{(x_1 - x_0)} y_1$$

and

$$y = \frac{(x - x_1)(x - x_2)}{(x_0 - x_1)(x_0 - x_2)} y_0 + \frac{(x - x_0)(x - x_2)}{(x_1 - x_0)(x_1 - x_2)} y_1 + \frac{(x - x_0)(x - x_1)}{(x_2 - x_1)(x_2 - x_0)} y_2$$

for three points.

Table 5.1 Empirical Cumulative Distribution Function

1	Scale the data so that the area of curve of the data is one.
2	Choose a number of partition intervals for x–axis, say $1/\Delta$.
3	Integrate by parts until the area under the curve is $1/\Delta$, $2/\Delta$, $3/\Delta$, etc., and mark the points on the x–axis.

Table 5.1 Empirical Cumulative Distribution Function (continued)

4	From this the cumulative density function is constructed as follows: Using the inverse transformation process we have a. Generate a y_i on the y–axis, extending this horizontally to meet the curve, then down vertically to the x–axis to determine x_i. b. Select $y_k < y_i \le y_{k+1}$ such that the interval y_k to y_{k+1} is $1/\Delta$ and that x_k correspond to y_k, and x_{k+1} corresponds to y_{k+1}. c. The x–axis and y–axis are therefore scaled proportionately. d. Random variables y_i's can now be used to generate x_i's (the variables we are looking for) by $$x = \frac{(y_i - y_{k+1})x_k}{y_k - y_{k+1}} + \frac{(y_i - y_k)x_{k+1}}{y_{k+1} - y_k}$$

Suppose the following ten values were recorded for the random variable **X**: 1, –1, 2, 0, 4, 6, 7, 9, 10, 3. The points (x_n, y_n) are

n	1	2	3	4	5	6	7	8	9	10
x_n	–1	0	1	2	3	4	6	7	9	10
y_n	0.1	0.2	0.3	0.4	0.5	0.6	0.7	0.8	0.9	1.0

Plotting these points and joining the sequential observations with straight lines gives the empirical distribution function illustrated in Figure 5.1.

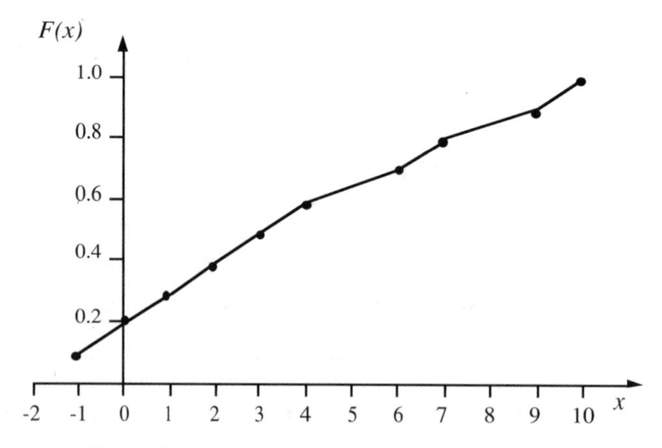

Figure 5.1. An empirical distribution function.

A distribution function constructed in this fashion has the same problems that any quantity based on sample observations has. If it is based on too few observations, it may not be representative of the true process. Thus it is desirable to collect as many observations on the random variable of interest as is possible, and to determine **normalcy** and **closeness of fit**.

Example 5.2 The exact procedure to determine normalcy and closeness of fit, itemized in Table 5.2, is as follows:

Table 5.2 Normalcy and Closeness of Fit Using the χ^2

1	Tabulate raw data x_i in a frequency and cumulative table. ***Interval*** — Equal spacing ***Frequency*** — Number of data points in given interval ***Cumulative Frequency*** — Running summation ***C%*** — $C\,F$ in %
2	Test data by χ^2 test to determine normalcy and closeness of fit.
3	If data are normal, plot the tabular points from the cumulative percentage on normal probability graph paper in which the mid–interval are the ordinate and $C\%$ the abscissa.
4	Draw best fit straight line.
5	Read the mean off the ordinate (the mean is at the intersection of the best fit straight line and the 50% percentile).
6	Determine the standard deviation. Read the 84.5% off the ordinate and subtract the mean. (The difference is the 1σ standard deviation).

Step 1: Tabulate the raw data x_i in a frequency table.

Step 2: Test the data by the χ^2 test to determine normalcy and closeness of fit.

χ^2 Test (see Table 5.3):

1. Tabulate the raw data x_i in frequency table.

2. Count the observed frequency θ_i in each k categories. That is, $\theta_1, \theta_2, \ldots, \theta_n$, where $\sum_{i=1}^{k} \theta_i = n$.

3. Determine the expected (or theoretical) frequencies $\hat{\theta}_i$ in each category.

$$\sum_{i=1}^{k} \hat{\theta}_i = n \quad \text{and} \quad \hat{\theta}_i \geq 5$$

4. The values $\hat{\theta}_i$ are determined from a table of the normal frequency function.

5. Compute $\chi^2 = \sum_{i=1}^{k} \frac{(\theta_i - \hat{\theta}_i)^2}{\hat{\theta}_i}$.

6. Compare χ^2 with published χ_0^2.

7. Locate χ_0^2 in the standard table for appropriate degrees of freedom (*df*),

$$df = k - 1$$

   Note: The percentile at the top of the column is the probability of χ^2 being less than χ_0^2.

8. If $\chi^2 \leq \chi_0^2$, the sample is considered consistent with the hypothesis. The better the fit, the smaller the χ^2 value.

Step 3: If the data is normal compute,

$$\bar{x} = \frac{1}{n}\sum_{i=1}^{n} x_i$$

$$S = \left[\frac{1}{n-1}\left(\sum_{i=1}^{n} x_i^2 - n\bar{x}^2\right)\right]^{1/2}$$

Table 5.3 Example χ^2 Test

Interval	F	CF	$C\%$	K	θ	$\hat{\theta}$	$\frac{(\theta-\hat{\theta})^2}{\hat{\theta}}$
0 – 5	0	0	0	0	0	0	0
5 – 10	2	2	1	1	2	2	0
10 – 15	2	4	2	2	2	3	0.33
15 – 20	6	10	5	3	6	7	0.14
20 – 25	10	20	10	4	10	12	0.33
25 – 30	20	40	20	5	20	20	0
30 – 35	28	68	34	6	28	28	0
35 – 40	32	100	50	7	32	32	0
40 – 45	32	132	66	8	32	30	0.13
45 – 5 0	28	160	80	9	28	26	0.15
50 – 55	18	178	89	10	18	20	0.20
55 – 60	12	190	95	11	12	10	0.40
60 – 65	6	196	98	12	6	6	0
65 – 70	4	200	100	13	4	4	0
70 – 75	0	200	100				

$$\chi^2 = \sum_{i=1}^{k} \frac{(\theta - \hat{\theta}_i)^2}{\hat{\theta}_i} = 1.68$$

$$df = k - 1 = 12$$

$$\chi^2 \leq \chi_0^2 (1.68 \leq 3.07)$$

for 99.5% and $df = 12$

5.3 Method of Least Squares

Construction of the cumulative distribution function by connecting sequential observations with straight–line segments as was done in Example 5.1 is generally not desirable. If it is done in this fashion, each of the ordered pairs must be retained. If the number of observations is large, this can be inconvenient. A more satisfactory way is to **fit a low–order polynomial**

$$y = a_0 + a_1 x + \ldots + a_k x^k$$

to the points. Then only the coefficients of the polynomial need be stored. In many cases $k = 2$ or $k = 3$ is sufficient to describe the distribution.

A number of techniques are available for fitting a curve to the points. One of the most common techniques is the **method of least squares**. This technique is designed to fit a polynomial of kth degree

$$P(x) = a_0 + a_1 x + a_2 x^2 + \ldots + a_k x^k$$

to a set of $N \geq k$ points. This procedure yields a set of predicted values $y_1, y_2, \ldots, y_N$ corresponding to each of the tabulated x_i, $i = 1, 2, \ldots, N$. The difference between the ith predicted value $\hat{y}_i$ and the ith observed value y_i is referred to as the ith **residual** and is given by $\gamma_i = \hat{y}_i - y_i$. The basis of the least–squares method is to find the kth degree polynomial such that the sum of the squares of these residuals is a minimum.

Formally, let S be the sum of the squared residuals. Then

$$S = \sum_{i=1}^{N} \gamma_i^2 = \sum_{i=1}^{N} \left(y_i - \sum_{j=0}^{k} a_j x_i^j \right)^2$$

Since S is always positive, its minimum can be obtained by setting

$$\partial S / \partial a_j = 0 \quad \text{for } j = 0, 1, 2, \ldots, k$$

and then solving the resulting set of $k + 1$ linear equations for the coefficients a_0, $a_1,\ldots,a_k$. In general, when $N = k + 1$, the polynomial (if obtainable) passes through each point (x_i, y_i) and all residuals are zero. If $N > k + 1$ and if there are $k + 1$ different values of x_i among points (x_i, y_i), then a solution can be expected. This technique will be illustrated by an example.

Example 5.3 Suppose we wish to fit the points of Example 5.1 with a first degree polynomial $P_1(x) = a_0 + a_1x$. Then

$$S = \sum_{i=1}^{10} \gamma_i^2 = \sum_{i=1}^{10} (y_i - a_0 - a_1x_i)^2$$

and

$$\frac{\partial S}{\partial a_0} = -2\sum_{i=1}^{10} (y_i - a_0 - a_1x_i)$$

while

$$\frac{\partial S}{\partial a_1} = -2\sum_{i=1}^{10} x_i(y_i - a_0 - a_1x_i)$$

Setting these derivatives equal to 0 and dividing by –2 yields the following equations

$$\sum_{i=1}^{10} (y_i - a_0 - a_1x_i) = 0$$

$$\sum_{i=1}^{10} x_i(y_i - a_0 - a_1x_i) = 0$$

These equations may be rewritten as

$$10a_0 + a_1\sum_{i=1}^{10} x_i = \sum_{i=1}^{10} y_i$$

$$a_0\sum_{i=1}^{10} x_i + a_1\sum_{i=1}^{10} x_i^2 = \sum_{i=1}^{10} x_iy_i$$

Solving these two equations for a_0 and a_1 gives

$$a_1 = \sum_{i=1}^{10} x_i y_i \bigg/ \sum_{i=1}^{10} x_i^2$$

$$a_0 = \bar{y} - a_1 \bar{x}$$

where $\bar{x}$ and $\bar{y}$ are the arithmetic means of the x's and y's respectively. Tabulation of the data necessary to compute these quantities for Example 5.1 is given in Table 5.4.

Table 5.4 Data for a Regression Example

i	x_i	y_i	x_i^2	$x_i y_i$	$\hat{y}_i$
1	–1	0.1	1	–0.1	–0.01
2	0	0.2	0	0	0.10
3	1	0.3	1	0.3	0.21
4	2	0.4	4	0.8	0.32
5	3	0.5	9	1.5	0.43
6	4	0.6	16	2.4	0.54
7	6	0.7	36	4.2	0.76
8	7	0.8	49	5.6	0.87
9	9	0.9	81	8.1	1.09
10	10	1.0	100	10.0	1.2
	41	5.5	297	32.8	

The respective means are $\bar{x} = 4.1$ and $\bar{y} = 0.55$, which yields

$$a_1 \doteq \frac{32.8}{297} = 0.11, \qquad a_0 \doteq 0.1$$

Thus the prediction equation is $\hat{y} = 0.1 + 0.11x$. The values of $\hat{y}_i$ for i = 1, 2, ..., 10 are given in the last column of Table 5.4. Note that the curve does not appear to fit well. This is normally an indication that a higher–order polynomial is needed.

5.4 Estimation

Whether a random variable of interest in a simulation study is represented by an empirical distribution or is known to follow a particular distribution, the analyst encounters the problem of estimating the appropriate parameters of the distribution. For example, it is not generally sufficient to know that a random variable **X** follows a normal distribution without having some estimate of its expected value (mean) or standard deviation.

We will not concern ourselves with a lot of theoretical properties of estimators, but will merely outline a few approaches.

1. Method of Least–squares

Let

$$x = f(\theta_1, \theta_2, \ldots, \theta_k) + \varepsilon_i$$

where x_i are the n observations in a random sample, θ_i are the parameters of some function describing the x_i (possibly a density function) and ε_i are the random errors. Next, with

$$L = \sum_{i=1}^{n} \varepsilon_i^2 = \sum_{i=1}^{n} (x_i - f(\theta_1, \theta_2, \ldots, \theta_k))^2$$

we want to select the θ_i such that L is minimized. This can be found by solving the equations

$$\frac{\partial L}{\partial \theta_i} = 0 \,, \qquad i = 1, 2, \ldots, k$$

2. Method of Moments

In this method we estimate the parameters of a distribution by equating the theoretical moments to the estimation of the moments for the data. If we were operating with a single parameter distribution, the single equation

$$\mu = \bar{x} = \frac{1}{n} \sum_{i=1}^{n} x_i$$

would be adequate, if we could express μ as a function of the distribution parameters. If the distribution had more parameters we would use higher moments to generate additional equations.

3. Maximum Likelihood Estimation (MLE)

Assume we are dealing with a single parameter distribution with parameter θ and that $x_1, x_2, \ldots, x_n$ is a random sample, then the likelihood function is defined as

$$L(\theta) = f(x_1)f(x_2)\ldots f(x_n) = \prod_{i=1}^{n} f(x_i)$$

We desire to choose θ such that $L(\theta)$ is maximized. In simple terms, we are choosing the parameter value that maximizes the probability of observing the data contained in the sample. The form of $L(\theta)$ is such that it is usually terribly nonlinear and thus difficult to maximize. This can sometimes be lessened by taking the log of $L(\theta)$ and maximizing this function (they have the same maximizing value of θ). We will see examples of where this technique can be applied later.

It is interesting to note that the different methods of estimation often produce the same result. For example, all methods yield that the best estimate of the true mean μ, of a pdf is the simple mean $\bar{x}$. Also, the best estimator of σ^2 is

$$\frac{1}{n}\sum_{i=1}^{n}(x_i - \bar{x}).$$

The method of MLE is the most difficult, but possesses the best theoretical properties, while the method of moments is a good compromise and most popular.

There are a number of desirable properties that any estimate of a population parameter should have. Some of these are as follows

1. ***Unbiasedness***: An estimate $\hat{\theta}$ of some parameter θ is said to be unbiased if $E(\hat{\theta}) = \theta$. For example, the sample mean $\mathbf{X} = n^{-1}\sum_{i=1}^{n} x_i$ is an unbiased estimate of the population mean μ. On the other hand, the sample standard deviation given by

$$S = \left(\sum_{i=1}^{n} (x_i - \bar{x})^2 / (n-1) \right)^{1/2}$$

is not an unbiased estimate of the population standard deviation σ.
2. ***Minimum Variance*** : An estimate $\hat{\theta}$ of some population parameter θ is said to be a minimum variance estimate if $\sigma^2_{\hat{\theta}} \le \sigma^2_{\theta^*}$ for any other estimate θ^*.
3. ***Sufficiency*** : An estimate $\hat{\theta}$ of some parameter θ is said to be sufficient if it utilizes all information in the sample.
4. ***Consistency*** : An estimate $\hat{\theta}$ is said to be consistent if the estimate approaches the value of the parameter as the size of the sample increases.

It is not always possible to obtain an estimate that possesses all these (desirable) traits. For example, the sample standard deviation is normally used as an estimate of the population standard deviation even though it is not an unbiased estimate.

A point estimate of some population parameter provides some indication of the true value of that parameter. This indication may or may not be meaningful, since no indication of the possible degrees of accuracy in the estimate is present. For this reason, estimates of population parameters are normally stated in the form of **confidence interval estimates**.

5.5 Confidence Interval Estimates

This **interval estimate** is normally stated in the form $\hat{\theta} \pm \varepsilon$, where ε is some positive quantity related to the size of the sample and to the standard deviation of the estimate. A **confidence interval** estimate normally carries some degree of assurance that the actual parameter falls within the specified interval.

DEFINITION 5.4

*A **confidence interval** is defined by*

$$P(X_L \leq \theta < X_U) = 1 - \alpha$$

*where X_L and X_U are lower and upper bound **confidence limits** for a 100(1 – α) percent of time.*

Example 5.4 Suppose that x follows some unknown pdf with unknown μ, but known σ^2. We have estimated μ by taking a sample of size n and computed $\bar{x}$ (whose sampling distribution is normal with mean μ and variance σ^2 / n, often written as $N(\mu, \sigma^2 / n)$). Thus, $z = \dfrac{\bar{x} - \mu}{\sigma / \sqrt{n}}$ is $N(0, 1)$.

If we let $z_{\alpha/2} \geq \dfrac{\bar{x} - \mu}{\sigma / \sqrt{n}}$, we have $\mu \geq \bar{x} - \left(\dfrac{\sigma}{\sqrt{n}}\right) z_{\alpha/2}$.

If we let $-z_{\alpha/2} \geq \dfrac{\bar{x} - \mu}{\sigma / \sqrt{n}}$, we have $\mu \geq \bar{x} + \left(\dfrac{\sigma}{\sqrt{n}}\right) z_{\alpha/2}$.

Thus, $P\left(\bar{x} - \left(\dfrac{\sigma}{\sqrt{n}}\right) z_{\alpha/2} \leq \mu \leq \bar{x} + \left(\dfrac{\sigma}{\sqrt{n}}\right) z_{\alpha/2}\right) = (1 - \alpha)$.

This defines the (1 – α) 100% confidence interval around the mean.

It is important to note that the three key features involved in an interval estimate, namely the **confidence limits,** the **confidence interval**, and the **probability** that the parameter is contained within the interval, are all interrelated with one another.

A number of techniques are available for constructing interval estimates; the choice of a technique depends on the distribution being studied as well as the parameter being estimated. For this reason, techniques will be introduced as needed.

5.6 Exercises

5.1. The following observations were recorded for a random variable **X**: { - 2, 0, 6, 4, - 1, -5, 12, 13, 3, 11}. Compute the sample mean and standard deviation.

5.2 Construct an empirical cumulative distribution function for the data of Exercise 5.1 by connecting sequential observations with a straight line.

5.3 Construct an empirical cumulative distribution function for the data of Exercise 5.1 by fitting the data with a second-degree polynomial $F(x) = a_0 + a_1x + a_2x^2$ using the method of least squares.

5.4 Assuming $\sigma = 6$ and $\alpha = 0.05$, test the following hypotheses for the data of Exercise 5.1.

a. H_0: $\mu = 4$, $A : \mu = 4$
b. H_0: $\mu = 6$, $A : \mu = 6$
c. H_0: $\mu \leq 5$, $A : \mu > 5$

5.5 Show that $S^2 = \sum_{i=1}^{n} (x_i - \bar{x})^2 / (n-1)$ is an unbiased estimate of the population variance.

6

STATISTICAL TESTS

6.1 Tests of Hypotheses

An alternative to constructing an empirical distribution is to hypothesize that the sample points of a random variable **X** come from some known distribution. Statistical methods can then be used to assess the validity of the hypothesis. In this chapter we review some of the fundamental concepts involved in making and testing statistical hypotheses. In later sections we examine particular tests and explore their application to simulation.

A **statistical hypothesis** is an assumption about the population being sampled. It could consist of theorizing that the population follows a given distribution or that a particular parameter of the population is a certain value. A **test** of a hypothesis is simply a rule by which a hypothesis is either accepted or rejected. This decision is normally based on statistics obtained from an examination of a sample or set of samples from the population. These sample statistics are referred to as **test statistics** when used to test hypotheses. The **critical region** of a test statistic consists of the values of the test statistic that result in rejection of the hypothesis.

Example 6.1 A simulation experiment is being designed to examine the operation of a service station. It is thought that customers arrive at the station at an average rate of 12 per hour. To test this assumption, the hypothesis could be stated as

$$H_0 \colon \mu = 12.$$

The statement H_0 (called the **null hypothesis**) implies another hypothesis (called the **alternate hypothesis**) which is true if H_0 is determined to be false. For this experiment the alternate hypothesis can be stated as

$$A: \mu \neq 12.$$

Suppose that the researcher decides to reject the null hypothesis if a sample collected yields an average arrival rate less than 10 per hour or greater than 14 per hour. The test statistic is then the sample mean $\mathbf{X}$, while the critical region is $|\mathbf{X} - 12| > 2$.

Hypothesis testing is based on statistics gleaned from a sample. Anything based on sample statistics involves some likelihood of making an erroneous decision. The sample collected may not be representative of the true population; thus basing a decision on the sample could result in an incorrect decision. There are two types of errors that can be made. If the sample statistics lead to rejection of a null hypothesis that is actually true, a Type I error has been committed. The probability of making this type of error for a given test situation is normally denoted by α and is called the **level of significance**. If the sample statistics fail to reject a null hypothesis when it is actually false, a Type II error has been made. The probability of making this type of error is normally designated β. The test situation and possible decisions are summarized in Figure 6.1.

Test Result	True Situation	
	H_0 true	H_0 false
H_0 accepted	Correct decision	Type II error (β)
H_0 rejected	Type I error (α)	Correct decision

Figure 6.1. Decisions made in hypothesis testing.

In order to compare the risks of making errors for various tests, a plot of $Pr\{\text{Accepting}\}$ versus the true state of nature is used. This is called an **operating characteristic curve** or "OC" curve (see Figure 6.2).

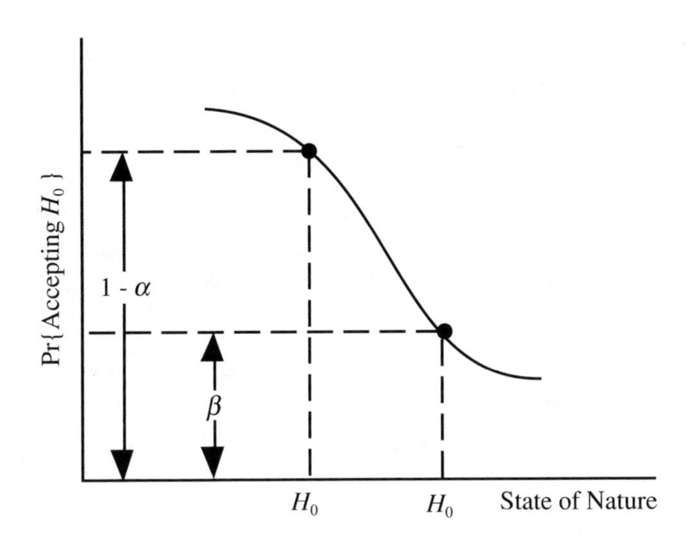

Figure 6.2. The operating characteristic curve.

Frequently, rather than plot β versus μ, one plots the complement $(1 - \beta)$ versus μ or $(1 - \alpha)$. The quantity $(1 - \beta)$ is known as the **power of the test**, $(1 - \alpha)$ the **extent of falsity**, and the plot of $(1 - \beta)$ versus μ or $(1 - \alpha)$, is known as the **power curve**. The typical power curve is "**S–shaped**", rising from coordinates near $(0, \alpha)$ to (a large value, almost 1).

Obviously one of the objectives in hypothesis testing is to minimize α and β, the probabilities of making an incorrect decision. Unfortunately if one probability is reduced, the other is increased. In fact, the only way to simultaneously decrease both risks is to base the decision on a sample statistic obtained from a larger sample. In most testing situations α is set at some predetermined acceptable level and the decision rule is formulated to minimize β.

The steps taken in hypothesis testing can be summarized as follows:

1. State the hypothesis to be tested (establish H_0 and A).
2. Determine an acceptable risk of rejecting a true null hypothesis (set α at some level).
3. Choose some suitable test statistic by which to test H_0. The choice of the test to be used will depend on what is known about the distribution of the population from which the sample is obtained. For example, a

nonparametric (or distribution free) test should be used if there is no knowledge about the distribution. On the other hand, parametric tests can generally be expected to give smaller β for fixed α and μ. Once the test has been chosen, fix α, then decide upon a suitable degree of difference (extent of falsity), fix β, and select an appropriate sample size. The statistical test is "often" viewed as a step in a decision process, and parameters of the test are so chosen as to optimize the cost consequences of the decision.

4. Assume H_0 is true and determine the sampling distribution of the test statistic.
5. Set up a critical region in which H_0 will be rejected $100\alpha\%$ of the time when α is true.
6. Collect a random sample of some predetermined size, compute the test statistic, and test the hypothesis. It is "assumed" that the observations are based on a random sample from the total population. That is, the sampling procedure must guarantee that each element in the population has the same chance of being included in the sample.

Example 6.2 Consider an example where a passing grade is of question. Suppose we want to decide if the mean passing grade is 72 or not.

1. H_0: $\mu = 72$, $A : \mu \neq 72$ (Note: mutually exclusive and exhaustive)
2. State sample size (n) and acceptance region (AR).

Consider three cases:

Test Statistic	n	*Acceptance Region (AR)*
$\bar{x}$, mean	4	$70.7 \le \bar{x} \le 73.3$
$\tilde{x}$, median	3	$70 \le \tilde{x} \le 74$
$x_{\max}$	4	$70 \le x_{\max} \le 74$

3. For illustrative purposes, consider the different OC curves of Figure 6.3. This figure illustrates the shape of the type of OC curve for which $\tilde{x}$ is best at 72, $x_{\max}$ is best for $70 \le \mu \le 74$, and $\bar{x}$ otherwise.

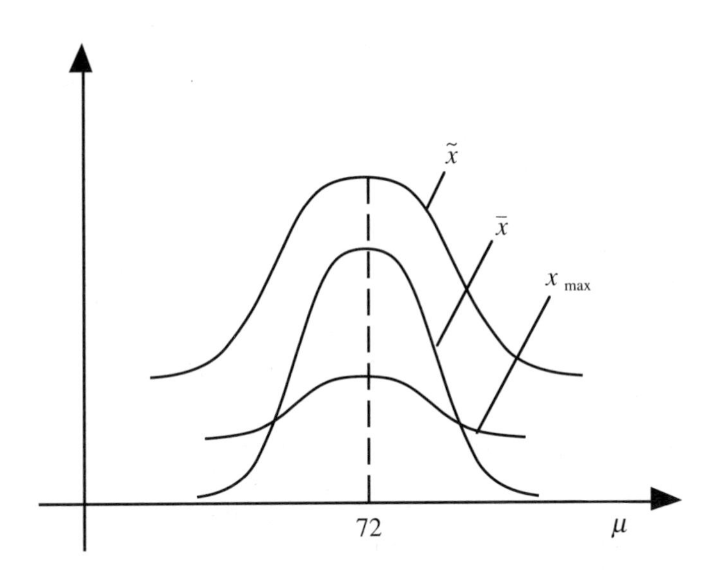

Figure 6.3. OC curves for Example 6.2.

However, in order to compare OC curves, we must decide on a reasonable α, and adjust the acceptance regions so that all tests have that same value (see Figure 6.4).

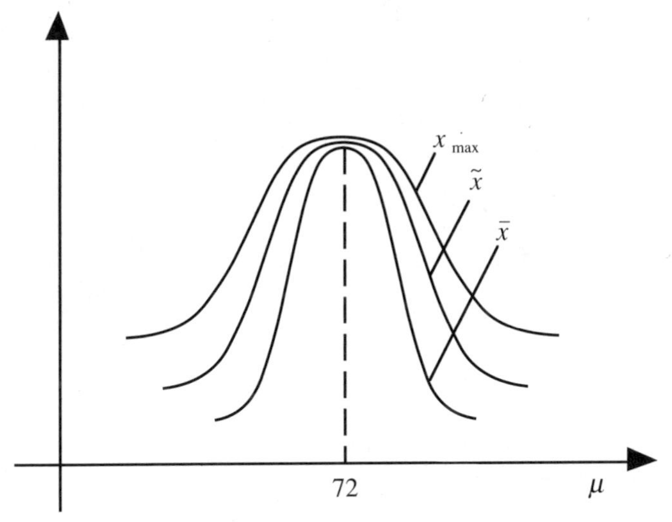

Figure 6.4. OC curves with adjusted acceptance regions.

Once this adjustment has been made, we can clearly see that $\bar{x}$ is the superior test statistic.

Example 6.3 Consider the experiment described in Example 6.1. A test of the stated hypothesis following the procedure just described is as follows:

1. H_0: $\mu = 12$, $A : \mu \neq 12$
2. Set $\alpha = 0.05$. That is, assume that a risk of rejecting a true null hypothesis 5% of the time is acceptable.
3. The test statistic to be used is $\overline{X}$, the average arrival rate determined from some sample. This statistic can be standardized as

$$\overline{X}: \; Z = \frac{\overline{X} - \mu}{\sigma / \sqrt{n}}$$

   where σ is the true standard deviation of the population and n is the size of the sample.
4. $\overline{X}$ is normally distributed with mean μ and standard deviation $\sigma / \sqrt{n}$. This result follows from the central limit theorem if n is large enough.
5. The critical region on Z for a risk of α =0.05 is depicted in Figure 6.5. Then H_0 will be rejected if

$$\left| \frac{\overline{X} - \mu}{\sigma / \sqrt{n}} \right| > 1.96$$

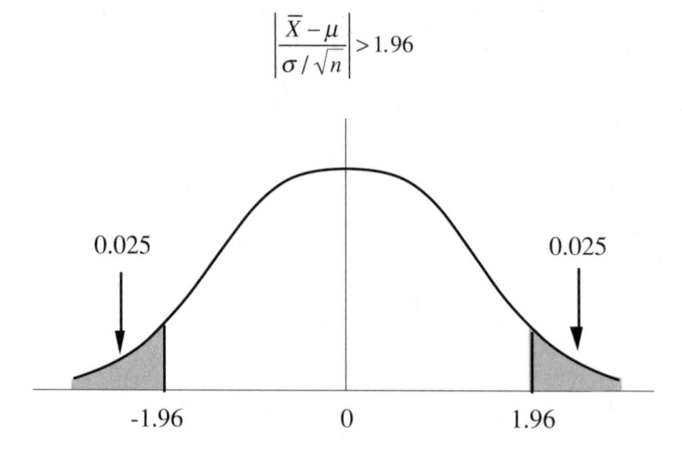

Figure 6.5. The critical region.

6. Assume that a sample of size 16 was collected, that $\sigma = 2$, and that the sample yielded $\overline{X} = 14$. Then

$$Z = \frac{14 - 12}{1} = 2$$

   So this sample would lead to rejection of H_0. The conclusion can then be drawn that $\mu \neq 12$.

The procedure outlined for testing a hypothesis does not consider β, the probability of a Type II error. To compute β, one must assume H_0 false and some specific alternative true. For example, if H_0 in Example 6.1 was assumed false and the alternate hypothesis A: $\mu = 14$ was assumed true, the situation would be as shown in Figure 6.6. The shaded area is β and can be calculated from the normal distribution function as

$$\begin{aligned}\beta &= P(\overline{X} \leq 13.96) \\ &= P\left(Z \leq \frac{13.96 - 14}{1}\right) \\ &= P(Z \leq -0.04) = 0.4840\end{aligned}$$

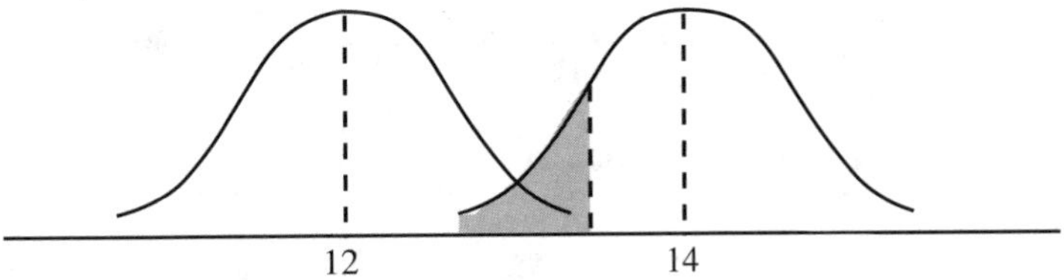

Figure 6.6. The value for β.

By computing β for various alternate hypotheses and plotting β versus μ, one obtains the **operating characteristic (OC) curve**. The OC curve for the data of Example 6.1 is shown in Figure 6.7.

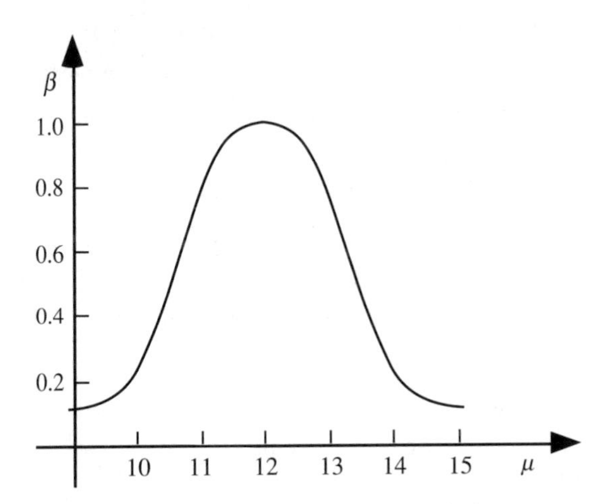

Figure 6.7. An operating characteristic curve.

The **power curve** for the data of Example 6.1 is shown in Figure 6.8.

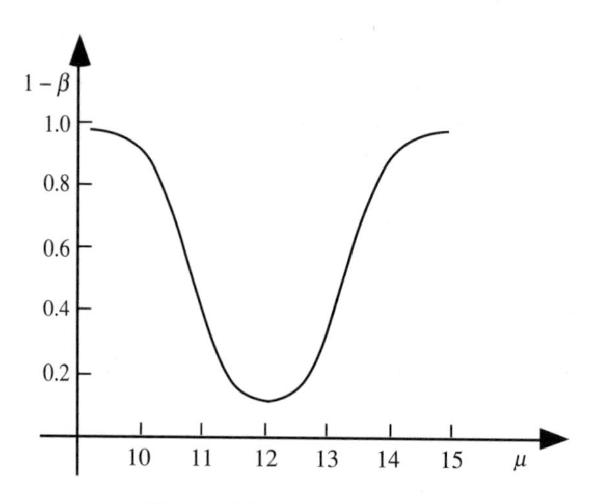

Figure 6.8. A power curve.

Note that rejection of a null hypothesis is a positive action, whereas acceptance of the null hypothesis is no action. This is pointed out through examination of the OC or power curves. In the previous example, if the true mean was 12.5, the null hypothesis H_0: $\mu = 12$ would be accepted with nearly as high a probability as if the mean was actually 12. Thus the test is not likely to detect small differences between the hypothesized and true parameter values. It will, however, detect large differences with a high probability. This is the reason for the statement that rejection of the null hypothesis is a positive action. Acceptance of the null hypothesis can be viewed as inconclusive. The hypothesis may be false, but we do not have sufficient evidence to prove it.

Many tests have been used in testing hypotheses. Which test to use depends on what one is trying to establish as well as how much is known about the underlying population. For example, Table 6.1 provides some guidance for testing hypothesis about means, while Table 6.2 illustrates the usefulness of various parametric and non parametric tests.

The next four sections will discuss four of the more useful tests, indicating the assumptions underlying their use as well as their utility.

Table 6.1 Tests of Hypothesis for Means

Null Hypothesis (Condition)	Test Statistic	Test Statistic Distribution	Degrees of Freedom	Alternative Hypothesis	Null Hypothesis Rejection Rule
$\mu = \mu_0$ (Known $\sigma_{\bar{X}}$)	$z = \dfrac{\bar{X} - \mu}{\sigma_{\bar{X}}}$	Standard Normal	—	$\mu > \mu_0$ $\mu < \mu_0$ $\mu \neq \mu_0$	$z > z_\alpha$ $z < -z_\alpha$ $\lvert z \rvert > z_{\alpha/2}$
$\mu = \mu_0$ (Unknown $\sigma_{\bar{X}}$)	$t = \dfrac{\bar{X} - \mu}{S_{\bar{X}}}$	Student *t*	$N - 1$	$\mu > \mu_0$ $\mu < \mu_0$ $\mu \neq \mu_0$	$t > t_\alpha$ $t < -t_\alpha$ $\lvert t \rvert > t_{\alpha/2}$
$\mu_{\bar{X}} = \mu_{\bar{Y}}$ $\mu_{\bar{X}} = \mu_{\bar{Y}}$ (Known $\mu_{\bar{X}}$ and $\mu_{\bar{Y}}$)	$z = \dfrac{X - Y}{\sqrt{\sigma_{\bar{X}}^2 + \sigma_{\bar{Y}}^2}}$	Standard Normal	—	$\mu_{\bar{X}} > \mu_{\bar{Y}}$ $\mu_{\bar{X}} < \mu_{\bar{Y}}$ $\mu_{\bar{X}} \neq \mu_{\bar{Y}}$	$z > z_\alpha$ $z < z_\alpha$ $\lvert z \rvert > z_{\alpha/2}$
$\mu_{\bar{X}} = \mu_{\bar{Y}}$ (Unknown $\mu_{\bar{X}}$ and $\mu_{\bar{Y}}$)	$t = \dfrac{\bar{X} - \bar{Y}}{\sqrt{S_{\bar{X}}^2 + S_{\bar{Y}}^2}}$	Student *t*	Nearest integer to: $\dfrac{\left(S_{\bar{X}}^2 + S_{\bar{Y}}^2\right)^2}{\dfrac{S_{\bar{X}}^4}{N_X} + \dfrac{S_{\bar{Y}}^2}{N_Y}} - 2$	$\mu_{\bar{X}} > \mu_{\bar{Y}}$ $\mu_{\bar{X}} < \mu_{\bar{Y}}$ $\mu_{\bar{X}} \neq \mu_{\bar{Y}}$	$t > t_\alpha$ $t < -t_\alpha$ $\lvert t \rvert > t_{\alpha/2}$

Table 6.2 Parametric and Nonparametric Tests

Name	Type	Used to Test
Student's *t*–test	Parametric	Differences in levels of two samples from a normal dis–tribution with unknown μ, and differences in regression coefficient.

Table 6.2 Parametric and Nonparametric Tests (continued)

F– test	Parametric	Dispersion of two samples from a normal population; or levels of several samples from normal population with equal variance.
Poisson parameter test	Parametric	Differences between Poisson parameter and a prescribed constant.
Kolmogorov–Smirnov test	Nonparametric	Goodness of fit to any cumulative distribution.
Mann–Whitney test	Nonparametric	Differences in distribution of two samples from any population.
Runs test	Nonparametric	Patterns in a sequence of observations from any distribution.
Chi–square test	Nonparametric	Goodness of fit to any density function.

6.2 Student's *t*- test

The ***t* – test** is used to test the hypothesis that two normal populations have the same mean. As the basis for the test, assume that we have two normally distributed populations, the first with mean μ_1 and variance σ_1^2 and the second with mean μ_2 and variance σ_2^2. In general, the variances are unknown, so they must be estimated from sample data. Assume further that a sample of size n_1, denoted $x_1, x_2, \ldots, x_{n_1}$, has been collected from the first population, while a sample of size n_2, denoted $y_1, y_2, \ldots, y_{n_2}$, has been collected from the second. The specific hypotheses we wish to test are

$$H_0: \mu_1 = \mu_2, \qquad A: \mu_1 \neq \mu_2$$

or alternately,

$$H_0 : \mu_1 - \mu_2 = 0 \qquad A : \mu_1 - \mu_2 \neq 0$$

If the variables x_i, y_i are independent, a further assumption required to yield an exact test is that the variances σ_1^2 and σ_2^2, while unknown, are equal. If this assumption is not valid, no exact test exists. However, there are approximate tests that may be used. These approximate tests will not be discussed here but are available in most standard statistical texts. Under the assumption of equal variances, the test statistic for testing the equality of mean is

$$t = \frac{(\overline{X} - \overline{Y})}{[S^2(1/n_1 + 1/n_2)]^{1/2}}$$

Where $\overline{X}$ is the sample mean from the first population, $\overline{Y}$ is the sample mean from the second population, and S^2 is an estimate of the variance given by

$$S^2 = \frac{[(n_1 - 1)S_1^2 + (n_2 - 1)S_2^2]}{(n_1 + n_2 - 2)}$$

S_1^2 is the sample variance from the first population, and S_2^2 is the sample variance from the second population. The sample variance S^2 is sometimes referred to as a **pooled S^2.** If the null hypothesis is true, t is distributed according to Student's t with $n_1 + n_2 - 2$ degrees of freedom. The critical region for the test is then

$$|t| > t_{(1-\alpha/2),(n_1+n_2-2)}$$

Example 6.4 Suppose that ten observations were taken on a random variable $\overline{X}$ and that the results were {1, 2, – 1, 3, 7, 8, 9, 4, 3, 2}. Assume further that eight observations were taken on a random variable $\overline{Y}$ and that the results were {0, 3, 6, –2, 4, 0, 7, 8}. Assume that the observations come from normal populations and that the variances are equal. Then

$$\overline{X} = \sum_{i=1}^{10} \frac{x_i}{10} = 3.8 \qquad \overline{Y} = \sum_{i=1}^{8} \frac{y_i}{10} = 3.25$$

$$S_1^2 = \sum_{i=1}^{10} \frac{(x_i - \overline{x})^2}{9} = 10.4 \qquad S_2^2 = \sum_{i=1}^{8} \frac{(y_i - \overline{y})^2}{7} = 13.36$$

$$S^2 = \frac{(9(10.4) + 7(13.36))}{16} = 11.695$$

The hypothesis for testing for equality of mean are

$$H_0\text{: } \mu_1 - \mu_2 = 0 \qquad A\text{: } \mu_1 - \mu_2 \neq 0$$

The test statistic is

$$t = \frac{(\overline{X} - \overline{Y})}{[S^2(1/n_1 + 1/n_2)]^{1/2}}$$

$$= \frac{(3.8 - 3.25)}{[(11.695)(1/10 + 1/8)]^{1/2}} = 0.34$$

If $\alpha = 0.05$, then $t_{0.05,\,16} = 2.120$. Since $t = 0.34 < 2.120$, we cannot reject the hypothesis that the means are equal.

Now suppose that the observations from the two populations cannot be considered independent but that corresponding observations can be paired. That is, they were both taken when conditions were similar (at the same time, and so on). This of course assumes that the size of each sample is the same, say n. Under these assumptions, rather than consider each random variable separately, we consider the differences D_i, $i = 1, 2, ..., n$, as a random sample of size n from a normal distribution with mean $\mu = \mu_1 - \mu_2$ and variance σ^2. The test for equality of means is then

$$H_0\text{: } \mu = 0 \qquad A\text{: } \mu \neq 0$$

The appropriate test statistic for testing this hypothesis is $t = n^{1/2}\overline{D} / S_D$ where $\overline{D}$ is the average difference and S_D is given by

$$S_D = \left[\frac{\sum_{i=1}^{n} (D_i - \overline{D})^2}{(n-1)} \right]^{1/2}$$

If the null hypothesis is true, this statistic is distributed according to Student's t within $n - 1$ degrees of freedom. The critical region is then

$$|t| > t_{\alpha, n-1}$$

Example 6.5 Suppose the following paired observations were taken on the random variables $\overline{X}$ and $\overline{Y}$.

i	1	2	3	4	5	6	7	8
x_i	1	0	3	6	2	1	7	6
y_i	–2	4	–2	1	4	7	3	4
$d_i = x_i - y_i$	3	–4	5	5	–2	–6	4	2

Then

$$\overline{D} = \frac{\sum_{i=1}^{8} d_i}{8} = 0.875, \qquad S_D^2 = \frac{\sum_{i=1}^{8} (d_i - \overline{D})^2}{7} = 18.41$$

and the test statistic is $t = n^{1/2}\overline{D} / S_D = (8^{1/2})\,(0.875)/\,4.29 = 0.577$. If $\alpha = 0.05$, then $t_{0.05,7} = 2.365$. Again we are not able to reject the hypothesis of equal means.

The two tests examined in this section are two–tailed, meaning that we are allowed rejection for extreme values above or below the hypothesized values. If one wished, for example, to test the hypothesis H_0: $\mu_1 \le \mu_2$ versus A: $\mu_1 > \mu_2$ where the samples are independent, a similar test is used, except that now rejection is possible only for large values of the test statistic. Rejection occurs only on one side of the hypothesized value. The only difference in the procedure is the calculation of the critical region. In this case the critical region is

$$t \geq t_{\alpha,(n_1+n_2-2)}$$

A similar procedure exists for paired values.

To summarize, the t–test is designed to test for equality of means when samples have been drawn from two normal populations. If the samples are independent, an exact test exists only if it can be assumed that the variance of the two populations are equal. In the case of correlated samples the differences rather than the individual observations are considered.

6.3 The *F* - test

The t–test discussed in Section 6.2 is used to test the hypothesis that two normal populations have the same mean. The ***F*–test** is used to test for equality of variances. Again an underlying assumption is that the two populations being sampled are normally distributed. Assume, just as in the previous section, that samples are taken from two normal populations with means μ_1 and μ_2 and variances of σ_1^2 and σ_2^2 respectively. Let the sample taken from the first population be represented as $x_1, x_2,...,x_{n_1}$ and the sample from the second population be represented as $y_1, y_2,...,y_{n_2}$. The hypotheses that we wish to test in this case are

$$H_0: \ \sigma_1^2 = \sigma_2^2 \qquad A: \ \sigma_1^2 \neq \sigma_2^2$$

The test statistic used to test these hypotheses is

$$F = S_1^2 \ / \ S_2^2$$

where S_1^2 is the sample variance of the first sample and S_2^2 is the sample variance of the second sample. If the null hypothesis is true, this test statistic is distributed according to the F–distribution, with $v_1 = n_1 - 1$ and $v_2 = n_2 - 1$. The critical region for the test is then

$$F \le F_{(1-\alpha/2),(n_1-1,n_2-1)} \quad \text{or} \quad F \ge F_{\alpha/2,(n_1-1,n_2-1)}$$

Example 6.6 Assume that samples are taken from two normal populations and that the following observations are recorded.

i	1	2	3	4	5	6	7	8	9	10
x_i	–5	4	–8	14	21	16	0	1	–	0
i	1	2	3	4	5	6	7	8		
y_i	0	1	–1	2	–2	0	1	2		

Then $S_1^2 = 73.73$ while $S_2^2 = 1.98$. The test statistic is

$$F = \frac{S_1^2}{S_2^2} = \frac{73.73}{1.98} = 37.2$$

If $\alpha = 0.05$, the critical region is

$$F \le F_{(0.975),(9,7)} = 4.82 \quad \text{or} \quad F \ge F_{0.025,(9,7)} = 0.238$$

Then the hypothesis of equal variances is rejected. Note that with these samples, the hypothesis would be rejected even if α were chosen as small as 0.00001. Thus one can say with a great deal of confidence that the population variances are not the same.

Recall from the previous section that to have an exact test for comparing the means of two normal populations when the samples are independent, we had to assume equal variances. The F–test is useful in testing this assumption.

Just as with the t–test, there are also one–sided F–tests. The test statistic used is the same as that for the two–sided test. Definition of the critical region does change, however. The appropriate hypotheses and critical regions for the one–sided tests are as follows.

Hypotheses	*Critical region*
$H_0:\ \sigma_1^2 \le \sigma_2^2$ $A:\ \sigma_1^2 > \sigma_2^2$	$F \le F_{\alpha,(n_1-1,n_2-1)}$
$H_0:\ \sigma_1^2 \ge \sigma_2^2$ $A:\ \sigma_1^2 < \sigma_2^2$	$F \ge F_{(1-\alpha),(n_1-1,n_2-1)}$

Example 6.7 The data in Example 6.6 resulted in rejection of the hypothesis of equal variances. Using that data, let's test the hypotheses

$$H_0:\ \sigma_1^2 \le \sigma_2^2, \qquad A:\ \sigma_1^2 > \sigma_2^2$$

If $\alpha = 0.05$, the critical region is$F \ge F_{0.05,(9,7)} = 3.68$. Thus, with the computed F of 37.2, we would also reject this null hypothesis.

To summarize, the F–test is useful in comparing the variances of two normal populations. Both two–tailed and one–tailed tests exist. The test statistic is the same in each case, but the critical regions differ.

6.4 The Chi–square Goodness–of–Fit Test

One of the major difficulties facing a researcher designing a simulation experiment is the characterization of the random variables of interest. Before the operation of a service station can be modeled, for example, the distribution of customers arriving at the station must be characterized. In many cases the random variable of interest is assumed to follow a particular distribution. Of course, the results obtained by the simulation study are usually very sensitive to this assumption. Thus there must be a method by which the assumption of a particular distribution can be checked. The **chi–square goodness–of–fit test** has proven useful in this regard.

This test makes a comparison between the actual and expected number of observations for various values of the random variable. The hypothesis that the observed and assumed distribution are the same is tested using the test statistic

$$\chi^2 = \sum_{i=1}^{r} \frac{(O_i - E_i)^2}{E_i}$$

where O_i is the observed frequency of observations in the ith interval and E_i is the frequency of observation if the assumed distribution is correct. A step–by–step procedure for conducting this test is as follows.

1. Construct a frequency table of the observed values of the random variable. The number of intervals is somewhat arbitrary; however, experience has shown that intervals with fewer than three to five observations tend to distort the test results.
2. Calculate the expected or theoretical frequencies for each interval under the assumption that the hypothesized distribution is correct. The parameters of the hypothesized distribution are usually estimated from the sample data.
3. Calculate the quantity $(O_i - E_i)^2/E_i$ for each interval, $i = 1, 2, ..., r$, where O_i is the actual frequency of observations expected under the hypothesized distribution.
4. Calculate the chi–square statistic using the formula

$$\chi^2 = \sum_{i=1}^{r} \frac{(O_i - E_i)^2}{E_i}$$

The degrees of freedom for this statistic are $r - p - 1$, where r is the number of intervals and p is the number of parameters estimated for the hypothesized distribution. As an example, suppose that there are ten intervals and that the distribution is assumed to be normal. Since both μ and σ^2 have to be estimated, the degrees of freedom for χ^2 are $10 - 3 = 7$. On the other hand, if the Poisson distribution is assumed, a single parameter needs to be estimated, so the degrees of freedom for χ^2 are $10 - 2 = 8$.

5. Choose a value for α and test the hypothesis, rejecting the hypothesis that the assumed and actual distributions are the same if

$$\chi^2 \geq \chi^2_{\alpha,r-p-1}$$

Example 6.8 A researcher is attempting to characterize the arrival pattern of customers to a service station. Through observations of the station over an extended period of time, the researcher constructed the following frequency table, where the quantity of interest is the number of customers arriving per hour.

Customers	*Observed frequency*
0 – 2	13
3 – 5	17
6 – 8	21
9 – 11	42
12 – 14	16
15 – 17	11
18 – 20	5

The researcher hypothesizes that the arrival pattern is Poisson. An estimate for λ, the average number of customers per hour, computed from this frequency table is $\overline{\mathbf{X}} = 9.0$. From the cumulative Poisson table with $\lambda = 9.0$, the following table of observed versus expected frequencies can be tabulated.

Customers	*Observed*	*Expected*
0 – 2	13	0.75
3 – 5	17	13.75
6 – 8	21	42.5
9 – 11	42	43.375
12 – 14	16	19.5
15 – 17	11	4.0
18 – 20	5	1.125

Now combining the first two intervals and the last two intervals gives

$$\chi^2 = \sum_{i=1}^{5} \frac{(O_i - E_i)^2}{E_i}$$

$$= \frac{(30-14.5)^2}{14.5} + \frac{(21-42.5)^2}{42.5} + \frac{(42-43.375)^2}{43.375} + \frac{(16-19.5)^2}{19.5} + \frac{(16-5.125)^2}{5.125}$$

$$= 51.2$$

Then since $\chi^2 = 51.2 > \chi^2_{.01,3} = 11.3$, the hypothesis that the arrival pattern is Poisson can be rejected with $\alpha = 0.01$.

Example 6.9 The scores on an exam in a large mathematics course are thought to be normally distributed. The actual versus expected frequencies (calculated from the cumulative normal with estimated mean 73.3 and standard deviation 10.05) are as follows.

Score	*Observed*	*Expected*	$(O_i - E_i)^2 / E_i$
50–59	16	11.98	1.35
60–69	31	39.39	1.79
70–79	72	58.84	2.94
80–89	26	34.65	2.16
90–99	9	8.33	0.05

Then $\chi^2 = 8.29$. If $\alpha = 0.01$, the critical region is $\chi^2 \leq \chi^2_{.01,2} = 9.21$. Thus we are unable to reject the hypothesis that the scores are normally distributed.

6.5 The Kolmogorov–Smirnov Test

An alternative to the chi–square goodness–of–fit test is the **Kolmogorov–Smirnov** test. Like the chi–square test, it is used to test the hypothesis that a sample follows some hypothesized distribution. It is somewhat more powerful than the chi–square test. Thus it would be more likely to detect small differences in the actual and hypothesized distributions. The techniques used in this test are as follows.

1. Let $S(x)$ be the empirical cumulative distribution function constructed from a sample of N observations using the technique outlined in Chapter 5.
2. Let $F(x)$ be the theoretical cumulative distribution function assuming that the null hypothesis is true.
3. For each of the N sample points, compute $F(x_i) - S(x_i)$. Let

$$D = \max_i |F(x_i) - S(x_i)|$$

4. Choose some value of α, and if the calculated value of D is greater than the tabulated critical value at that level of significance, reject the hypothesis.

Example 6.10 Consider the data of Example 6.8. Using the right endpoint of each interval as the value of the sample, one can construct the following table.

x	$S(x)$	$F(x)$	$\lvert F(x) - S(x)\rvert$
2	0.104	0.006	0.098
5	0.240	0.116	0.124
8	0.408	0.456	0.048
11	0.744	0.803	0.059
14	0.872	0.959	0.087
17	0.960	0.995	0.035
20	1.000	1.000	0.000

In this table the hypothesized distribution was the Poisson distribution with $\lambda = 9.0$. The true sample mean for this sample was 9.02. It was rounded down to correspond to the table entries.

Now if $\alpha = 0.05$, the critical value from the table is $1.36/\sqrt{125} = 0.122$. Since $D = 0.124 > 0.122$, the hypothesis that the distribution is Poisson with mean 9.0 is rejected.

Example 6.11 Consider the data of Example 6.9. Using the right endpoint of each interval as the value of the sample and the normal distribution (mean 73.3, standard deviation 10.05) as the hypothesized distribution, one can construct the following table.

x	$S(x)$	$F(x)$	$\lvert F(x) - S(x) \rvert$
59	0.10390	0.07780	0.0261
69	0.30519	0.33360	0.0284
79	0.77273	0.71566	0.0571
89	0.94156	0.94062	0.0009
99	1.00000	0.99477	0.0052

If $\alpha = 0.01$, the critical value is $1.63/\sqrt{154} = 0.1313$. Since $D = 0.0571 < 0.1313$, we are unable to reject the hypothesis that the distribution is normal with mean 73.3, standard deviation 10.05.

6.6 Summary

In this chapter we have reviewed the fundamental concepts of estimation and hypothesis testing. We have also examined four of the more common statistical tests used in simulation studies. As in the previous chapter, the coverage was brief. The reader desiring more thorough coverage should consult any of many standard statistical texts.

6.7 Exercises

6.1 Independent samples were drawn from two normal populations and recorded as **X** and **Y**.

X : {2, 4, 6, 7,9 , 14, 16, 3}
Y : {- 1, 4, 1, 6, 12, 12, 1, 0}

Under the assumption of equal variances, test the hypothesis of equal means, with $\alpha = 0.05$.

6.2 Test the assumption of equal variances made in Exercise 6.1.

6.3 If the observations in Exercise 6.1 are paired rather than independent, test the hypothesis of equal means.

6.4 Messages arriving at a communications center are thought to follow a Poisson distribution. This of course implies that the interarrival distribution is exponential. Test this hypothesis using the chi-square test given the following frequency table of interarrival times.

Time	0-1	1-2	2-3	3-4	4-5	5-6	6-7
Frequency	2	6	12	16	21	11	6

6.5 Test the hypothesis of Exercise 6.4 using the Kolmogorov-Smirnov test.

6.6 The following frequency table was recorded for some random variable **X**. Test the hypothesis that **X** comes from a normal distribution using the

a. Chi-square test
b. Kolmogorov-Smirnov test

Interval	0-5	5-10	10-15	15-20	20-25
Frequency	0	2	2	6	10
Interval	25-30	30-35	35-40	40-45	45-50
Frequency	20	28	32	32	28
Interval	50-55	55-60	60-65	65-70	70-75
Frequency	18	12	6	4	0

7

GENERATION OF RANDOM NUMBERS

In many simulations events appear to occur at random or to involve attributes whose values must be assigned somewhat by chance. This occurs because most simulations are based on knowledge expressed as general or historical relationships. For instance, in many cases the duration of an event is known to fall within a certain range. Simulation of the event requires that a particular value be assigned. Consider the simulation of a general–purpose computer system. One event that must be modeled is the retrieval of a record from a direct–access storage device. The duration of this event can be determined to fall within a certain interval; the actual value, however, is influenced by chance variables such as the position of the record relative to the read head when the request is made. Another instance in which chance appears to play a part is in the widespread use of decision logic in simulation. For example, suppose that in the operation of a system, a given path is known to be taken a certain percentage of the time. Simulation of the system requires a method for selecting this path over others so that the long–run behavior of the simulator is similar to that of the actual system. Since in most cases these decisions are nondeterministic, the choice is normally based on probabilistic relationships.

For these reasons and others, one of the requirements of almost any simulation model is some facility for generating random numbers. One must be able to assign a particular value to seeming random events on any given simulation run. In the previous example of the simulation of a computer system, one must assign a particular value to the time required to retrieve a record from a direct–access storage device each time a simulated request is received. In the case of a decision block one must be able to direct the simulator to take a given path on certain runs and other paths on other runs.

Modeling of random events, as described in Chapters 2 and 3, is done using the uniform distribution. Thus random numbers can be generated by sampling from this distribution. Some sources of true random numbers that have been used are (1) a sack of unnumbered beads that can be sampled with replacement, (2) low–order digits on a microsecond clock, (3) a random electronic noise source whose output is quantized periodically. This last technique was used by the RAND Corporation to generate its widely published table of 1 million random digits (7.5). These techniques appear to generate random numbers, but they all have a common disadvantage when used in simulation studies; sequences generated by these techniques are generally not reproducible, and reproducibility is a requirement in most cases.

7.1 Pseudorandom Numbers

A number of techniques have been applied to overcome the inherent nonreproducibility of random sequences. Before considering some of these, we might find it useful to discuss some of the requirements of a random number generator.

1. Numbers produced must follow the uniform distribution, because truly random events follow this distribution. Any simulation of random events must therefore follow it at least approximately. For goodness of fit, a chi–square is recommended with a large sample of N between 1000 – 10000 numbers.

2. Numbers produced must be statistically independent. The value of one number in a random sequence must not affect the value of the next number. Tests appropriate for independence include the serial test, the runs test, the poker test, the spectral test, and the autocorrelation test. None of the previous tests actually test independence. They test for properties consistent with independence. Their rejection supports lack of independence, but their acceptance does not prove independence.

3. The sequence of random numbers produced must be reproducible. This allows replication of the simulation experiment.

4. The sequence must be nonrepeating for any desired length. This is not theoretically possible, but for practical purposes a long repeatability cycle is adequate. The repeatability cycle of a random number generator is known as its **period**.

5. Generation of the random numbers must be fast. In the course of a simulation run a large number of random numbers are usually required. If the generator is slow, it can greatly increase the time and thus the cost of the simulation run.

6. The method used in the generation of random numbers should use as little memory as possible. Simulation models generally have large memory requirements. Since memory is usually limited, as little as possible of this valuable resource should be devoted to the generation of random numbers.

In considering pseudorandom number generators, the test for uniformity is always applied. Next one considers tests of independence depending on how strongly independence need be assured. The need for independence may not be as great if "randomization" is enhanced by one or more of the following: multiple generators or streams, sort and retrieve strategies, and programmatic design. A quick test is usually a "runs" test, while a discerning (and more time consuming) is either a serial test or a auto–correlation test.

With these requirements, it is now possible to evaluate the approaches taken to compensate for the lack of reproducibility of random sequences. The first approach is to generate the sequence by some means and to store it, say on tape. This approach is generally unsatisfactory because of the time involved. Each time a random number is required, a read operation must be initiated, and this is a time–consuming operation. This technique also potentially suffers from a short repeatability cycle unless a large sequence is stored. The second approach is to generate a random sequence and hold it in memory. This approach would overcome the speed problem of the previous technique; however, to store a list large enough to satisfy the requirements of many simulation studies would require an inordinate amount of core. The third and most common approach is to use a specified input value to generate a random number using some algorithm. This technique overcomes the problems of speed, and memory requirements but suffers from potential problems with independence and repeatability.

The use of an algorithm to generate random numbers seems to violate the basic principle of randomness. For this reason numbers generated by an algorithm are called **synthetic or pseudorandom numbers**. These numbers meet certain criteria for randomness but always begin with a certain initial value called the **seed** and proceed in a completely deterministic, repeatable fashion. Extreme care must be taken when using pseudorandom sequences to insure that a fair degree of randomness is present (that the uniform distribution is followed) and that the repeatability cycle is long enough. Random numbers are also important to simulation studies that much work has been done in devising and testing algorithms that produce pseudorandom sequences of numbers.

7.2 Algorithms for Generating Pseudorandom Numbers

A lot of work has been done in designing and testing algorithms to produce pseudorandom number sequences. The algorithms differ not only in technique but in speed of generation, length of the repeatability cycle, and ease of programming. Some of the more common algorithms are surveyed in this section.

7.2.1 The Midsquare Method

The **midsquare technique** was developed in the mid 1940s by John von Neuman (7.3). The technique starts with some initial number, or seed. The number is then squared, and the middle digits of this square are used as the second number of the sequence. This second number is then squared, and the middle digits of this square are used as the third number of the sequence. The algorithm continues in this fashion.

Example 7.1 Suppose that one wishes to generate a sequence of four–digit random numbers using the midsquare method. Let the first number of the sequence be 3187. Then

$$
\begin{aligned}
x_0 &= 3187 \\
(3187)^2 &= 10156969 \Rightarrow x_1 = 1569 \\
(1569)^2 &= 02461761 \Rightarrow x_2 = 4617 \\
(4617)^2 &= 21316689 \Rightarrow x_3 = 3166 \\
(3166)^2 &= 10023556 \Rightarrow x_4 = 0235 \\
(0235)^2 &= 00055225 \Rightarrow x_5 = 0552 \\
(0552)^2 &= 00304704 \Rightarrow x_6 = 3047 \\
(3047)^2 &= 09284209 \Rightarrow x_7 = 2842
\end{aligned}
$$

This process could be continued to produce 0769, 5913, 9635, 8332, 4222, 8542,. . . .

This technique has a number of shortcomings. First, sequences generated by this technique generally have short repeatability periods. Second, of the longer sequences, the numbers may not pass the statistical tests for randomness. Third, whenever a zero is generated, all succeeding numbers will be zero. This phenomenon, if it occurred in the midst of a large simulation study could drive even the most conscientious analyst to distraction; it is illustrated by the following example.

Example 7.2 Suppose that one wishes to generate two–digit random numbers beginning with 44. Then

$$
\begin{aligned}
x_0 &= 44 \\
(44)^2 &= 1936 \Rightarrow x_1 = 93 \\
(93)^2 &= 8649 \Rightarrow x_2 = 64 \\
(64)^2 &= 4096 \Rightarrow x_3 = 09 \\
(09)^2 &= 0081 \Rightarrow x_4 = 08 \\
(08)^2 &= 0064 \Rightarrow x_5 = 06 \\
(06)^2 &= 0036 \Rightarrow x_6 = 03 \\
(03)^2 &= 0009 \Rightarrow x_7 = 00 \\
(00)^2 &= 0000 \Rightarrow x_8 = 00
\end{aligned}
$$

The problems with the midsquare method led to development of alternate algorithms to provide a more reliable source of random numbers.

7.2.2 The Linear Congruential Method

Most of the random number generators used are modifications of the linear congruential scheme devised by Lehmer (7.4). In this algorithm successive numbers in the sequence are generated by the recursion relation

$$x_{n+1} = (ax_n + c) \bmod m, \qquad n \geq 0$$

The initial value x_0 is known as the **seed**, the constant a is the **multiplier**, the constant c the **increment**, and the m the **modulus**. The selection of values for these constants has a dramatic effect on the length of the period of the generated sequence of random numbers.

Example 7.3 Let $a = 2$, c = 3, $m = 10$, and $x_0 = 0$. Then

$$\begin{aligned}
x_0 &= 0 \\
x_1 &= (2 \times 0 + 3) \bmod 10 = 3 \\
x_2 &= (2 \times 3 + 3) \bmod 10 = 9 \\
x_3 &= (2 \times 9 + 3) \bmod 10 = 1 \\
x_4 &= (2 \times 1 + 3) \bmod 10 = 5 \\
x_5 &= (2 \times 5 + 3) \bmod 10 = 3 \\
x_6 &= (2 \times 3 + 3) \bmod 10 = 9 \\
x_7 &= (2 \times 9 + 3) \bmod 10 = 1 \\
x_8 &= (2 \times 1 + 3) \bmod 10 = 5
\end{aligned}$$

Example 7.4 Let $a = 2$, $c = 0$, $m = 10$, and $x_0 = 1$. Then

$$\begin{aligned}
x_0 &= 1 \\
x_1 &= (2 \times 1) \bmod 10 = 2 \\
x_2 &= (2 \times 2) \bmod 10 = 4
\end{aligned}$$

$$x_3 = (2 \times 4) \bmod 10 = 8$$
$$x_4 = (2 \times 8) \bmod 10 = 6$$
$$x_5 = (2 \times 6) \bmod 10 = 2$$
$$x_6 = (2 \times 2) \bmod 10 = 4$$
$$x_7 = (2 \times 4) \bmod 10 = 8$$
$$x_8 = (2 \times 8) \bmod 10 = 6$$

Example 7.4 illustrates the case in which $c = 0$. This algorithm is called a **multiplicative congruential technique**. If $c \neq 0$, the technique is called a **mixed congruential scheme**. Both examples illustrate the repeatability of any sequence generated by this scheme. The sequence 3, 9, 1, 5 in Example 7.3 repeats endlessly, while it is the sequence 2, 4, 8, 6 in Example 7.4. Knuth (7.3) has shown that judicious choices of the constants a, c, x_0 and m can make the period sufficiently long for most studies. The arguments for each of the constants are summarized as follows.

1. *Choice of m.* Since the period will always be less than m, a large value of m is desirable. Furthermore, a value of m that facilitates the solution of the congruence relation should be used. For machines that utilize a binary number representation, a value of $2^k - 1$, where k is the word size of the machine, has proven excellent.

2. *Choice of a and c.* A sequence generated by a linear congruential scheme has period m if and only if

   a. c is relatively prime to m.
   b. $a - 1$ is a multiple of every prime dividing m.
   c. $a - 1$ is a multiple of 4 if m is a multiple of 4.

These constraints yield multiplier values of the form $a = z^P + 1$, where z is the radix used in the number representation of the computer, k is the word size of the computer (number of bits per word), $m = z^k$, and $z \leq p < k$. In particular, choices of $a = 2^{16} + 5 = 65541$ or $2^{16} + 3 = 65539$ have proved successful. As for the choice of c, it must only satisfy the requirement that it is relatively prime to m.

3. *Choice of* x_0. If the period of the sequence is m, the choice of x_0 is immaterial, since the entire sequence will be generated. Some care must be taken, since a choice of $x_0 = 0$, for example, will yield a degenerate sequence if the multiplicative congruential scheme is used.

7.2.3 Additive Congruential Generator

The **additive congruential technique** requires as its seed, a sequence of n numbers $x_1, x_2, \ldots, x_n$. This sequence of numbers can be generated using some other technique. Application of the algorithm will produce an extension to the sequence $x_{n+1}, x_{n+2}, x_{n+3}, \ldots$. Specifically the algorithm is

$$x_j = (x_{j-1} + x_{j-n}) \bmod m$$

The main advantage of this technique is speed; no multiplications are necessary. It can yield periods greater than m. As Knuth mentions (7.3), the theoretical behavior of this technique is not as well understood as the behavior of the mixed congruential technique. Thus careful validation of any number sequence generated by this technique is necessary.

Example 7.5 Let $m = 10$, and extend the sequence 1, 2, 4, 8, 6 generated in Example 7.4.

$$\begin{aligned}
x_1 &= 1\\
x_2 &= 2\\
x_3 &= 4\\
x_4 &= 8\\
x_5 &= 6\\
x_6 &= (x_5 + x_1) \bmod 10 = (6+1) \bmod 10 = 7\\
x_7 &= (x_6 + x_2) \bmod 10 = (7+2) \bmod 10 = 9\\
x_8 &= (x_7 + x_3) \bmod 10 = (9+4) \bmod 10 = 3\\
x_9 &= (x_8 + x_4) \bmod 10 = (3+8) \bmod 10 = 1\\
x_{10} &= (x_9 + x_5) \bmod 10 = (1+6) \bmod 10 = 7\\
x_{11} &= (x_{10} + x_6) \bmod 10 = (7+7) \bmod 10 = 4\\
x_{12} &= (x_{11} + x_7) \bmod 10 = (4+9) \bmod 10 = 3\\
x_{13} &= (x_{12} + x_8) \bmod 10 = (3+3) \bmod 10 = 6
\end{aligned}$$

$$x_{14} = (x_{13} + x_9)\bmod 10 = (6+1)\bmod 10 = 7$$
$$x_{15} = (x_{14} + x_{10})\bmod 10 = (7+7)\bmod 10 = 4$$
$$x_{16} = (x_{15} + x_{11})\bmod 10 = (4+4)\bmod 10 = 8$$
$$x_{17} = (x_{16} + x_{12})\bmod 10 = (8+3)\bmod 10 = 1$$
$$x_{18} = (x_{17} + x_{13})\bmod 10 = (1+6)\bmod 10 = 7$$
$$x_{19} = (x_{18} + x_{14})\bmod 10 = (7+7)\bmod 10 = 4$$
$$x_{20} = (x_{19} + x_{15})\bmod 10 = (4+4)\bmod 10 = 8$$

7.2.4 Quadratic Congruence Generator

The **quadratic congruence method**, proposed by Coveyou (7.1), may be used when m is a power of 2. It is nearly equivalent to the double–precision midsquare method but has a longer period. The recursion relation for this method is given by

$$\boxed{x_{n+1} = (x_n(x_n + 1))\bmod m, \qquad n > 0}$$

The seed x_0 must satisfy the relation $x_0 \bmod 4 = 2$.

Example 7.6 Let $x_0 = 2$, $m = 16$, and generate a random number sequence using the quadratic congruence generator.

$$x_0 = 2$$
$$x_1 = (2(3))\bmod 16 = 6$$
$$x_2 = (6(7))\bmod 16 = 10$$
$$x_3 = (10(11))\bmod 16 = 14$$
$$x_4 = (14(15))\bmod 16 = 2$$
$$x_5 = (2(3))\bmod 16 = 6$$
$$x_6 = (6(7))\bmod 16 = 10$$
$$x_7 = 14$$
$$x_8 = 2$$

7.2.5 Pseudorandom Number Generator

The **pseudorandom number (PRN) technique** is useful for generating pseudorandom numbers that fall in the interval (0, 1). Let

$$x_{n+1} = \langle 10^p c x_n \rangle$$

where

$\langle a \rangle$ denotes the fractional part of a

p is the number of digits in the pseudorandom number

c is a constant multiplier, $0 < c < 1$

We shall not go into details of how c is to be picked, but it has been established that a value of c equal to 10^{-p} (200A ± B) will provide satisfactory results, where A is any non–negative integer and B is any number from {3, 11, 13, 19, 21, 27, 29, 37, 53, 59, 61, 67, 69, 77, 83, 91}. The seed should be chosen to be $x_0 = 10^{-p} k$, where k is any integer not divisible by 2 or 5 such that $0 < k < 10^p$.

Example 7.7 Let $x_0 = 0.33$, $A = 0$, and $B = 11$. Suppose that two–digit numbers are desired. Then

$$c = 10^{-2}(11.0) = 0.11$$

and

$$\begin{aligned} x_0 &= 0.33 \\ x_1 &= \langle 100(0.11)(0.33) \rangle = 0.63 \\ x_2 &= \langle 100(0.11)(0.63) \rangle = 0.93 \\ x_3 &= \langle 100(0.11)(0.93) \rangle = 0.23 \\ x_4 &= \langle 100(0.11)(0.23) \rangle = 0.53 \\ x_5 &= \langle 100(0.11)(0.53) \rangle = 0.83 \end{aligned}$$

The main disadvantage of this scheme is that it is slow because so many multiplications are required to generate the sequence of random numbers.

7.2.6 GPSS Random Number Generator

GPSS/360 incorporates eight pseudorandom number generators specified as RN1–RN8. Each will produce the same sequence of numbers, or it may be altered to provide up to eight unique sequences. The description of these generators that appeared in Felder (7.2) is summarized here.

Three eight–by–one arrays are maintained: a base number array containing the seed, a multiplier array containing the multiplier for each generator (all are initially 1), and an index array containing the index for each generator (all are initially 0). A pseudorandom number is generated by the following procedure.

1. The appropriate word from the index array points to the base number array. Since each index is initially zero, the first base number is used as the seed for all generators.
2. The appropriate multiplier is multiplied by the base number selected in step 1.
3. If the higher–order bit of the low–order 32 bits of this product is a 1, the low–order 32 bits are replaced by their two's complement.
4. The low–order 31 bits of this possibly transformed product is stored in the multiplier array for future use.
5. Three bits of the high–order 16 bits of the product are stored in the index array, also for future use.

6a. If a fractional number is required, the middle 32 bits of the product are divided by 10^6, and the remainder produced is the desired six–digit number.

6b. If an integer is required, the middle 32 bits are divided by 10^3 and the remainder produced is the desired three–digit number.

This generator starts as if it were a multiplicative congruential generator but then diverges from that procedure because no recursion relation is used to produce successive numbers. The scrambling operation is designed to reduce nonrandomness, but it precludes analysis of the generator. For a further description of the generator see Felder (7.2).

7.3 Testing and Validating Pseudorandom Sequences

The generation of pseudorandom numbers is used to simulate the sampling from a continuous uniform distribution. Testing and validation of a sequence of pseudorandom numbers normally entails the comparison of the sequence with what would be expected from the uniform distribution. A multitude of tests are in use. Knuth (7.3) describes ten such tests. In this section we will review some of the more common tests. For more detail see Knuth.

7.3.1 Frequency Test

The **frequency test** is designed to test the uniformity of successive sets of numbers in the sequence. A procedure for this test is as follows.

1. Generate a sequence of M (say 10) consecutive sets of N (say 100) random numbers each.
2. Partition the range into intervals (say 10).
3. Tabulate the frequency within each interval for each of the M groups.
4. Compare the results of the M groups with each other and with the expected values (continuous uniform distribution) using the chi–square goodness–of–fit test described in Chapter 6.

7.3.2 Serial Test

The **serial test** measures the degree of randomness between successive numbers in a sequence. A procedure for this test is as follows.

1. Generate a sequence of M (say 10) consecutive sets of N (say 100) random numbers each.
2. Partition the number range into k intervals (say 10).
3. For each group construct an array of size k x k. The values of the array are initially 0. Examine the sequence of numbers from left to right, pairwise. Do not examine any number twice. If the left member of the pair is in interval i while the right member is in interval j, increment the (i, j) element of the array by 1.
4. When an array has been constructed for each group, compare the results of the M groups with each other and with the expected value (each pair should be equiprobable) using the chi–square test.

This test can be extended to examine triples, quadruples, and so on. The number of random numbers required for a valid test tends to become very large in these cases.

7.3.3 Kolmogorov–Smirnov Test

The **Kolmogorov–Smirnov** test, described in Chapter 6, is used to test the degree of randomness. It can be applied to the overall distribution of the sequence, resulting in a frequency or equidistribution test. This test compares the distribution of the set of numbers to a theoretical (i.e., standard uniform) distribution. The unit interval is divided into a number of subintervals and the cdf of the sequences of the pseudorandom numbers is calculated up to the endpoint of each subinterval. For example, if x_i is the ith endpoint and f_i is the cumulative relative frequency, and $F(x) = x$, then all that is required for this test is to find the maximal value of

$$\max_i \left| F(x_i) - S(x_i) \right| .$$

$F(x_i)$ is the theoretical value, and $S(x_i)$ the empirical cdf. Its utility in testing the hypothesis that a sequence of numbers are from a uniform distribution should be obvious.

7.3.4 Runs Test

The **runs test** is used to test the randomness of oscillation of numbers in the sequence. A procedure for this test is as follows.

1. Generate N random numbers (say 10,000).
2. Build a binary sequence such that for any two consecutive numbers (x_j, x_{j+1}) of the sequence, the jth bit is 0 if $x_j < x_{j+1}$ and 1 otherwise.
3. Tabulate the frequency of occurrence of runs (consecutive 0's or 1's) of each length (2, 3, 4, 5, . . .).
4. Compare the tabulated frequencies with expected values. As Knuth points out (7.3), it is not possible to apply the chi–square statistics directly because consecutive runs are not independent. Knuth has devised a

special statistic for use in this test, and any interested reader is referred to his work for the details.

7.3.5 Poker Test

The **poker test** examines successive groups of five random numbers. The number of distinct values in the set of the five numbers is counted, giving five categories:

1. Five different
2. Four different (one pair)
3. Three different (two pairs or three of a kind)
4. Two different (full house or four of a kind)
5. One different (five of a kind)

The theoretical frequency of each category is calculated under the assumption that each number is equally likely, and the chi–square statistic is used to determine whether the random sequence conforms to the expected distribution.

7.3.6 Permutation Test

The **permutation test** is a generalization of the poker test. The random sequence is broken into n subsequences, each with k elements. Each of the possible $k\,!$ permutations of the elements of each group is considered equiprobable, and the theoretical frequency of each obtained ordering is calculated. The chi–square statistic is then used to compare the actual frequency with the expected frequency.

7.3.7 Distance Test

The **distance test** considers successive pairs of random numbers to be the coordinates of points in the unit square. For example, if the random sequence is r_1, $r_2, r_3, \ldots, r_n$, the points $(r_1, r_2), (r_3, r_4), \ldots, (r_{n-1}, r_n)$ would be plotted in an $x-y$ plane. The square of the Euclidean distance D^2 is then calculated between each of the points. If the points are distributed randomly in the unit square (which they should be in the case of a truly random sequence), the probability that the observed value of D^2 is less than or equal to some value k is given by

$$F(k) = \begin{cases} \pi k - \frac{8}{3}k^{3/2} + \frac{k^2}{2}, & \text{for } k \le 1.0 \\ \frac{1}{3} + (\pi - 2)k + 4(k-1)^{1/2} + \frac{8}{3}(k-1)^{3/2} - \\ \frac{k^2}{2} - 4k \text{ arc sec}\sqrt{k}\,, & \text{for } 1.0 \le k \le 2.0 \end{cases}$$

That is

$$\begin{aligned} F(0.0) &= 0 \\ F(0.25) &= 0.483 \\ F(0.50) &= 0.753 \\ F(0.75) &= 0.905 \\ F(1.0) &= 0.975 \\ F(1.5) &= 0.999 \\ F(2.0) &= 1.0 \end{aligned}$$

Using this distribution function, one can calculate the theoretical frequencies for the valued of D^2 in any prescribed interval. Comparisons of the actual frequencies with the theoretical frequencies can be made with the use of the chi–square statistic.

7.3.8 Spectral Test

The **spectral test** is used to check for a flat spectrum by checking the observed estimated cumulative spectral density function with the Kolmogorov–Smirnov test, and thus measuring the independence of adjacent sets of numbers.

7.3.9 Gap Test

Define some gap $\alpha - \beta$, scan the sequence of pseudorandom numbers until n gaps are recorded. The length of the gaps is recorded, and the frequency of the gap lengths is compared to the expected frequency using a chi–square test.

7.3.10 Autocorrelation Test

Using the **auto covariance function** estimator $\hat{R}_k$, a normal significance test is possible. The function $\hat{R}_k$ is defined as

$$\hat{R}_k = \frac{1}{n-k}\sum_{i=1}^{n-k}\left[\left(r_i - \frac{1}{2}\right)\left(r_{i+k} - \frac{1}{2}\right)\right]$$

for $k > 0$, has mean of 0 and variance of $(1/144)(n-k)$. For moderately large n, the distribution $(n-k)^{1/2}\hat{R}_k$ is normal $(0, 1/144)$.

7.3.11 Serial Correlation Test

For this test consider the generated numbers to be in a cycle. That is, compare r_i to r_j where i is modulo r. Calculate

$$\alpha = \frac{\sum_{i,j=1}^{N} r_i r_j - \frac{1}{N}\sum_{i=1}^{N} r_i \sum_{j=1}^{N} r_j}{\left[\left(\sum_{i=1}^{N} r_i^2 - \frac{\left(\sum_{i=1}^{N} r_i\right)^2}{N}\right)\left(\sum_{j=1}^{N} r_j^2 - \frac{\left(\sum_{j=1}^{N} r_j\right)^2}{N}\right)\right]^{1/2}}$$

where N is the number of random numbers. The closer the value of α approaches 1.0, the more r_i and r_j are related (i.e., non–random).

7.4 Generation of Nonuniform Variates

All the generators used thus far are designed to generate sequences of numbers following a uniform distribution. However, other theoretical distributions such as the normal, exponential, Poisson, and gamma distributions are encountered more often in simulation studies than the uniform distribution. In many cases no appropriate theoretical distribution can be found, and an empirical distribution is used. Thus it is necessary to have a technique for generating random numbers that simulate the sampling of any arbitrary distribution. Note that all the techniques discussed start by generating one or more pseudorandom numbers from the uniform distribution. A transformation is then applied to this uniform variable to generate the nonuniform pseudorandom number.

7.4.1 The Inverse Transformation Method

The **inverse transformation technique** is useful for transforming a standard uniform deviate into any other distribution. It is particularly useful whenever the density function $f(x)$ can be integrated to find the cumulative function $F(x)$, or $F(x)$ is an empirical distribution.

Suppose that we wish to generate a pseudorandom number from a distribution given by $F(x)$, where F satisfies all the properties of a cumulative distribution function outlined in Chapters 2 and 3. This generation is done in two steps.

1. Generate a standard uniform random number using one of the techniques outlined in Section 7.2.
2. If r is the standard uniform number generated in step 1, then

$$x_0 = F^{-1}(r)$$

is the desired nonuniform variate.

This process is depicted graphically in Figure 7.1.

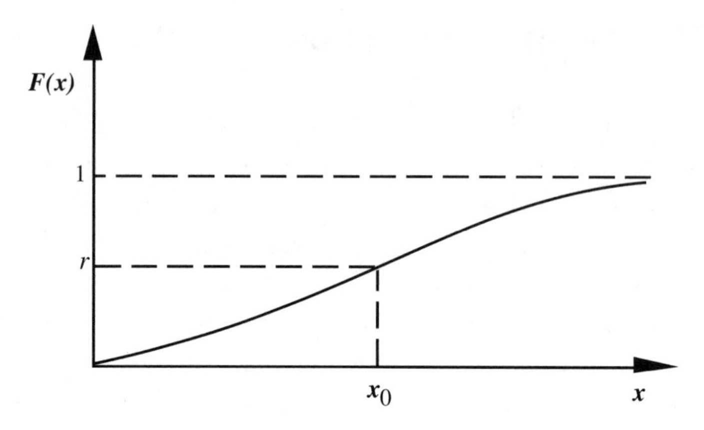

Figure 7.1. The inverse transformation method.

Example 7.8 Suppose that the standard uniform random numbers 0.1021, 0.2162, and 0.7621 have been generated and that we wish to transform them into a distribution given by

$$F(x)=\begin{cases}0, & x<0\\ x, & 0\le x<1/4\\ (3x+1)/7, & 1/4\le x<2\\ 1, & x>2\end{cases}$$

This distribution function is illustrated in Figure 7.2. Then

$$F^{-1}(a)=\begin{cases}a, & 0\le a<1/4\\ (7a-1)/3, & 1/4\le a\le 1\end{cases}$$

So

$$F^{-1}(0.1021)=0.1021$$
$$F^{-1}(0.2162)=0.2162$$
$$F^{-1}(0.7621)=1.4449$$

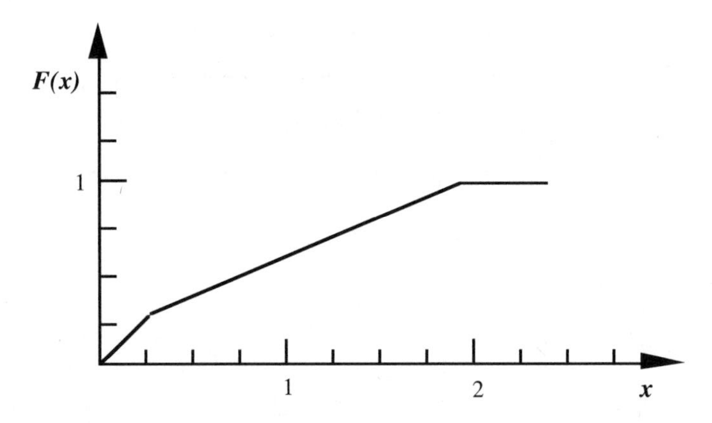

Figure 7.2. The distribution function for transforming variates.

These numbers then are random numbers from the distribution given by $F(x)$.

The inverse transformation method is particularly useful when the cumulative distribution function is tabulated. The inverse function in this case is obtained simply by reversing the roles of the abscissa and the ordinate values of each tabulated point.

Example 7.9 Suppose a distribution has been tabulated as follows.

x	0.0	0.5	1.0	1.5	2.0	2.5	3.0
$F(x)$	0.0	0.1	0.2	0.5	0.7	0.9	1.0

Suppose further than the standard uniform random number is 0.25 has been generated. The corresponding transformed random number, obtained by using linear interpolation between 1.0 and 1.5, is 1.0833.

Just as with the above continuous case, we need to develop a relationship between r and $F(x)$ in the general discrete case. Unfortunately, in the discrete case this is not an equation, but rather a "band" resulting from an inequality set. That is,

$$F(x) = \sum_{y \le x} P(y)$$

and the random variable **X** takes on the specific value x when

$$F(x-1) < r \le F(x)$$

For example, let $P(x) = \dfrac{1}{k+1}$, $\quad x = 0, 1, 2, \ldots, k$

then $F(x) = \dfrac{x+1}{k+1}$.

Therefore, $\dfrac{x}{k+1} < r \le \dfrac{x+1}{k+1}$, or $\quad (k+1)r - 1 < x \le (k+1)r$.

7.4.2 Generation of Nonstandard Uniform Random Numbers

Most standard routines to generate pseudorandom numbers are designed to simulate sampling from the standard uniform distribution. They produce random numbers in the range from 0 to 1. The nonstandard uniform distribution is given by

$$F(x) = \begin{cases} 0, & x < a \\ (x-a)/(b-a), & a \le x \le b \\ 1, & x > b \end{cases}$$

To produce nonstandard uniformly distributed random numbers from standard uniformly distribution random numbers, it is only necessary to scale the numbers. If r is a random number from a standard uniform distribution, then

$$x = a + (b-a)r$$

is a random number from a nonstandard uniform distribution with the range from a to b.

Example 7.10 Let us consider the uniform distribution

$$f(u) = \begin{cases} 0\,, & u < a \\ 1/(b-a)\,, & a \le u \le b \\ 0, & u > b \end{cases}$$

the uniform cdf

$$F(u) = \int_{-\infty}^{u} f(x)dx = \begin{cases} 0\,, & u < a \\ (u-a)/(b-a), & a \le u \le b \\ 1\,, & u > b \end{cases}$$

a and b are most often chosen such that $a = 0$ and $b = 1$. Variates from other distributions can be obtained by applying a suitable transformation.

Consider a variate x with cdf $F(x)$ such that

$$\alpha = F(x)$$

Applying the given transformation show that the result has a uniform distribution with $a = 0$ and $b = 1$. That is, show that

$$U(\alpha) = \begin{cases} 0, & \alpha < 0 \\ \alpha, & 0 \le \alpha \le 1 \\ 1, & \alpha > 1 \end{cases}$$

Proof: Consider those distributions $F(x)$ which have

1. $\lim_{x \to -\infty} F(x) = 0$

2. $\lim_{x \to +\infty} F(x) = 1$

3. $F(x_2) > F(x_1)$ for all x_1 and x_2, and $x_1 > x_2$

4. $F(x)$ is continuous

Statements 1 and 2 are true of all cdf's. Statements 3 and 4 require that $\alpha = F(x)$ has a unique inverse $x = F^{-1}(\alpha)$ such that$F^{-1}(\alpha_2) = x_2 > x_1 = F^{-1}(\alpha_1)$. Since $F(x)$ cannot be negative, α cannot be negative. Since by definition $U(R)$ is defined to be Prob ($\alpha \le R$), $U(R) = 0$ for $R \le 0$, and $U(R) = 1$ for $R \ge 0$ (since $\alpha = F(x) \le 1$).

Next, we need to show that $U(R) = R$ for $0 < R < 1$. Using Prob ($\alpha \le R$) = $U(R)$ and Prob ($x \le \mathbf{X}$) = $F(\mathbf{X}) = R$, then for $-\infty < x \le \mathbf{X} < \infty$ we have the following

$$\begin{aligned} U(R) &= \text{Prob}(\alpha \le R) = \text{Prob}(F(x) \le R) \\ &= \text{Prob}\left\{F^{-1}(F(x)) \le F^{-1}(R)\right\} \\ &= \text{Prob}\left\{x \le F^{-1}(R)\right\} \\ &= \text{Prob}\left\{x \le \mathbf{X}\right\} = F(\mathbf{X}) = R \end{aligned}$$

For example, consider variate x from the Rayleigh probability density function

$$f(x) = \begin{cases} xe^{-x^2/2}, & x > 0 \\ 0, & x \le 0 \end{cases}$$

mapped into variates α from the uniform distribution by

$$\alpha = F(x) = \int_{-\infty}^{x} f(y)dy = 1 - e^{-x^2/2}$$

The inverse cdf is

$$x = F^{-1}(\alpha) = (-2\ln(1-\alpha))^{1/2} \quad ; \quad 0 \le \alpha < 1 .$$

Letting $\alpha = r_i$, $i = 1, 2, \ldots$ generated by the computer, the desired pseudorandom numbers are generated from $x_i = (-2\ln(1-r_i))^{1/2}$.

7.4.3 Generation of Normal Random Numbers

Random variables following a **normal distribution** are commonly encountered in simulation studies. A number of techniques are used in transforming standard uniform random numbers into normal random numbers. Two of the more common techniques are given here.

The first technique is to use the inverse transformation technique. In Chapter 3 we showed that there is no closed–form functional representation of the cumulative distribution function for the normal distribution; thus a tabular representation must be used. A sufficient number of points must be included to give the desired degree of detail.

Example 7.11 Suppose that we want to obtain random numbers from a normal distribution with mean 3 and variance 4. Suppose further the standard uniform random numbers 0.2163, 0.3241, 0.1021, and 0.7621 have been generated. A 13–point tabular representation of the standard normal (mean 0, variance 1) is given in Table 7.1.

The standard normal random numbers corresponding to the standard uniform random numbers in Table 7.1, obtained by using linear interpolation, are –0.8077, –0.4594, –1.3070, and 0.7357.

To transform a standard normal variable z into a normal random variable x with mean μ and standard deviation σ, the relationship

$$x = \mu + \sigma z$$

is used. The final transformed numbers are 1.3846, 2.0812, 0.3842, and 4.4714.

Table 7.1 Tabular Representation of the Standard Normal Distribution

z	$F(z)$	z	$F(z)$
–3	0.00135	0.5	0.69146
–2.5	0.00621	1.0	0.84134
–2.0	0.02275	1.5	0.93319
–1.5	0.06681	2.0	0.97725
–1.0	0.15866	2.5	0.99379
–0.5	0.30854	3.0	0.99865
0.0	0.50000		

A second, more common technique for generating normally distributed random numbers is to use the **central limit theorem**. This theorem states that the sum of identically distributed independent random variables $\mathbf{X}_1, \mathbf{X}_2, \ldots, \mathbf{X}_n$ has approximately a normal distribution with a mean $n\mu$ and variance $n\sigma^2$, where μ and σ^2, are respectively the mean and the variance of $\mathbf{X}_i$. If the variables $\mathbf{X}_i$, $i = 1, 2, \ldots, n$, follow the standard uniform distribution, then $\mu = 0.5$ and $\sigma^2 = 1/12$. Thus summing n standard uniform variates gives an approximate normal distribution with mean $0.5n$ and variance $n/12$.

The choice of n is largely up to the analyst. Of course, the larger the value of n chosen, the better the approximation to the normal distribution. Studies have shown that with $n = 12$ the technique provides fairly good results while at the same time maintaining calculation efficiency. This is because it yields $\sigma = 1$; so in the transformation from a nonstandard normal to the standard normal, a division operation is saved.

Example 7.12 Suppose the standard uniform random numbers 0.1062, 0.1124, 0.7642, 0.4314, 0.6241, 0.9443, 0.8121, 0.2419, 0.3124, 0.5412, 0.6212, 0.0021 have been generated. Generate a normal random number from a distribution with mean 25 and variance 9.

Summing the 12 standard uniform numbers gives $Y = 5.5135$. This number is from an approximate normal distribution with a mean of 6 and a variance of 1. The corresponding standard normal number is $Z = Y - 6 = -0.4865$. Now transforming this number to a normal distribution with mean 25 and a variance 9 generates the desired result.

$$X = \mu + \sigma Z = 25 + 3(0.4865) = 23.5405$$

7.4.4 Generation of Binomially Distributed Random Numbers

The **binomial distribution** is used to model n successive trials of some experiment having two possible outcomes on each trial. The binomial random variable **X** counts the number of successes in each of these n trials, where the probability of a success on any given trial is p. To generate binomially distributed random numbers, one simulates the outcome of a trial of the experiment by generating a standard uniform random number. After all n trials have been simulated (n standard uniform numbers have been generated), the value of the binomially distributed random variable is simply a count of those standard uniform numbers that are less than or equal to p. This procedure is useful for small to moderate values of n. For large values of n, the normal approximation to the binomial should be used. Rather than generate binomially distributed numbers, generate numbers from a normal distribution with mean np and variance $np(1-p)$.

Example 7.13 Generate a random number from a binomial distribution with $n = 7, p = 0.3$. To simulate seven trials, generate seven standard uniform numbers. Assume that the numbers are 0.02011, 0.85393, 0.97265, 0.61680, 0.16656, 0.42751, and 0.69994. Now two of these numbers are less than $p = 0.3$, so the desired binomial random number is $\mathbf{X} = 2$.

7.4.5 Generation of Exponentially Distributed Random Numbers

The generation of **exponentially distributed random numbers** is easily accomp-lished with the use of the inverse transformation technique described in Section 7.4.1. Recall that the cumulative distribution function for an exponentially distributed random variable **X** is

$$F(x) = 1 - e^{-\alpha x}, \quad x > 0$$

The inverse of F is then

$$F^{-1}(a) = -\frac{1}{\alpha}\ln(1-a)$$

If a is uniformly distributed, however, then $1 - a$ is also uniformly distributed, and the desired random numbers can be generated by using

$$F^{-1}(a) = -\frac{1}{\alpha}\ln(a)$$

Thus generating a standard uniform number r permits the formation of an exponentially distributed random number by

$$\mathbf{X} = -\frac{1}{\alpha}\ln(r)$$

Note that this method is simple to program, yet it is very time–consuming because it involves the calculation of the natural logarithm function. If computation time is a problem, tabulated values of the exponential distribution function can be used in the manner illustrated in Section 7.4.1.

Example 7.14 Suppose we want to generate random numbers from an exponential distribution with $\alpha = 1$. Then

$$F(x) = 1 - e^{-x}, \quad x > 0$$
$$F^{-1}(a) = -\ln(1-a)$$

Now if a standard uniformly distributed random number, say 0.02104, has been generated, the desired random number from the exponential distribution would be

$$\mathbf{X} = -\ln(0.02104) = -(-3.8613) = 3.8613$$

7.4.6 Generation of Poisson Distributed Random Numbers

Generation of random numbers from a **Poisson distribution** with a mean of λ can be accomplished by multiplying successively generated standard uniform random numbers. Specifically, multiply N standard uniform random numbers U_i until

$$\prod_{i=1}^{N} U_i < e^{-\lambda}$$

Then the value of the Poisson random variable $\mathbf{X}$ is $N - 1$.

Example 7.15 Suppose that we wish to generate a random number from a Poisson distribution with $\lambda = 2.5$. Suppose further that we have generated the standard uniform random numbers 0.91646, 0.89198, 0.64809, 0.16376, 0.91782, 0.45624, 0.31641. Then e = 0.08208, and

$$(0.91646)(0.89198) = 0.81746 > 0.08208$$
$$(0.91646)(0.89198)(0.64809) = 0.52979 > 0.08208$$
$$(0.91646)(0.89198)(0.64809)(0.16376) = 0.08675 > 0.08208$$
$$(0.91646)(0.89198)(0.64809)(0.16376)(0.91782) = 0.07963 < 0.08208$$

Then $N = 5$ and hence the value of the Poisson–distributed random number is $\mathbf{X} = N - 1 = 4$.

7.4.7 The Rejection Method

When the inverse transformation method is impractical, other methods must be employed. One such technique is called the **rejection method.**

The basis for this approach is that the probability of r being less than or equal to $c \cdot f(x)$ is $c \cdot f(x)$ itself. That is,

$$\text{Prob}[r \leq c \cdot f(x)] = c \cdot f(x)$$

Thus, if x is generated randomly in the interval (a, b), and x is rejected if $r > c \cdot f(x)$, then the accepted x's will satisfy the density function $f(x)$. In order to use this approach, $f(x)$ has to be bounded and x valid over some range $(a \leq x \leq b)$. The steps of this procedure can be summarized as follows:

1. Normalize the range of $f(x)$ such that

$$c \cdot f(x) \leq 1 \quad ; \quad a \leq x \leq b$$

2. Define x as a uniform continuous random variable,

$$x = a + (b - a)r \ .$$

3. Generate a pair of random numbers (r_1 , r_2).

4. If the pair satisfies the property

$$r_2 \leq c \cdot f(x)$$

then set the random deviate to

$$x = a + (b - a)r_1 \ .$$

5. If the test in step 4 fails, return to step 3 and repeat steps 3 and 4.

Example 7.16 Suppose we wish to generate deviates from

$$f(x) = \begin{cases} 2x, & 0 \le x \le 1 \\ 0, & \text{otherwise} \end{cases}$$

Clearly $f(x)$ satisfies the necessary conditions required for the rejection procedure.

Step 1. $c \cdot f(x) \le 1$
Step 2. $x = a + (b - a)r = r$
Step 3. Generate r_1 and r_2
Step 4. Test to see whether $r_2 \le \frac{1}{2} \cdot 2(r_1)$
if yes, $x = r_1$; else repeat step 3.

The rejection method has been popularly applied to generate beta and gamma variates with non–integral parameters and binomial variates with large N.

7.4.8 The Composite Method

In this procedure $f(x)$ is expressed as a probability on x of selected density functions g(x),

$$f(x) = \sum_{n=1}^{N} P_n g_n(x)$$

The selection of the $g_n(x)$ are motivated on the basis of best fit and effort to produce $f(x)$.

Example 7.17 Let

$$F(x) = \sum_{i=1}^{n-1} a_i F_i(x) + \left(1 - \sum_{i=1}^{n-1} a_i\right) F_n(x)$$

with $a_1, a_2, \ldots, a_{n-1} > 0$ and $\sum_{i=1}^{n-1} a_i < 1$.

A random number r_1 is generated from a standard uniform distribution. Then if

$$S_j = \sum_{i=1}^{j} a_i , \qquad j = 1,2,\ldots,n$$

there exists a value m such that

$$\boxed{S_{m-1} < r_1 < S_m}$$

Suppose $n = 2$, then

$$F(x) = pF_1(x) + (1-p)F_2(x)$$

If $r < p$, then x is chosen from $F_1(x)$, while if $r \geq p$, x is chosen from $F_2(x)$.

This procedure can be extremely useful if one of the values of a_i is near 1 and if F_i is an easily accommodated distribution.

7.5 Summary

In this chapter some of the more common techniques for generating pseudorandom numbers and tests for validating the corresponding random number generators have been reviewed. The generation of pseudorandom numbers can be viewed as the simulated sampling of a given distribution. This simulated sampling technique is generally referred to as the **Monte Carlo technique**. It is of great value when sampling is desirable but either impossible or impractical. Monte Carlo simulation has been applied to the solution of waiting–line problems, inventory control problems, and purchasing problems, as well as to the approximate solution of differential equations and integral equations. It is a valuable tool when the analytic model of the system under consideration is complex or unwieldy.

7.6 Exercises

7.1 Generate a sequence of ten pseudorandom numbers using the mid–square method and beginning with $x_0 = 0.6677$.

7.2 Use the linear congruential scheme to generate ten pseudorandom numbers beginning with $x_0 = 21$, $a = 4$, $c = 1$, and $m = 100$.

7.3 Repeat Exercise 7.2 with $c = 0$.

7.4 Assess the choices of a, c, and m according to the criteria established in Section 7.2.2. Could the generator be expected to have a full period?

7.5 Use the additive congruential generator of Section 7.2.3 to extend the sequence generated in Exercise 7.3 to a length of 25.

7.6 Generate a sequence of ten pseudorandom numbers using the quadratic congruence generator of Section 7.2.4 with $x_0 = 10$ and $m = 128$.

7.7 Generate a sequence of ten two–digit random numbers using the pseudorandom number generator scheme of Section 7.2.5, with $A = 0$ and $B = 13$. Start with $x_0 = 0.49$.

7.8 Use the inverse transformation method to generate a random number from a distribution with the probability density function given by

$$f(x) = \begin{cases} 1/4, & 0 \le x < 1 \\ 3/4, & 1 \le x \le 2 \end{cases}$$

7.9 Generate a random number from a normal distribution with mean 10 and variance 16.

7.10 Generate a random number from a binomial distribution with $n = 10$, $p = 0.35$.

7.11 Generate a random number from an exponential distribution with $\alpha = 1.5$.

7.12 Generate a random number from a Poisson distribution with $\lambda = 4.0$.

7.13 Write a general purpose FORTRAN subroutine to implement the linear congruential scheme of Section 7.2.2. Scale the output so that standard uniform random numbers are produced.

7.14 Write a FORTRAN subroutine that generates normal random numbers and uses the output of the subroutine of Exercise 7.13. Input parameters should include the desired mean and variance.

7.15 Write a FORTRAN subroutine to generate exponentially distributed random numbers.

7.16 Write a FORTRAN subroutine to generate Poisson–distributed random numbers.

7.17 Using one of the generators developed in Exercises 7.13–7.16, generate a list of 100 pseudorandom numbers. Use the Kolmogorov–Smirnov test to verify that these numbers have the desired distribution.

7.7 References

7.1 Coveyou, R. R., "Serial Correlation in the Generation of Pseudo–Random Numbers." *J. ACM 7* (1960): 72–74.

7.2 Felder, H., "The GPSS/360 Random Number Generator." *Digest of the Second Conference of Applications of Simulation*, New York, December 1968.

7.3 Knuth, D. E., *The Art of Computer Programming, Vol 2: Seminumerical Algorithms*. Reading, MA: Addison–Wesley, 1969.

7.4 Lehmer, D. H., *Proceedings of the Second Symposium on Large–Scale Digital Computing Machinery*. Cambridge, MA: Harvard University Press, 1951.

7.5 RAND Corporation, *A Million Random Digits with 100,000 Normal Deviates*. New York: Free Press, 1955.

8

DISCRETE SYSTEM SIMULATION

An understanding of the basic concepts in probability and statistical methods that were introduced in previous chapters is necessary for the full understanding of simulation methods. This chapter and the next few discuss some of the factors that influence the development of a system simulation model. Let us review some of the simulation terminology introduced earlier.

8.1 Simulation Terminology

Simulation models of systems can be classified as discrete change or continuous change models. **Discrete simulation** implies that the dependent variables (i.e., state indicators) change discretely at points in time referred to as events. On the other hand, **continuous simulation** implies that the state (dependent) variables change in a continuous manner in time. It is important to remember that the terms discrete and continuous refers to changes in the state of the system, and not of time management. Time, in fact, may be modeled in both discrete or continuous manner in either type of simulation. Some simulation models as well as languages combine both concepts, called the combined or **hybrid simulation**, and thus provide the ultimate in modeling capabilities. We, however, will concentrate only on the development of discrete simulation models. For these models, it should be remembered that the state of the system changes only when events occur, and simulated time advances in occurrence with those events.

An **object** is called an **entity**. The characteristics of an entity are referred to as **attributes**. **Files** are groups of entities with a common attribute or common relationship to other entities in the group. The collection of entities and attributes for a given system is referred to as the **system state**, and it is generally expressed in a time frame of reference. Any process that changes the system state is referred to as an **activity**. The occurrence of such a change at a point in time is referred to as an **event**. An event may be a change in the value of some attribute, the creation or destruction of an entity, or the initiation or termination of an activity. A **process** is the time ordered collection of events which may include several activities.

The key concept in discrete event simulation is the scheduling of events and the changes for events (additions, deletions, modifications, etc.). Events have to be synchronized with one or more of the following approaches: event orientation, activity scanning, and process coordination.

Event orientation describes the change in a system state that occurs at each event, and the scheduling of events occur in proper time order. In this orientation, the user must identify the events that may occur. The simulationist must also describe the activities associated with each of the events, and how the system states may be altered by the events, whether conditioned on the current system states or not. Thus, the result of events are very much conditioned on prior events and system states. In summary, entities have attributes, interact with activities that may be conditioned, creating events that change the state of the system.

Activity scanning describes the event in which entities of the system interact with respect to time, and time is managed in discrete time steps. The world model view differs significantly from that of the previous approach, and is best suited for continuous change simulation models. With this approach the specific conditions which initiate or end an activity also describe the dynamics of the activity. The state of the activities is checked at each small increment of time, and whenever predefined conditions are satisfied the initiation or termination of these activities are instantiated, with the simulation model responding accordingly. A typical example that would incorporate such a world view includes the simulation of pumping oil from various sources into a tank, monitoring corresponding tank levels, and releasing that same oil by way of several output pipelines. This example clearly corresponds to a natural continuous process. This technique, however, may also be applied to discrete phenomena such as population models.

Finally, **process orientation** describes the processes of the system in a way whereby the entities of the system flow in time, and thus force the progress of entities. This world view, also referred to as **transaction orientation** view, operates with the view that entities will be subjected to repetitive event patterns that may be repeated in loops, or by using macros. **Macros** are modules of code that repeat statements, with associated parameters, and may be graphically represented as networks or flow graphs. A process oriented approach, which uses the flow of entities or transactions, models the simulation components by following the set of activated processes. One example that can be used to demonstrate this model view approach is to use a simple queueing system. One can generate an arrival every t time units, await the server, advance time, free the server, and then when done terminate the server.

Each of the above procedures is referred to as a **world view** of the simulation model.

The **state of a system** – continuous or discrete – is usually expressed as a function of time. Two time references are involved in the simulation of a system: simulation time and run time. The term **simulation time** is used to refer to the period of time simulated by the model – 30 minutes, 5 years, 100 years, whatever the interval the researcher is interested in. This simulation time is usually set to 0 at the beginning of the simulation run and acts as a counter to the number of simulation time units. The **duration** of a simulation run is the difference between the initial clock setting and the setting at the termination of the run. The **run time** is simply the time that it takes the computer to simulate the period of interest. There is generally very little if any correlation between the two measures of time. The run time is influenced by factors such as the complexity of the simulation model and the number of events simulated, as well as by the duration of the run. Run time in most cases is considerably shorter than simulation time. For example, one can generally simulate the operation of this country's economic system over a period of years in minutes on a computer. In some cases, however, the run time is considerably longer than the simulation time. Consider, for example, the simulation of a computer system. Simulation time in this case is measured in microseconds and nanoseconds. But because of the complexity of any realistic computer system model and the number of events that generally occur in any period of interest, the run time would probably be several thousand times greater than the simulation time. Unless otherwise indicated, references to time in this chapter and succeeding chapters are to simulation time. Run time, although crucial from an economic standpoint, is generally not as easily controlled by the analyst as simulation time. References to run time are made only to compare different simulation approaches.

8.2 Time Management Methods

Simulation models have been used to model both static (time–independent) and dynamic (time–dependent) situations. A **static model** shows the relationships between entities and attributes when the system is in a state of equilibrium. In this case, when the value of an attribute is changed to allow observation of its effect on the rest of the system, new attribute values can be derived for the rest of the system. No information can be obtained, however, on the manner in which this change occurred. The **dynamic model** allows the changes in attribute values to be derived as a function of time. The manner in which a change occurs can then be studied.

Most simulation models are dynamic models. Many static models can be solved analytically and provide better results than simulation. Analytic solutions of dynamic models often require many simplifying assumptions to make the model tractable and thus rarely yield useful results. Thus a simulation is generally a numerically solved dynamic mathematical model.

Dynamic models are time dependent. For that reason a simulator must generally include a means for depicting a time change in the system. This has resulted in the use of the phrase **time management**. There are a number of different ways of managing time in a simulator. Two of the more common ways are periodic scan and even scan.

The **periodic scan,** or **fixed–time increment technique** adjusts the simulation clock by one predetermined uniform unit and then examines the system to determine whether any events occurred during that interval. If any occurred, the event or events are simulated; otherwise no action is taken. The simulation clock is then advanced another unit, and the process is repeated. In the periodic scan (Interval–oriented; Time–sliced; Unit–advance; Fixed time increment) method, the unit interval must be smaller than the smallest inter–event interval, or else the method will become very inefficient. An example of this time management procedure is illustrated in Figure 8.1.

In Figure 8.1 no event occurs in the first unit of simulated time, so the clock is immediately advanced and the system scanned. Then event E_1 occurs in the second time increment. This event would be simulated and the clock advanced again. Since there is no event to simulate during the third interval, the clock is again advanced. During the fourth interval two events are to be simulated, E_2 and E_3. Following their simulation the clock is again advanced. This process of advancing the clock, scanning the system, and simulating events if necessary is repeated until the duration of the simulation run is reached. With this method the exact time of the occurrence of particular events is largely ignored. All events that occur during a given interval are treated as if these events occurred at the end of that interval.

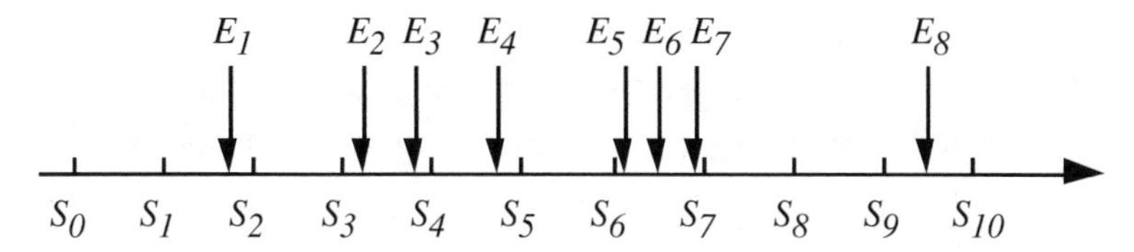

Figure 8.1. The periodic scan approach.

Example 8.1 Consider the simulation of a single–bay service station for a period of ten minutes. Assume that the system is empty at the beginning and at the end of the period. Assume also that four customers are serviced, that arrivals occur at simulation times 1.8, 3.2, 6.1, and 7.4, and that service completions occur at simulation times 2.6, 4.8, 7.3, and 8.1. If a time interval of one minute is used, arrival events are denoted by A_1, A_2, A_3, and A_4 and completion events by C_1, C_2, C_3, and C_4. This system is shown in Figure 8.2.

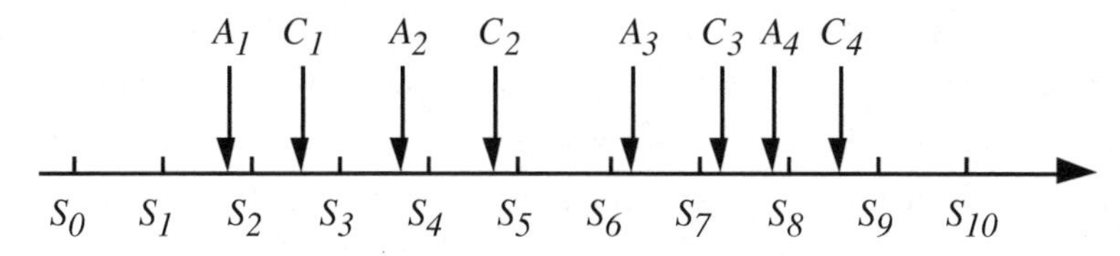

Figure 8.2. Arrival and completion of service events.

In this example events C_3 and A_4 are both considered to have occurred at clock time 8.0, although C_3 actually occurred before A_4. Thus a problem with the periodic scan approach is that events separated in time appear to occur simultaneously. To an observer the system state would not appear to have changed with the simultaneous arrival and departure events, although the actual system was idle for a short period of time.

The initial consideration in using the periodic scan approach is the determination of the length of the interval to be used. As Example 8.1 shows, if the time unit selected is too large, events separated in time appear to occur simultaneously, and information on the operation of the system is lost. Thus the time advance increment should generally be small relative to the likelihood of occurring events. More precisely, the time increment should be small enough that the probability of multiple events occurring during a single time interval is small. If the time unit selected in Example 8.1 had been 0.1 minute rather than 1 minute, the time separation between events C_3 and A_4 would be detected.

From this discussion it appears that the time increment used in the periodic scan approach should be as small as possible to minimize the possibility of lost information, but a moment's reflection should reveal that this is not always the case. The smaller the time unit, the larger the number of calculations necessary to complete the simulation. Thus the actual run time of the simulation and hence the cost of the simulation is increased. There is obviously a trade–off between the need for the precise, detailed model that can be obtained by decreasing the time advance interval and the increasing cost of the computer simulation run. Conversely, the cost of the computer simulation run can be decreased by increasing the size of the time advance unit.

Even if the time advance unit is small enough that the likelihood of multiple event occurrences within a single interval is small, the precise point at which an event occurs within that interval is still uncertain. In some models the exact time of occurrence may be unimportant. However, if a great deal of precision is desired, the time of the exact occurrence might be crucial. There seems to be no solution to this problem, short of making the time advance interval even smaller and thus increasing the cost of the computer simulation.

In many simulation models there are periods of high activity separated by periods of inactivity. The periodic scan approach is not well suited to this type of behavior because it is designed to give equal attention to each interval when more emphasis should be given to the periods of high activity. This fact, along with the problems of determining the time advance interval length, suggests an alternative approach to time management – the event scan approach.

In the **event scan** (Event–advance; Variable time increment) approach the clock is advanced by the amount necessary to trigger the occurrence of the next, most imminent event, not by some fixed, predetermined interval. Thus the time advance intervals are of variable lengths. This approach requires some scheme for determining when events are to occur. When events are discovered or generated, they are generally stacked in a list, or queue, in time order. The length of the required time advance interval can then be determined merely by scanning the event lists to determine the next earliest event. The simulation clock is then advanced to that time, and occurrence of the event is simulated.

Example 8.2 Consider the simulation model described in Example 8.1. If the event scan approach was used in lieu of the periodic scan approach, the simulation clock would initially be advanced to time 1.8 minute and event A_1 simulated. The clock would then be advanced to time 2.6 minutes and event C_1 simulated. The process would continue until all events have been executed or until the simulation duration had been met, whichever occurs first.

The event scan approach avoids some of the problems inherent in the periodic scan approach. The periodic scan approach requires that the analyst supply a fixed unit to be used to advance the simulation clock. The optimal size of this time unit is not known in advance and must be determined from a trade–off between precision and maximum run time. The event scan approach, on the other hand, does not require this artificial time increment; instead, the simulation clock is merely advanced to the next scheduled event occurrence time.

As pointed out in Example 8.1, some information can be lost in the periodic scan approach because events that are actually separated in time may be treated as if they occur simultaneously. The event scan approach avoids this problem because the clock is advanced only to the next occurrence time. Thus in an event–scheduling approach two events will be treated as occurring simultaneously only of they actually occur simultaneously.

The third problem with the periodic scan approach is that the exact occurrence time of events is not known, because each event is treated as if it occurred at the end of the interval in which it occurred. The resulting loss of precision in the model can be overcome only by decreasing the size of the time advance interval and thus increasing the cost of the simulation run. This problem is not encountered in the event scan approach, since the simulation clock is advanced to precisely the instant at which the next event occurs.

Another problem with the periodic scan is that during periods of inactivity the simulator is cycling, doing nothing useful, merely advancing the clock. This problem may seem minor, but if the system is characterized by relatively long periods of inactivity separated by short periods of high activity, a significant percentage of the run time associated with a given simulation run might be devoted to merely updating the clock. This problem is again avoided by the use of the event scan approach.

This comparison of the two techniques shows that the event scan approach appears to be more efficient in terms of run time, while the periodic scan approach is more advantageous in ease of implementation and simplicity in bookkeeping. Both approaches are illustrated in Figure 8.3.

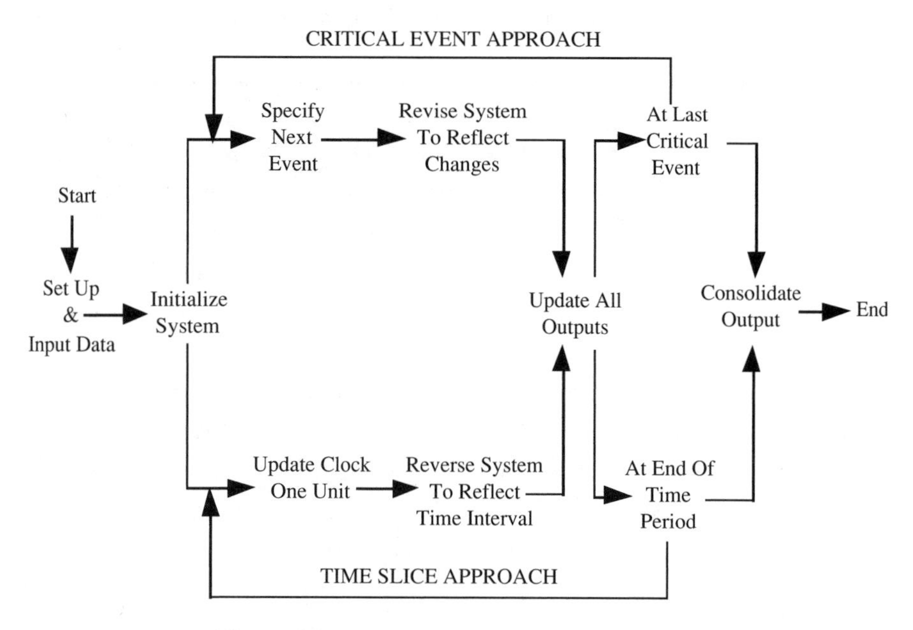

Figure 8.3. Techniques for time management.

8.3 Object Generation

A **system** as defined in Chapter 1 is a collection of objects with a well–defined set of interactions among them. Therefore when a system is being simulated, there must be some way of representing and introducing objects to the system.

An **object** or an **entity** of a simulation model must be uniquely identified in the program; in most cases this is done by the assignment of a serial number to the entity. For example, in a service station model customers might be numbered in the order of their arrival to the system. Along with the serial number, entities are characterized by a set of attributes. The **state of an entity** at any point in time is the current set of values of its attributes. The **state of the system** can be thought of as the collection of all present entities represented by their state vectors. Entities that will be present throughout the course of a simulation run are called **permanent**. Entities that will be introduced to the system, remain for a period of time, and then exit the system are called **transitory**.

Permanent entities are generally input to the simulation model during the initialization of that model. Transitory entities, on the other hand, are introduced at particular points in the course of the simulation run and then removed from the model. Thus the generation of a transitory entity involves the generation of an arrival time as well as the assignment of values to various attributes.

The arrival of a transitory entity to the system is an event. It alters the state description of the system by adding one more entity. Arrivals can be made to occur at regular fixed intervals or in some random fashion by sampling the interarrival distribution. A bootstrapping scheme is commonly used in generating the arrival times for successive transitory entities. The arrival time for the first entity is generated by sampling the appropriate distribution and placing the arrival event on the event list. When the simulation clock has advanced to that point at which the event is completed, the system state is updated and the arrival time for the second entity is generated. An alternative to this bootstrapping scheme is to generate all the arrival times at once. This scheme, however, requires the storage of arrival times, thus increasing the storage requirements of the model.

Example 8.3 Again consider the problem of simulating the operation of a message center. The transitory entities for this system are the messages themselves. Suppose that the messages arrive according to a Poisson distribution with a mean rate of λ. The interarrival times for the messages can then be obtained by generating exponentially distributed random numbers using the techniques outlined in Chapter 7.

Once the arrival time for the transitory entity is determined, values must be assigned to the various attributes. Attribute values can be generated either at the time the arrival time is generated or at the instant at which the arrival event actually occurs (8.2). The factor that determines which of these two times is to be used is whether the assignment of attribute values is state dependent. State–dependent attribute values should be assigned at the time the arrival actually occurs. And, as with the arrival times, attribute values are normally assigned by the simulated sampling of a given distribution, using the random number generation techniques outlined in Chapter 7

Example 8.4 Consider the message center example outlined in Example 8.3. Once the arrival time has been determined through simulated sampling from the exponential distribution, attribute values are assigned. Two attributes of interest are the message precedence and the message length. Suppose that messages are equally likely to be one of three precedences – routine, priority, and emergency. Suppose also that the message length is approximately normally distributed with mean μ and variance σ^2. The assignment of message precedence can be accomplished by the generation of a standard uniform random variate. If the random number is between 0 and 0.33, routine precedence is assigned; between 0.33 and 0.66, priority precedence; and between 0.66 and 1.00, emergency precedence. Assignment of message length can be accomplished by the generation of an appropriate random normal deviate.

8.4 Events and Event Synchronization

An event is the occurrence of a change in the system state at some point in time. In this sense an event has no duration but exists as an instant in time. An important aspect of a discrete system simulation is the scheduling and synchronization of events.

Events can be categorized by type. Events that involve similar changes to the system state are said to be of the same type. There are two broad event types: system events and program events. A **system event** is an event that represents a simulation of a comparable event in the real system. A **program event** has nothing to do with the system being simulated but deals only with the simulation program itself. Program details such as when the collected statistics are to be printed out are program events. Although these program events may seem artificial, they must be scheduled just as system events must be scheduled.

System events can be further broken down. The scheduled occurrence time for certain types of system event as well as the effect of their occurrence on the system state may be altered by some event that occurs after the event is scheduled but before it actually occurs. Such events are referred to as **contingent events**. System events that cannot be modified by any intervening event once they are scheduled are referred to as **non–contingent events**.

Example 8.5 Consider the simulation of a single–bay service station. Suppose that an arrival event to the service station is scheduled for some time in the future. Suppose furthermore that after the arrival event was scheduled but before it actually occurred, the event "station capacity reached" occurs. The arrival event then would be canceled. The arrival event in this case could be thought of as a contingent event.

System events can also be categorized as decision or nondecision events. A **decision event** is one in which some decision must be made about the event's effect on the system state. A **nondecision event** is one for which no decision is required to determine the effect of the event's occurrence. This breakdown of event types is mentioned merely to point out that special bookkeeping routines may be necessary to handle the various types.

Event sequencing and **synchronization** are important aspects of a discrete system simulation. Events are represented as instances in time. Once an event is scheduled and the simulation clock is advanced to the scheduled time, an event execution routine for the particular type of event is invoked to update the system state, thus simulating the effect that event would have on the real system. As long as scheduled events are separated by some interval of time, there is no problem with event sequencing. If two or more events are by chance scheduled to occur simultaneously, the simulation model must decide which event is to be executed first. The decision logic of the simulation model should reflect as much as possible the decision rule in effect in the real system. In other cases, in which either the simultaneity does not occur in the real system or the events are actually handled simultaneously by the real system, some ordering of the simultaneous events must be made so that the execution routines can be invoked. Possible orderings of simultaneous events range from elaborate priority schemes to simple random selection schemes.

The proper sequencing of events is crucial because the effects of certain events can have consequences on other events. For example, some events generate the occurrence of other scheduled events, while yet others may cause a cancellation of some already scheduled event. Events not sequenced in their proper order can have drastic effects on the system's performance.

Events are scheduled in most discrete system simulation programs by an **event calendar** or **event chain**. The events generated by the model join a time–ordered list. As the simulation clock is advanced using the periodic scan approach, the event calendar is scanned for events that should have occurred in the last time unit. Execution routines for those events are then invoked, and after they are completed, the simulation clock is again advanced. On the other hand, if the event scan approach is used, the simulation clock is actually controlled by the event calendar. The event calendar is scanned for the earliest scheduled occurrence of an event. The simulation clock is then advanced to this time, and the appropriate execution routine invoked. After this routine is completed, the event calendar is again scanned and the process repeated.

The number of events and the complexity of the corresponding execution routines are the primary determinants of the run time and consequently the cost of the simulation. For example, if there are a large number of contingent events, just updating the event calendar can be time–consuming.

While it is important to have a description of the simulation modeling process, it is also important to have a mathematical representation of this same process. A mathematical description is useful in determining the completeness of the simulation model, establishing evaluation criteria, and in assessing the number and importance of the simulation model parameters. The reader may skip the remainder of this section without loss if not interested in this level of detail.

The two types of information used in such a mathematical description can be characterized as being descriptive and designational. **Descriptive information** is about an event occurrence or an object (unit) in terms of parameters and variables of the simulation model. **Designational information** identifies the event or object as distinct from all other such occurrences or objects.

Event occurrences can be described in terms of four information elements, sequence number (k), the time of occurrence (t), the type of event (T_e), and the simulation run (l). Thus the kth event of the lth simulation is represented as follows:

$$E_k \equiv E_k(l, T_e, t)$$

The analogous representation for an object is

$$\boxed{O_i \equiv O_i(l, T_o, t)}$$

where O_i represents the ith object in the lth simulation, T_o the object's type or subtype, and i the object's sequence number.

Next we must consider the **representation for parameters** and random variables. Let x_n, $n = 1, 2, \ldots, N$ be N random variables. Furthermore, let

$$\bar{b} = (b_1, b_2, \ldots, b_N)$$

be a **binary usage vector** for the description of objects and event occurrences. b_i will take on the value 0 when x_i is not used in the description of the object or event, or the value 1 otherwise. Thus, $\bar{b}_{T_e}$ will indicate which random variables are used in the description of the T_e–th type of event, and $\bar{b}_{T_o}$ for the analogous T_o-th type of object.

The above random variables, usually obtained from distribution functions, can be expressed as polynomials

$$x_n = \sum_{\alpha=0}^{k} a_{n,\alpha} \hat{x}^{\alpha} = x_n(\bar{a}_n, \hat{x})$$

where k is the degree of the polynomial, $\hat{x}$ the random number, and $\bar{a}_n = (a_{n,0}, a_{n,1}, \ldots, a_{n,k})$ is the coefficient vector for $\hat{x}$.

We also must have a vector describing the parameters involved in the simulation model. Let P_m be such a **generic parameter descriptor**. Then

$$\bar{P}_i \equiv (P_{i,1}, P_{i,2}, \ldots, P_{i,m})$$

can be used as part of the ith object or kth event occurrence, where $P_{i,m}$ would take on the actual parameter value if relevant to the description of the ith object.

To complete the description of event occurrences, let $\bar{c} \equiv (c_1, c_2, \ldots, c_k)$ be a binary vector, where c_i takes on the value of 1 if the kth object is involved in the particular event occurrence, and 0 otherwise.

Let $\overline{E} \equiv (E_1, E_2, \ldots, E_{T_e})$ indicate which **type of event** occurrences can be added (A) or deleted (D) by the T_e–th type of event. In fact, the **event addition** and **deletion process** can be represented by

$$\overline{E}_A(T_e) = (E_{T_e, A_1}, E_{T_e A_2}, \ldots)$$

and

$$\overline{E}_D(T_e) = (E_{T_e, D_1}, E_{T_e D_2}, \ldots)$$

where E_{T_e, A_1} (or $E_{T_e D_1}$) take on the value of 1 if the ith type of occurrence may be added (deleted) by an event of that type, and 0 otherwise. Furthermore, let $E_{q'}$ be the event occurrence which added E_q. The complete representation of an event occurrence encompasses

$$E_q = E_q(l, T_e, t; E_{q'}, \overline{c}, \overline{b}, \overline{P}, \overline{E}_A(T_e), \overline{E}_D(T_e))$$

where each vector is assumed to have components relevant to the event occurrence equations, while the complete representation of an object is

$$O_i = O_i(l, T_0, t; \overline{b}, \overline{P})$$

It is now possible to determine the mathematical relationship between the occurrence of an event and its effect on the state of an object. Suppose that an event occurred at time t_β. The effect of this event occurrence on the ith object is as follows:

$$O_i(l, T_0, t_k; \overline{b}, \overline{P}) = O_i(l, T_0, t_{k-1}; \overline{b}, \overline{P}) + \Delta O_i(E_q(l, T_e, t_k); t_k - t_{k-1})$$

The first term is the state of an object at given time t_k, while the third term reflects changes in the components $\overline{P}$ resulting from the evaluation of event E_q and the lapsed time $t_k - t_{k-1}$. Furthermore, the state of the modeled system is given $\overline{S}(l,T_0,t_k)$ with components $S_i(l,\ T_0,\ t_k\ ;\ \overline{b},\ \overline{P})$. It should be noted that the state of the system also is changed, in a manner similar to the equation above, by the occurrence of an event.

The **event status calendar** of added/deleted events can be maintained each time an event occurrence is processed. Assume that the kth event E_k is processed, the event calendar at time t_k during the lth simulation is then ordered by time

$$\begin{aligned}\overline{E}(l,t_k) = [&E_{k_1}(l,T_{e_1},t_1;\ldots),\\ &E_{k_2}(l,T_{e_2},t_2;\ldots),\ \ldots\\ &E_{k_N}(l,T_{e_N},t_N;\ldots)]\end{aligned}$$

where $t_k \le t_1 \le t_2 \le \ \ldots\ \le t_N$. As a result of processing event E_k, the calendar is pudated by adding [deleting] $E_A(t_k)$ $[E_D(t_k)]$. Recall $\overline{E}_A(t_k)$ are the occurrences which are generated by the actual event occurrence E_q, and $E_D(t_k)$ the actual event occurrences to be deleted from the calendar. Thus,

$$\overline{E}(l,\ t) = \overline{E}(l,\ t_k)\ \text{AND}\ \overline{E}_A(t_k)\ \text{MINUS}\ \overline{E}_D(t_k)$$

where t, the simulation time, has a value not less than t_k and not greater than the earliest occurrence time in the calendar, and AND, MINUS are Boolean addition, deletion operators.

Finally let us consider how to evaluate the **performance of the simulation model**. Since simulation models and experiments involve many random variables, many simulation runs may have to be made so that average values, variances, and other statistics for the performance can be obtained. The measures of performance F are evaluated with respect to l runs.

An important part of the evaluation of a simulation model is to determine the value of l which assures an acceptable description of each performance measure F_i. Let $F_{i,l}$, $i = 1, 2, \ldots, M$, be the ith performance measure, evaluated at any time t during the lth simulation. Suppose $t_k \le t \le t_{k+1}$, then the functional representation of the ith performance measure is

$$F_{i,l}(t) = F_{i,l}[\overline{S}(l,t_1),\ \overline{S}(l,t_2),\ldots,\overline{S}(l,t_k)]$$

which may involve all or part of the state histories up to time t. The mean and the variance, for l simulation runs, are then respectively

$$\overline{F}(t;l) = \frac{1}{l}\sum_{j=1}^{l} F_j(t_i)$$

and

$$S^2[F_i(t);l] = \frac{1}{l-1}\sum_{j=1}^{l}[F_{i,j}(t) - \overline{F}_i(t;l)]^2$$

The analysis of simulation model runs may indicate that certain variables appear to be mathematically related. Such a hypothesized relation could be checked using the equations of the simulation model. An attempt may be made to fix some parameters, and then to test any proposed parametric relationships, by either additional simulation runs or by parametric analysis of the equations of the simulation model. However, from the equations of this discussion, it should be clear that the formal descriptions of objects, event occurrences, and distributions are very cumbersome, and from a practical standpoint can easily overtax the analysts capabilities.

8.5 Queue Management and List Processing

Waiting lines or **queues**, are encountered in almost every system that is modeled because nearly all systems use some limited resource, whether it is the number of servers in a service station or the number of available I/O channels on a general–purpose computing system. Queues arise because of the competition for these limited resources. Thus a major consideration in developing a simulation model of a system is the representation and manipulation of queues. Queues can be distinguished by their corresponding queueing disciplines, that is by the manner in which the next customer gained control of the resource and was selected from the waiting line. The most common queueing discipline is the **first–in, first–out**

(FIFO) discipline, in which the customer who has been waiting the longest is given control. Other disciplines include **last–in, first–out (LIFO)**, in which the customer who has been waiting in the queue the shortest time is given control; **random (RAND)**, in which a customer is selected at random from the queue; and a myriad of **priority (PRI)** disciplines, in which some customer is selected based on a particular value associated with one attribute.

Whenever an arriving entity is generated and scheduled, the values of its attributes are also assigned, either through the simulated sampling from given distributions or through some other pre–established procedure. In many cases the entity is also assigned a distinguished serial number to allow unique identification. A convenient means of representing the entity arriving in the system is through an activation record such as illustrated in Figure 8.4. As the entity makes its way through the system model, the values of its attributes are adjusted to reflect event occurrences. Eventually the entity should traverse the entire system, at which time it ceases to be an active entity and becomes part of the summary statistics.

Serial number	Arrival time	Attribute 1	•••	Attribute n

Figure 8.4. An activation record.

Just as the activation record is a convenient way of representing an arriving entity to the system, a convenient way of representing a queue (nothing more than a collection of arriving entities waiting for a common resource) is a list. The organization of queues is a list supports such operations as inserting a new record (adding another entity to the queue), deleting a record (selecting an entity to gain control of the resource and thus eliminating it from those waiting), and accessing a particular record from a list to examine the value of one of its attributes.

With list processing, objects are represented in a computer in a manner that shows the relationship among these objects in an ordered set of entries. It is this ordering that makes the list so convenient for the representation of queues. The manner in which list processing is performed depends largely on the structure of the list.

There are a number of considerations in the construction of lists. First, the **dimension of the list** must be defined. A list comprising single entries is a one–

dimensional list; a list of sublists is a two–dimensional list, and so on. Second, the **density of the list** must be considered. Objects in the list may be stored in contiguous or widely separated memory locations. Third, the **sequence of the list** should be determined. For example, a one–dimensional list can be sequenced from top to bottom or from bottom to top.

A number of factors influence the **construction of a list**. One is the type of application. The list must be constructed in the manner that best fits the application. Inappropriately constructed lists can lead to high processing overhead. A second factor important for the construction of a list is the **search technique** to be used. The search technique should be matched to the application and construction scheme. A third factor is the **alteration** of the list. Consideration must be given to changes likely to result from adding, deleting, or altering elements.

A **list** is a collection of records in which the extent of the list must be delineated. Objects may be located in a variety of ways depending on the construction of the list. First, the item may be **serially locatable**. In this scheme the list is sequentially searched until the desired record is found. Second, the item may be **directly locatable**. A key is assigned to each record, and this key is used to extract the desired record directly from the list. A third scheme is **indirect locatability**. In this scheme items of the list are linked or chained by pointers. There may be single or multiple pointers to enhance locatability.

The **linked**, or **threaded**, list is commonly used to implement a queue structure. The linking or chaining in this queue structure is accomplished by appending a pointer to each record in the list. The pointer indicates the next item in the chain. If multiple chaining is desired, multiple pointers can be appended. Linked lists are therefore composed of records and pointers to the next item on a chain. The main advantage of the linked list is that the list can be altered easily. Items may be added or deleted from the list simply by manipulating pointers rather than entire records. An example of a **singly linked list** is given in Figure 8.5. A **doubly linked list** can be obtained from the singly linked list by appending a pointer that links in the opposite direction. That is, record 109 would point to 108, 108 to 107, and so on. A **circular linked list** could be obtained by linking the tail of the chain (record 109) back to the head. In a singly linked list there is normally a separate pointer that points to the first record in the chain, while there may or may not be a separate pointer to the last record in the chain.

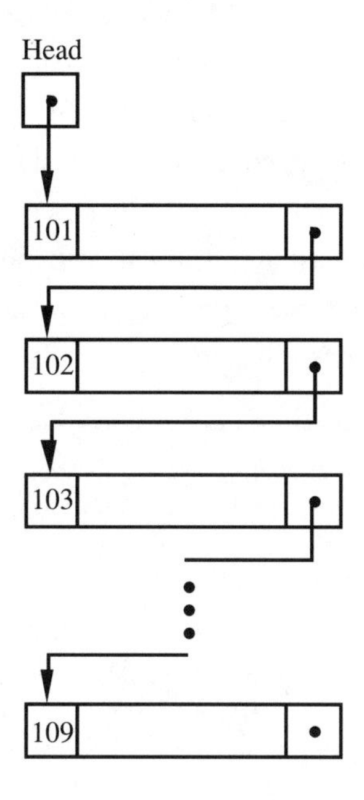

Figure 8.5. A singly linked list.

At this point it might be useful to indicate how the singly linked list of Figure 8.5 could be changed. Suppose that record 102 is to be deleted from the list. This could be accomplished merely by changing the pointer appended to record 101 to point to record 103, thus severing all ties to record 102. Note that information in record 102 is no longer retrievable. The storage used by record 102 should then be reclaimed by some "garbage collection" routine and reused. Suppose that instead of deleting record 102 we wish to add a new record, call it 102a, between records 102 and 103. This could be accomplished by appending a pointer to record 102a which points to the address of record 103, and then changing the pointer appended to record 102 to point to record 102a. Note that this scheme does not require record 102a to be located contiguous to record 102; record 102a can be located anywhere since it is locatable through the pointer appended to record 102. This avoids the need to shift entire records as would be necessary if a simple sequential arrangement were used.

Most of the common queueing disciplines can be readily implemented by using linked lists. To simulate the operation of a **FIFO queue**, a singly linked list would be sufficient. Arriving entities would join the tail of the queue. This of course would necessitate a separate pointer to the tail of the list. The processing of the queue (list) begins at its head, thus simulating first–in, first–out processing. A **LIFO queue** is also easily simulated using a singly linked list. Arriving entities are placed at the head of the list, and the queue is processed from its head. **Priority queues** are readily simulated using linked lists if the linking is based on the value of the attribute that determines the priority rather than on the time of arrival to the queue. An example may clarify this point.

Example 8.6 Consider a queueing system composed of a single server. Arriving entities are characterized by an assigned serial number by the time that they arrive in the queue, and by the estimated length of service that they require. Suppose that there are five objects in the queue awaiting service having the following attributes.

Serial number	101	102	103	104	105
Arrival time	1013	1009	1011	1015	1014
Service required(min)	2	4	6	1	5

These arriving entities could be represented to the system by an activation record consisting of the fields illustrated in Figure 8.6.

Serial number	Arrival time	Service required	Pointer to next record

Figure 8.6. Activation record for arriving customers.

If the queueing discipline to be simulated is a FIFO discipline, the following list arrangement is used.

Head
102

Tail
104

101	1013	2	105
102	1009	4	103
103	1011	6	101
104	1015	1	
105	1014	5	104

Processing then begins with record 102 and, following the chain, proceeds with 103, 101, 105, and 104; each record is deleted as it is processed. When a new arrival occurs (say record 106), it is added to the list by entering record 106 into the pointer field of record 104 and changing the tail pointer to point to record 106.

If a LIFO discipline is to be simulated, the following list arrangement must be used.

Head: 104

101	1013	2	103
102	1009	4	
103	1011	6	102
104	1015	1	105
105	1014	5	101

In this arrangement processing begins with record 104 and, following the chain, proceeds with records 105, 101, 103, and 102. If a new arrival (again say record 106) occurs, it is added to the queue by changing the head pointer to 106 and by placing 104 in the pointer field of record 106.

A common priority scheme used in service demand systems is the **shortest–job–first discipline**, in which the priority is based on the amount of resources requested. The highest priority is given to the arrival that requires the least service. Only a moment's reflection is necessary to be convinced that this discipline minimizes the average waiting time. This simple priority scheme can be simulated for our examples by using the following list arrangement.

Head: 104

101	1013	2	102
102	1009	4	105
103	1011	6	
104	1015	1	101
105	1014	5	103

Processing begins with record 104 and proceeds to records 101, 102, 105, and 103 in that order. If a new arrival occurs, it is placed in the proper place in the list based on its requested service time. For example, if record 106 arrives requesting three minutes of service time, it is placed in the chain between records 101 and 102 by placing the address of record 102 in the pointer field of record 106 and by changing the pointer field of record to 101 to point to record 106.

This example illustrates how linked lists can be used to simulate three of the more common queueing disciplines. More complex queueing disciplines can be simulated by including more pointers. For example, the **RAND (random)** discipline requires that every record be locatable from every other record. When the linked list is used care should be taken so that the overhead (storage requirements for the pointers) does not become excessive. As items are deleted from the list, some mechanism must be included to reclaim the storage occupied by that record (garbage collection).

8.6 Collecting and Recording Simulation Data

A simulation model that produces no output is not very useful to the researcher. Thus an important aspect in the design, development, and use of the simulation model is the means by which data are collected and summarized. One of the major advantages of using a special–purpose simulation language rather than a general–purpose language such as FORTRAN is that a special–purpose language facilitates data collection. Depending on the system being simulated, different data of the system's performance must be recorded so that the system can be analyzed under varying conditions. This section describes some of the more general types of data that are usually collected and used to characterize a simulation run.

The most common type of data collected from a simulation run is count data – for example, the number of occurrences of a given type of event, the number of entities in each queue of the model at each interval of time, and the total clock time that an entity remains in the system. These count data are then manipulated at the completion of the simulation run to provide summary measures of the system's performance. Count data are easily collected through the use of counter variables initialized at the beginning of the simulation run and updated whenever an event occurs that affects the item of interest.

Example 8.7 Suppose that a simulation model has been developed for a single–bay service station and that one item of interest is the number of arrivals that occurs over the duration of the run. This information can be recorded by initializing a count variable NRARR to zero at the beginning of the simulation run and incrementing it each time that the occurrence of an arrival event is processed. Another item of interest might be the total time that entities remain in the system. This information could be collected by initializing a count variable NTIME to zero at the beginning of the simulation run. Then each time the simulation clock is incremented, the amount of the increment is multiplied by the total number of entities in the system and added to NTIME. Of course, the total number of entities in the system would be collected by initializing a count variable NRENT to zero at the beginning of the run, incrementing it whenever an arrival occurs and decrementing it whenever a departure occurs.

Particular attention must be given to the place in the simulation model where the data are collected. The most obvious place to collect the data is during the processing of an event occurrence that affects the variable of interest. For interest, in Example 8.7, the count variable NRARR should be incremented either when the routine is invoked to process the arrival event occurrence or in the arrival event occurrence routine itself. Care should be taken to ensure that the data are collected at each point in the simulation model where an event occurrence affects the variable of interest. If arrival events are processed at two points in the model, data on the number of arrivals must be collected at both points.

The type of summary statistics to be calculated influences the type of data collected. For example, the mean and standard deviation of some variable may be desired. To calculate these two items of interest, one would need to accumulate the sum of the observations of the variable, the sum of the squares of the observations, and the number of observations on which these accumulated sums are based.

Example 8.8 Suppose that we want to calculate the mean and standard deviation of the length of the queue in the queueing system described in Example 8.7. This could be accomplished by collecting the following data. Three count variables SUM, SUMSQ, and NTIM are initialized to zero. Then when the simulation clock is incremented, the number in the queue, call it NRQUE, is added to SUM, the square of NRQUE is added to SUMSQ, and NTIM is incremented by 1. At the end of the simulation run the mean and standard deviation of the queue length are calculated using the standard formulas

$$\text{MEAN} = \text{SUM} / \text{NTIM}$$

$$\text{STDDEV} = \frac{[(\text{SUMSQ})^2 - (\text{NTIM})\,(\text{MEAN})^2]^{1/2}}{(\text{NTIM} - 1)}$$

In addition to count data, frequency data are often found recorded during a simulation run. The recording of these data makes it possible to summarize certain aspects of the system's performance through the use of a **frequency histogram**. Frequency data are usually recorded for events such as interarrival times and queue lengths.

Example 8.9 Suppose that the analyst wishes to summarize the interarrival distribution to the single–bay service station discussed in Example 8.7. This could be done as follows. A number of classes are established for the interarrival times. In this example suppose that the times between successive arrivals to the system have been recorded using the count variable INTARR. Suppose also that the classes have been established as

Class 1	$0 \leq \text{INTARR} < 1$
Class 2	$1 \leq \text{INTARR} < 2$
Class 3	$2 \leq \text{INTARR} < 3$
Class 4	$3 \leq \text{INTARR} < 4$
•	•
•	•
•	•
Class 10	$9 \leq \text{INTARR} < 10$

An array composed of ten elements, ARRDIST, is then initialized to 0. Each time an arrival event is processed, INTARR is checked to determine the event's class, and the appropriate array element is incremented. Note that INTARR must be reinitialized after each arrival. After the simulation run has been completed, the interarrival distribution is summarized in array ARRDIST in the form of a frequency table. This can then be subjected to goodness–of–fit testing or presented as a summary of that aspect of the system's performance. If a relative frequency table is desired, each element of the array ARRDIST is divided by the total number of arrivals, which could be accumulated with another count variable NRARR.

The data collected and the analysis methods used on the data are nearly as numerous as the types of simulation models. In this section we have surveyed some of the more common types of data and collection techniques. The analyst has to tailor the data collection methods to the requirements of the system being modeled.

8.7 Summary

In the course of this chapter we have surveyed some of the factors and techniques that have been considered and employed in the development of computer simulation models. In particular we have addressed not only all the major components that make up a simulation model, but also the major problem areas (see Table 8.1) as well as typical solutions. The reader will find in Table 8.2

suggested guides concerning model complexities. By this time the reader should have some idea how to develop a simulator. In a subsequent chapter we will survey some of the more common discrete–event simulation languages. These languages were developed largely to ease the analyst's burden of developing the simulation model. Some additional considerations in the development of a simulation model will be introduced as features of the various languages are discussed.

Table 8.1 Major Problem Areas of Discrete System Simulation

1	Time Management
2	Event Synchronization <Event – Scan Management> • Simultaneous events • Dependent events
3	Queue Management • Queue manipulation • Queue record keeping
4	Queueing System Interaction • Interaction determined by time • Interaction determined by historical record • Arrival patterns • Queue sharing
5	Collection and Recording Simulation Data • Correlate data to entities • Correlate data to events • Correlate data to time
6	Evaluation of Simulation Data

Table 8.2 Model Complexity Guidelines

1	The greater the number and types of different objects, the greater the model complexity.
2	Different descriptions (routines) are required for different types of objects and events.

Table 8.2 Model Complexity Guidelines (continued)

3	The more that objects are able to interact with each other, the more complicated the model is likely to be.
4	The greater the number of different types of events, the greater are the possibilities for complex instructions.
5	The greater the proportion of contingent events (these can be delayed, prevented, or otherwise be altered by other events), the more complex the model.
6	The number and complexity of decisions made by events (routines) affect the overall complexity of the model.

8.8 Exercises

8.1 Develop a FORTRAN subroutine ARR that simulates the arrival of a customer to a queueing system. The arrival process to be simulated is Poisson. Parameters to be passed to ARR are the time of the last arrival and the mean arrival rate. The returned value should be the time of the next arrival.

8.2 Develop a FORTRAN subroutine SERVE that assigns required service times to the arrivals to a queueing system. The service–time distribution to be simulated is the exponential distribution. The mean service rate should be passed to the sub routine as a parameter. The returned value should be the service time required by the next arriving customer.

8.3 Develop a FORTRAN subroutine POSTQ that enters arriving customers into the queue. The maximum queue length should be passed as a parameter. Customers attempting to enter a full queue should be turned away. The attributes to be entered for each customer are customer number, arrival time, and required service time. Assume that the queueing discipline is FIFO.

8.4 Develop a FORTRAN subroutine REMOVQ that removes the first customer from the queue and moves the remaining customers up accordingly (simulate a FIFO queue). An attempt to remove a customer from an empty queue should be signaled by setting a flag.

8.5 Develop a logical flowchart of a single–channel queueing system (M/M/1/K/FIFO) using the periodic scan approach to time management.

8.6 Develop a logical flowchart of a single–channel queueing system (M/M/1/K/FIFO) using the event scan approach to time management.

8.7 Code the model depicted in the flowchart developed in Exercise 8.5. The routines developed in Exercises 8.1–8.4 should be used.

8.8 Code the model depicted in the flowchart developed in Exercise 8.6. The routines developed in Exercises 8.1–8.4 should be used.

8.9 Run the simulation models developed in Exercises 8.7 and 8.8 with an average service rate of $\mu = 18$ per minute and an average arrival rate of $\lambda = 6$ per minute. Collect statistics on the number in the queue. Assume a maximum queue capacity of 25.

8.10 Compare and contrast the models developed in Exercises 8.7 and 8.8 in terms of complexity and results obtained.

8.9 References

8.1 Emshoff, J. R., and Sisson, R. L., *Design and Use of Computer Simulation Models*. New York: Macmillan, 1970.

8.2 Gordon, G., *System Simulation*. Englewood Cliffs, NJ: Prentice–Hall, 1978.

8.3 Martin, F. F., *Computer Modeling and Simulation.* New York: John Wiley and Sons, 1968.

9

MODEL VALIDATION

Once a simulator has been designed and coded, the analyst must evaluate the validity of the model. Evaluating a model is determining how well the simulation model predicts the real system's performance. The ability of a simulation model to accurately predict the performance of the real system depends on both the validity of the model and the reliability of the performance measures produced. In this chapter we discuss problems in these areas and survey some techniques that have been used to enhance reliability and to demonstrate validity.

9.1 Evaluation of the Simulation Model

The validity of a simulation model depends on the accuracy of the model representing the real system. Also of concern to the analyst is the detail of the model. The model must be sufficiently detailed to provide the analyst with information on the aspects of the system's performance that are of primary interest. The truly satisfactory method of validating a simulation model is to judge its performance. If the inferences drawn from the analysis of the simulator's output allow correct conclusions to be drawn about the system or the situation being modeled, then the simulation model can be assumed to be valid for that particular situation. However, using this test for the validation procedure has some drawbacks. Courses of action not taken or decisions not made by the simulator could provide some added insight into the true validity of the model. Thus it may not be sufficient to assess the validity of the simulation model only from observations of the model's performance.

A number of methods for assessing the validity of a model use techniques other than direct observation of its performance. These methods, just like the observation technique, do not guarantee that the model is valid, but they do provide a basis for assuming that the model is valid. The first of these methods deals with **validating the design** of the model. This form of validation is simply a checking problem in which the design of the model is verified at different stages of its development. The process of modeling a system is broken into two phases, the conceptual phase and the implementation phase (9.8).

In the **conceptual phase** the logical flow of the system being modeled is determined, and the relationships between the various subsystems are formulated. During this phase the factors likely to influence the performance of the model are isolated and tentatively selected for inclusion. There are two procedures by which the model is validated at this stage. The first is to have the model reviewed by disinterested qualified observer. If this observer confirms the decisions made by the model designer, the judgment of the model's validity is reinforced. Another method is to trace through the model in reverse order. This technique is somewhat analogous to verifying the accuracy of an arithmetic result by applying the inverse operation to the result.

The **implementation phase** of the model includes selecting and quantifying procedures for the model, coding the model, and actually using the model. Martin (9.8) suggests that the only practical method of ascertaining the validity of the model during this phase is to check the model at pre–established milestones by comparing it with the previous stages of the model development. The following milestones are suggested.

1. Following the development of the logical flow chart of the model
2. Following the program coding of each model subsystem
3. Following the integration of the subsystem modules into a complete coded model

At the second milestone each subsystem should be tested with sample input before it is incorporated into the model. When all subsystems have been separately tested, the model as a whole should be tested.

Emshoff and Sisson (9.4) have suggested other validity tests, including internal validity tests and variable–parameter validity tests. The **internal validity** test consists of performing several simulations using the same model and the same input parameters and then comparing the outputs to detect variability. If the variability of results is high, the model will probably prove of little value as a predictor, since it will be difficult to assess whether changes in output are due to changes in input parameter settings, the model's inherent variability, or a combination of the two. The limiting effect of this internal variability on the usefulness of the model can also be viewed in light of the real system's possible behavior under similar circumstances. It is unlikely that any real system of interest, when presented with identical operating conditions, will produce radically different results.

The **variable–parameter validity** *test* consists of varying parameters and variables to determine their effect on the simulator and the subsequent output. If the impact of certain variables or parameters is large compared with the initial estimate of their impact, the validity of the model must be questioned.

Probably no analyst is ever completely sure of the validity of the simulation model. In many cases validity is assumed until the contrary is shown. This could happen long after the original analyst has completed the work. The analyst must, however, make a serious attempt to validate the model before using it.

9.2 Validation Description

Decisions about the structural characteristics of a simulation model need to be made in conjunction with decisions about how the simulation model is to be used. Some definitions are in order.

DEFINITION 9.1

*A **simulation run** is an uninterrupted recording of the simulation system's performance given a specified combination of controllable variables (e.g., range of values, values on the parameter, etc.).*

DEFINITION 9.2

*A **simulation duplication** is a recording of the simulation system's performance given the same or **replicated conditions** and/or combinations, but with different random variates.*

DEFINITION 9.3

*A **simulation observation** is a simulation run or a segment of a simulation run that is sufficient for estimating the value of each of the performance measures.*

DEFINITION 9.4

***Steady** or **stable state** of a simulation system is achieved when successive system performance measurements are statistically indistinguishable (i.e., no new information is obtained about the future behavior of the system).*

It should be noted that steady state is approximately the same as statistical stationarity with respect to the performance measure of the system simulation output. For example, $f(x)$ at x is approximately the as $f(x + \Delta)$, where $\Delta > 0$. This does exclude all systems that have perfectly predictable cyclical behavior.

Steady state is indicated by either one of the following two methods. The first is whenever in the number of observations, the output is greater than the average at a given point as often as the output is less. The second is to compute a moving average of the output and stop whenever the average no longer changes $\pm\varepsilon$ over time, where ε is a user provided variation.

DEFINITION 9.5

*The **transient state** of a system is that state in which a system operates, for a period of time, such that either the effects of the starting conditions become insignificant, or for some systems no steady state conditions are expected.*

Reasons for studying transient behavior include that steady state may not exist in the simulation system, or a desire to analyze problems associated with system initializations. Because the latter includes radically changing situations and conditions, such as exceptional or extreme conditions (e.g., maximum numbers, probabilities of exceeding safety levels, probability of failure of new components, etc.), analytical methods are frequently not applicable.

As with the analysis of other type of simulation models, stochastic transient models require numerous replicated restarts to obtain a distribution of results. Furthermore, if the output of interest is the possibility of an exception or an estimate of an extreme, a large number of replications may be required. This is to ensure a high probability of including the unlikely events in the sample.

Output data obtained during the simulation system's transient period may be bypassed by one of several methods. These include setting initialization conditions in such a manner as to assume that transients are insignificant; starting the simulation with initial conditions indicating *a priori* steady state; introducing an initial nonrecording period during the transient time and then resetting and/or initializing all counters once steady state has been detected; and using long simulation runs so that the data from the transient period becomes insignificant relative to the data in steady state.

Validation concerns itself with how accurate of a representation the simulation model is to the actual physical system that is being simulated. Validation of the design of the simulation model may be considered by investigating internal validity, face validity, variable–parameter validity, hypothetical validity, and event or time series validity (9.4, 9.5).

Internal validity may be measured as low variability in internal noise effects. That is to say, stochastic models with a high variance owing to internal processing will obscure changes in output resulting from changes in controlled or environmental variables. **Face validity** is measured by comparing actual output results from individuals familiar with the real system. **Variable–parameter validity** tests the sensitivity or interrelationship among one or more factors of the simulation model to determine the effect of their output, and how effectively certain dependencies more closely model historical data. **Hypothetical validity**, rarely properly investigated, is a test of negation. **Event** or **time series validity** is used to determine if the simulation model predicts observable events, event patterns, or the variations in the output variables.

9.3 Sampling Methods

Performance measures may be used to determine the level of confidence an analyst has in a simulation model. This requires information about the variance of the output data, which in turn requires information about the relationship between the observations in the sample, leading to the question of how samples are selected. Should the simulation run be replicated? How long should the observation period be? What can be done to eliminate or reduce interdependencies?

From statistical sampling theory we may use the concept of **repetitions** to answer questions concerning the number and length of simulation runs, and the concept of **blocking** to avoid the transient periods. Let us consider the notion of simulation run reliability. Let the average estimated performance measure be

$$\hat{\mu} = \sum_{i=1}^{n} \frac{x_i}{n}$$

where x_i are the individual observations, and n is the number of observations in the sample. If each of the x_i 's is independent, the **confidence** of this performance measure is the **estimate of the variance**

$$\boxed{\hat{\sigma}_\mu^2 = \frac{\sigma^2}{n}}$$

Let us examine the effect of replicating a simulation and introducing a **correlation** between the replications. X and Y can be averaged to produce an estimate of the mean and the variance of the performance measure. Suppose that each replication has an equal number of independent observations, $n/2$, then

$$\mu = \sum_{i=1}^{n/2} \frac{(x_i + y_i)}{2}$$

with a confidence given by

$$\sigma_{\mu}^{2} = \frac{\sigma^{2}}{n}(1-\alpha)$$

where α is the **replication correlation coefficient**. Obviously, if the two replications are independent, $\alpha = 0$ and the variance estimate is identical to one replication of twice its length.

If we introduce a negative correlation between the pairs of observations, the variance to the sum of the observations $(x_1 + y_1), (x_2 + y_2), \ldots, (x_{n/2} + y_{n/2})$ will be less than the variance of one continuous simulation of n observations. Variables produced using a negative correlation are called **antithetic variates** (9.6, 9.7).

A procedure most commonly used to generate negatively correlated variates is to use uniformly distributed random numbers over the interval (0, 1) to generate probabilistic events for one simulation, and to use $1 - r$ for the equivalent event in the second simulation run.

The effects of correlated variates can be used for comparing two simulation runs with different controllable variables. For example, suppose we have independently estimated the mean and variance of two simulation runs using two sets of controllable conditions $\bar{\mu}_1, \hat{\sigma}_1^2$, and $\bar{\mu}_2, \hat{\sigma}_2^2$. The mean and variance of the differences is

$$\bar{\mu}_D = \hat{\mu}_1 - \hat{\mu}_2$$

and

$$\hat{\sigma}_D^2 = \hat{\sigma}_1^2 - \hat{\sigma}_2^2 \, .$$

We can then test to see if $\hat{\mu}_D^2$ is significantly different from zero by determining how small $\hat{\sigma}_D^2$ is.

To reduce the variance between x_1 and x_2, we can introduce a positive correlation such that

$$\hat{\sigma}_D^2 = \hat{\sigma}_1^2 + \hat{\sigma}_2^2 - 2\alpha\, \hat{\sigma}_1\, \hat{\sigma}_2$$

Thus, if we analyze two different courses of action using the same random numbers for equivalent events, we can reduce the variance at no increase in simulation run time. This is called **blocking** in physical experimental designs (9.3, 9.5).

If each condition of a simulation run is replicated once, we have

$$\hat{\sigma}_1^2 = \hat{\sigma}_1^2(1+\alpha_{11'})$$

$$\hat{\sigma}_2^2 = \hat{\sigma}_2^2(1+\alpha_{22'})$$

and

$$\boxed{\hat{\sigma}_D^2 = \hat{\sigma}_1^2 + \hat{\sigma}_2^2 - 2\,\alpha_{12}\;\hat{\sigma}_1\,\hat{\sigma}_2}$$

From the above equation we can see that the introduction of a negative correlation (i.e., $\alpha_{11'}$, $\alpha_{22'}$, negative) between **replicated** simulation runs results in a reduced variance for the within run estimate, and the introduction of a positive correlation ($\alpha_{1\,2}$, positive) between simulation runs under **different** conditions also results in a reduced variance, but for the difference between the simulation runs.

Thus, as we have also seen in Chapter 7, the **correlation coefficient** may be regarded as a measure of the degree of association between x and y, and in general is given as

$$\alpha = \frac{n\sum_{i=1}^{n} x_i y_i - \sum_{i=1}^{n} x_i \sum_{i=1}^{n} y_i}{\left[\left(n\sum_{i=1}^{n} x_i^2 - \left(\sum_{i=1}^{n} x_i\right)^2\right)\left(n\sum_{i=1}^{n} y_i^2 - \left(\sum_{i=1}^{n} y_i\right)^2\right)\right]^{1/2}}$$

Linear dependence is measured by the degree to which α approaches one, and **linear independence** by the degree to which α approaches zero.

From the above discussion the reader might conclude that the most efficient experimental design method might involve several replications of each simulation with correlations between replicated pairs. This however, would only be true if no initial transient phase existed in the simulation runs. Thus, it may be that the time required to pass over the transient phase starting conditions would negate any advantages of the correlated simulations , once steady state conditions occurred. Thus we deduce, as Conway (9.2) that "estimating mean effects using one long run is usually more efficient than several correlated replicated shorter runs". This conclusion, of course, depends on the stable conditions and the length of the transient time.

The reader may recall, from Chapter 7, that optimal run lengths are very sensitive to the amount of **interdependence** or **autocorrelation** within each simulation. There are two frequently used methods to analyze run lengths assuming autocorrelation. In the first, we estimate precisely the autocorrelation function and include its effects in the estimation of the mean and the variance of the state variables. In the second, we essentially group the time series output data into blocks of consecutive observations such that each block represents an independent observation. Let us consider each of these methods in more detail.

In the **autocorrelation function estimation** method, we first obtain test output data under steady state conditions during a simulation run. We then use this test data to estimate the correlation between observations at simulated time t and time $t + \Delta t$. That is,

$$\alpha(\Delta) = \frac{E\{[x_t - \mu][x_{t+\Delta} - \mu]\}}{\sigma^2}$$

where x_t is the individual output observation at time t, μ the average output over the simulation test run, and σ^2 the variance of the individual observations over simulation test run. Estimates made for all positive values of Δ for which $\alpha(\Delta)$ is significantly different from zero. Once the autocorrelation function has been estimated, the simulation model is run to compare performance characteristics using different operating characteristics. For example, if the simulation run using specific conditions results in n correlated observations, the estimated output would be

$$\hat{\mu} = \sum_{i=1}^{n} \frac{x_i}{n}$$

and

$$\hat{\sigma}_{\mu}^{2} = \frac{\sigma_i^2}{n}\left[1 + 2\sum_{\Delta=1}^{n-1}\left(1 - \frac{\Delta}{n}\right)\alpha(\Delta)\right].$$

Clearly, such observations are independent if $\alpha(\Delta) = 0$.

In the **blocking** method (see Figure 9.1), the simulation is run beyond the normal run length, some additional time, avoiding the transient period for this second data collection phase. We must note however, that the observations or blocking periods are not independent. In other words, the average performance measures of the second of two adjacent periods, depends in part, on the interdependence of what happened during the first period.

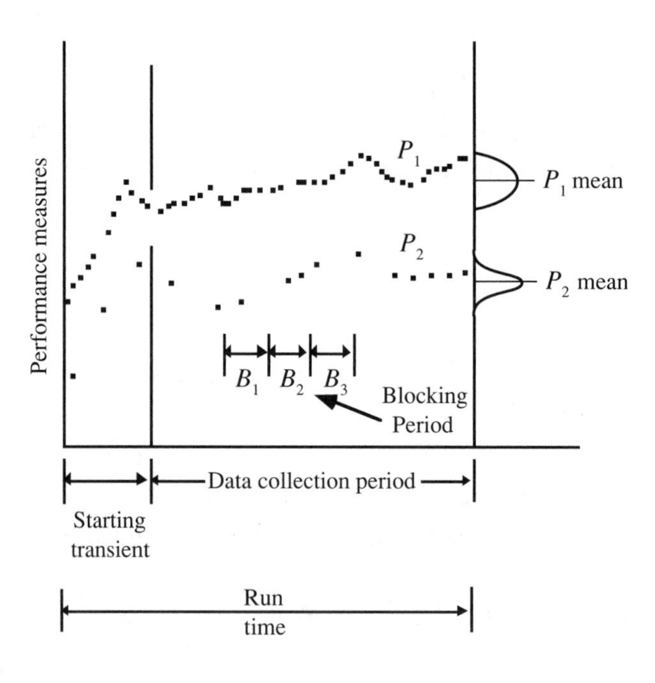

Figure 9.1. The blocking method (9.4).

Let us consider this procedure in detail. Suppose a simulation run consisted of n autocorrelated (interdependent) observations, $x_1, x_2, \ldots, x_n$, group these observations into k consecutive blocks, such that each block satisfies the following properties. An observation is redefined to be the average of m observations within a block. For example,

$$y_1 = \frac{x_1 + x_2 + \ldots + x_m}{m}$$

$$y_2 = \frac{x_{m+1} + x_{m+2} + \ldots + x_{2m}}{m} \quad , \quad \ldots \quad ,$$

$$y_3 = \frac{x_{m(k-1)+1} + x_{m(k-1)+2} + \ldots + x_n}{m}$$

where $m = n / k$, and the block size is chosen large enough so that the y_i's are independent. Then the estimated mean and variance are given by

$$\hat{\mu} = \sum_{i=1}^{k} \frac{y_i}{k} = \sum_{l=1}^{n} \frac{x_l}{n} \quad ,$$

and

$$\hat{\sigma}_{\mu}^{2} = \frac{\sigma_y^2}{k} \quad .$$

Both methods are efficient for estimated mean values, but not variance values. The method that explicitly uses the autocorrelation function always provides the lower (minimum) variance estimate. Its advantage over the blocking method is that the cost of running the simulation model to obtain individual observations of the system's behavior is relatively low, and the size of the variance of the estimated behavior decreases. The blocking method always loses at least some independent information because of the associated aggregation. This may be analyzed, in detail, using regression analysis.

The use of an autocorrelation function (or its estimate), in a simulation model, always guarantees a specified confidence interval in fewer observations than the blocking method and the ability to obtain variance estimates to within specified limits using the smallest number of observations. As the computer cost per simulation observation increases, the relative advantage of the autocorrelation function method increases over the blocking method (9.1, 9.7, 9.9).

9.4 References

9.1 Banks, J. and Carson, J. S., *Discrete–Event Simulation*, Englewood Hills, NJ: Prentice–Hall, Inc. 1984.

9.2 Conway, R. W., "Some Tactical Problems in Digital Simulation", *Management Science*, Vol. 10, No. 1, October, 1963.

9.3 Davies, O, ed., *Design and Analysis of Industrial Experiments, 2nd Ed.*, New York: Hafner, 1960.

9.4 Emshoff, J. R. and Sisson, R. L., *Design and Use of Computer Simulation Models*, New York: MacMillan, 1970.

9.5 Fishman, G. S., *Concepts and Methods in Discrete Event Digital Simulation*, New York: John Wiley and Sons, 1973.

9.6 Hammerly, J. M. and Handscomb, D.C., *Monte Carlo Methods*, New York: John Wiley and Sons, 1964.

9.7 Lewis, P. A. W. and Orav, E. J., *Simulation Methodology for Statisticians, Operations Analysts, and Engineers, Vol. I*, Pacific Grove, CA: Wadsworth & Brooks/Cole, 1989.

9.8 Martin, F. F., *Computer Modeling and Simulation*, New York: John Wiley and Sons, 1968.

9.9 Matloff, N. S., *Probability Modeling and Computer Simulation*, Boston: PWS–Kent Publishing. 1988.

10

THE DESIGN OF SIMULATION EXPERIMENTS

One of the most difficult problems encountered in a simulation study is validating the simulation model.

> DEFINITION 10.1
> *The* ***validation of a simulator*** *refers to the proof that the simulator accurately reflects the behavior of the system being modeled when confronted with identical conditions. One approach used in validation is to observe the system being modeled under a set of controlled or measurable conditions.*

The simulator is then run under identical conditions, and its output is compared to the results obtained from the system being modeled. When the compared results agree within some given level of tolerance, the simulator is considered to be validated.

Whether this validation procedure actually proves the validity of the simulation model is open to question. The degree of confidence attached to such a validation decision is tempered by the number of experiments on which it is based, as well as by the range of parameter settings (conditions) considered in each experiment. To measure the degree of confidence that one might attach to the validation decision, as well as to reduce the number of experiments required to achieve some preselected minimum level of confidence, the well–known technique for the statistical design and analysis of experiments can be applied.

Statistical experimental design methodology is also useful in the execution phase. Once the simulator has been validated, the behavior of the system being modeled can be studied under a variety of conditions by exercising (running) the simulator. Experimental design methodology is useful in this phase to indicate the significance of simulated effects through the tests of various hypotheses and to minimize the number of simulation runs necessary to measure given effects.

For the purpose of this chapter the simulation model can be visualized as a black box. The input to the model can be specified in terms of parameters set at various levels. The simulator is run under these input conditions, and a measure of the system's performance under these conditions is produced. This black–box approach is shown in Figure 10.1.

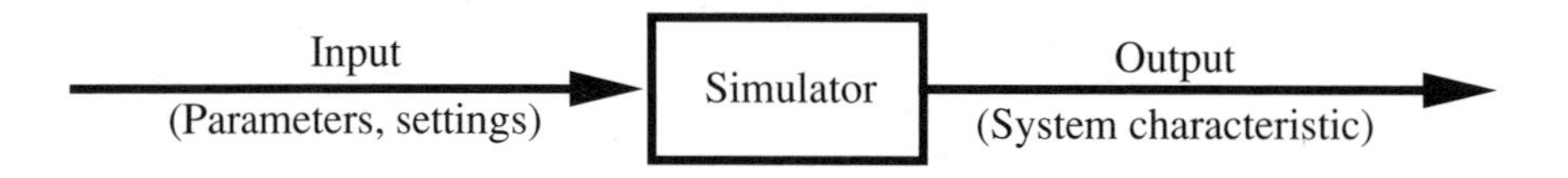

Figure 10.1. Simplified view of a simulator.

The parameters that are set before any simulation run correspond to the factors that have influence on the aspect of the system's performance being studied. As mentioned by Ferrari (10.2), the identification of the factors that influence a given performance measure is in general a very difficult task. The analyst has to rely on past experience, knowledge of the system, and in some cases intuition. Once the factors affecting a certain aspect of system performance have been determined, the allowable levels of each factor must be determined. In some cases the allowable levels are apparent, such as for an on–off type of parameter. In other cases the allowable levels must be determined carefully, so that the fewest possible parameter settings are used to explain the total range of variability in the performance measure.

The statistical design of experiments is a collection of principles that provide a procedure for designing experiments that maximize the information gained from each run of the simulator. This procedure also permits valid inferences about the effect on the performance variables of varying input parameters. Thus, experimental designs are to determine the best solution to the problem with the minimum expected computation costs. Please note, there exists no procedure available for determining how to design the best simulation experiment. There are, however, aids found to be useful in designing these experiments. The fundamental concepts of this methodology were formalized by R.A. Fisher (10.3). Most of this methodology was developed for agricultural and laboratory experiments, so the terminology that has evolved reflects those applications. The system variables that affect the measure of the system performance of interest are known as **factors**. The allowable settings of those factors are known as **levels**. A combination of a specific factor and level is called a **treatment**. The measure of system performance, stated as a function of the factors, is called the **responsive variable**. If the simulator is run repeatedly with the same treatment, the experiment has been **replicated**. We will use this terminology in this chapter when we review some of the common experimental designs.

10.1 Completely Randomized Design

The completely randomized design is applicable when the response variable is affected by a single factor that may be set at various levels. The order in which these levels are set for successive runs of the simulator is completely random, hence the name completely **randomized design**. The number of simulation runs (observations) for each level is determined by economic considerations (how many runs one can afford) and by the degree of confidence desired in the inferences. With this design the same number of observations is not required for each level. The key to this design is that the treatments (factor levels) are assigned at random. The randomization tends to average out the effect of factors that are not being considered or controlled in a given experiment.

Once a simulation experiment has been run, the data must be analyzed. To this end a mathematical model is developed to describe the experiment. The validity of the model is then tested. The model for the completely randomized design can be stated as

$$Y_{ij} = \mu + T_j + \varepsilon_{ij}$$

where Y_{ij} is the ith observation of the jth treatment, μ is the common "average" effect for all treatments, T_j is the additional effect due to the particular treatment, and ε_{ij} is the random error associated with the ith observation of the jth treatment.

These error terms are usually considered independent, identically distributed normal random variables with mean 0 and variance σ_ε^2. Furthermore, it is usually assumed that

$$\sum_{j=1}^{n} T_j = 0.$$

Example 10.1 Suppose that the system to be simulated is a single–channel queueing system in which arrivals occur according to a Poisson process with rate λ and service times are exponentially distributed with service rate μ. Suppose further that it is desired to test the effect that various queueing disciplines (FIFO, SIRO, LIFO) have on the number of customers in the system at the end of one hour. The response variable then is the number of customers in the system after one hour ; the single factor influencing this variable is the queueing discipline, and this factor can assume three levels. To collect data using a completely randomized design, it is necessary to randomly assign one of the queueing disciplines, run the simulation to simulate one hour of operation, and then count the number in the system. Then one of the two remaining disciplines is selected, and the experiment repeated. Finally, the third discipline is assigned and the experiment repeated. If multiple replications of the experiment are desired, the whole process is repeated as often as necessary. Care must be taken to randomize the treatment for each replication.

This example illustrates that the factors influencing a response variable need not be quantitative. The response variable, however, does have to be quantifiable.

To analyze the data obtained from a completely randomized design, one normally performs a **one–way analysis of variance** (see Table 10.1 for the itemized procedure). This analysis involves partitioning the variability of the recorded data into components due to the treatments and due to random error. In particular, suppose that there are k levels of the factor, and that n_j replications were obtained on the jth treatment, giving a total of

$$\sum_{j=1}^{k} n_j = N$$

observations.

Table 10.1 Outline of Steps for a One–Way Analysis of Variance

1	The raw data and group means.
2	Calculation of within groups sum of squared deviations (Note: The sum of squared deviations within each group, when divided by their degrees of freedom, will yield the normal table variance).
3	Calculation of between groups sum squared deviations (weighted by group sample size).
4	Calculation of total sum of squared deviations.
5	Summary analysis of variance table, including source of variation, sum of squared deviations, degrees of freedom, variance estimate, and *F*–test.

Source of Variation	Sum of Squared Deviations (*SS*)	*df*	Variance Estimates
Total	$SS_T = \sum_{j=1}^{k}\sum_{i=1}^{n_j} x_{ij}^2 - \frac{\left(\sum_{j=1}^{k}\sum_{i=1}^{n_j} x_{ij}\right)^2}{\sum_{j=1}^{k} n_j}$	$\sum_{j=1}^{k} n_j - 1$	
Between groups	$SS_{bg} = \sum_{j=1}^{k} \frac{\left(\sum_{i=1}^{n_j} x_{ij}\right)^2}{n_j} - \frac{\left(\sum_{j=1}^{k}\sum_{i=1}^{n_j} x_{ij}\right)^2}{\sum_{j=1}^{k} n_j}$	$k-1$	$s_{bg}^2 = \frac{SS_{bg}}{k-1}$
Within groups	$SS_W = SS_T - SS_{bg}$	$\sum_{j=1}^{k} n_j - k$	$s_w^2 = \frac{SS_w}{\sum_{j=1}^{k} n_j - k}$

Then we have the following:

Total sum of squares

$$SS_{TOT} = \sum_{j=1}^{k} \sum_{i=1}^{n_j} Y_{ij}^2$$

Treatment sum of squares

$$SS_{TRT} = \sum_{j=1}^{k} n_j (\bar{Y}_{\bullet j} - \bar{Y}_{\bullet\bullet})^2$$

where

$$\bar{Y}_{\bullet j} = \sum_{i=1}^{n_j} \frac{Y_{ij}}{n_j} \quad \text{and} \quad \bar{Y}_{\bullet\bullet} = \sum_{j=1}^{k} \sum_{i=1}^{n_j} \frac{Y_{ij}}{N}$$

Error sum of squares

$$SS_{ERR} = \sum_{j=1}^{k} \sum_{i=1}^{n_j} (Y_{ij} - \bar{Y}_{\bullet j})^2$$

Mean sum of squares

$$SS_{MEAN} = N\bar{Y}_{\bullet\bullet}^2$$

The dot subscript indicates that the response variable is summed over the entire range of that subscript. One can readily verify that

$$SS_{TOT} = SS_{MEAN} + SS_{TRT} + SS_{ERR}$$

Each of these sums of squares is divided by the appropriate degrees of freedom to produce sample variances. The sample variance corresponding to the treatment effect is compared with the sample variance of the error using an F–test to assess the statistical significance of the difference in treatment effect. The particular hypothesis that is normally tested is

$$H_0:\ T_j = 0, \qquad j = 1, \ldots, k$$

That is, **all treatment effects are the same**. The test statistic is

$$F = (SS_{TRT} \,/\, (k-1)) \,/\, (SS_{ERR} \,/\, (N-k))$$

and the hypothesis is rejected if

$$F > F_{\alpha,(k-1,N-k)}$$

Rejection of such a hypothesis indicates that there is a significant difference in the effects of one or more of the treatments. It does not tell which of the effects is different or how much it is different. Other tests are available to determine this difference. See for example, Hicks (10.4) for these tests.

Example 10.2 Suppose that in the experiment described in Example 10.1, the simulator was run four times for each of the queueing disciplines and that the following observations for the response variable were noted.

	j		
i	FIFO	SIRO	LIFO
1	50	32	80
2	48	18	73
3	61	64	54
4	58	72	63

Test the hypothesis that the queueing discipline has no effect on the number of customers in the system after one hour. From these data, it can be determined that

$$Y_{\bullet 1} = 54.25$$
$$Y_{\bullet 2} = 46.50$$
$$Y_{\bullet 3} = 67.50$$
$$Y_{\bullet\bullet} = 56.08 \sum_{i=1}^{4} \sum_{j=1}^{3} Y_{ij}^{2} = 41{,}131$$

and

$$SS_{TOT} = 41{,}131$$
$$SS_{MEAN} = 37{,}739.6$$
$$SS_{TRT} = 902.17$$
$$SS_{ERR} = 2489.23$$

with the result that $F = (902.17/2)/(2489.23/9) = 451.08/276.58 = 1.63$. If $\alpha = 0.05$, then $F_{0.05,(2,9)} = 4.26$, and the hypothesis that the queueing discipline has no effect cannot be rejected.

10.2 Randomized Complete Block Design

There are sometimes additional factors that cannot be controlled or whose effect on the response variable is of no interest to the analyst. If the effects of these secondary factors cannot be assumed to be constant or negligible, it is sometimes convenient to subtract these effects from consideration by using a **randomized**

complete block design. This design accounts for the effect of a secondary factor by including a block effect in the model. For data collection a simulation run must be made for each level of the primary factor and for each level of the secondary factor, or **block variable**. The order in which the various levels of the primary factor are set is again random.

Example 10.3 Suppose that in the simulation experiment described in Example 10.1 there are four different service rates, corresponding to shift changes in the real system. That is, one rate is in effect when the period between midnight and 6 a.m. is being simulated; a second rate for the period between 6 a.m. and noon; a third rate between noon and 6 p.m.; and the fourth rate between 6 p.m. and midnight. This difference in service rates will affect the number of customers in the system after one hour of elapsed simulation time. However, if we are interested only in assessing the effect of queueing discipline, a randomized complete block design, with four blocks corresponding to the different servers might be appropriate. In this case three simulation runs for each of the four blocks would be required, with the order of the setting of the queueing discipline made at random for each of the blocks.

The mathematical model for the randomized complete block experiment is

$$Y_{ij} = \mu + B_i + T_j + \varepsilon_{ij}$$

The B_i term accounts for the **block effect**, and the other terms are defined in the previous section. Analysis of the data obtained from a design of this type is usually accomplished by using a **two–way analysis of variance**. In addition to the components of variation listed in the previous section for the completely randomized design, a component due to the block effect can be isolated. If there are n blocked and k levels of the primary factor, the appropriate sums of squares can be calculated from

$$SS_{TOT} = \sum_{i=1}^{n} \sum_{j=1}^{k} Y_{ij}^{2}$$

$$SS_{MEAN} = nkY_{\bullet\bullet}^{2}$$

$$SS_{TRT} = \sum_{j=1}^{n} nY_{\bullet j}^{2} - nkY_{\bullet\bullet}^{2}$$

$$SS_{BLOCK} = \sum_{i=1}^{n} kY_{i\bullet}^{2} - nkY_{\bullet\bullet}^{2}$$

$$SS_{ERR} = SS_{TOT} - SS_{MEAN} - SS_{TRT} - SS_{BLOCK}$$

Thus the hypothesis is that there is no difference in treatment effect is tested using the test statistic

$$F = [SS_{TRT} / (k-1)] / [SS_{ERR} / (n-1)(k-1)]$$

and the hypothesis is rejected whenever

$$F > F_{\alpha,\ [(k-1),\ (n-1)(k-1)]}$$

The hypothesis can also be used to test that there is no difference in the effect of the blocks. This is generally not done, because it is frequently either of no interest or meaningless to the analyst.

Example 10.4 Consider the experiment described in Example 10.3, and assume that the data listed in Example 10.2 were collected for this experiment. That is, each row of the data listed as a replication in Example 10.2 should be considered a block. Then

$$\bar{Y}_{\bullet 1} = 54.25 \qquad \bar{Y}_{1\bullet} = 54.0$$
$$\bar{Y}_{\bullet 2} = 46.50 \qquad \bar{Y}_{2\bullet} = 46.33$$
$$\bar{Y}_{\bullet 3} = 67.50 \qquad \bar{Y}_{3\bullet} = 59.67$$
$$\bar{Y}_{4\bullet} = 64.33$$
$$\bar{Y}_{\bullet\bullet} = 56.08 \qquad \sum_{i=1}^{4}\sum_{j=1}^{3} Y_{ij}^{2} = 41.131$$

and

$$SS_{TOT} = 41{,}131$$
$$SS_{MEAN} = 37{,}739.6$$
$$SS_{TRT} = 902.17$$
$$SS_{BLOCK} = 544.38$$
$$SS_{ERR} = 1944.85$$

with the results that

$$\begin{aligned} F &= (902.17/2)/(1944.85/6) \\ &= 451.08/324.14 \\ &= 1.39 \end{aligned}$$

But $F_{0.05,\,(2,\,6)} = 5.14$, and we are still not able to reject the hypothesis that the treatment effects are the same.

It may not be possible to apply each treatment to each block because of time or money constraints. When this is the case, a randomized incomplete design should be used. This design is described in the work by Hicks (10.4).

10.3 Factorial Design

The two previous experimental designs concerned the effect of a single factor on the response variable. In most simulation experiments we must consider the effect of multiple factors on a given response variable. An experimental design known as the **factorial design** is then applicable. Consider an experiment in which a response variable Y is affected by two factors, A and B. For purposes of illustration, assume that factor A can assume three levels while factor B can assume two levels. Now the effect that factor A has on the response variable can be tested by holding factor B constant and making, say, n runs of the simulator at each of the three levels of factor A. The model used to test this effect then would be

$$Y_{ij} = \mu + A_j + \varepsilon_{ij}$$

where i would vary from 1 to n and j would vary from 1 to 3. Similarly, the effect of factor B on the response variable could be tested by holding factor A constant and making n runs at each level of factor B. The model used to test this effect would be

$$Y_{ij} = \mu + B_j + \varepsilon_{ij}$$

where again i would vary from 1 to n and j would vary from 1 to 2. Thus to test the main effects of factors A and B for this example would require $5n$ runs of the simulator.

In many cases the effect of multiple factors on a response variable is characterized not only by the effects of the individual factors but also by a synergistic or interaction effect caused by the simultaneous application of the factors. This means that we should test the effect of factor A, the effect of factor B, and the effect of the interaction of the two, AB. Unfortunately the one–at–a–time design does not allow testing of this interaction effect. This drawback, along with the inordinate number of runs required in one–at–a–time testing is the primary motivation for using the factorial design.

With the factorial design, $n/2$ simulation runs are made for each unique treatment combination, with the treatment combinations assigned in random order to each run. Therefore there are 3 x 2 = 6 unique treatment combinations, which require $3n$ simulation runs to acquire the same degree of precision as with the one–at–a–time design. As an illustration, assume that the levels of factor A are denoted 0, 1, and 2 while the levels of factor B are denoted 0 and 1. The unique treatment combinations are 00, 01, 10, 11, and 21, where the level of factor A is listed first in this notation. The $3n$ runs result in $3n/2$ observations of levels 0 and 1 for factor B, compared with n observations in each case with the one–at–a–time design. Thus we have achieved more information about the factors in fewer runs of the simulator. This simultaneous varying of the levels of the two factors allows us to isolate and to test for the effect of the interaction of the two factors.

Let us assume that the proper randomization has been done, and the data collected for $m = n/2$ replications of the experiment. The data can then be analyzed using the mathematical model

$$Y_{ijk} = \mu + A_i + B_j + (AB)_{ij} + \varepsilon_{ijk}$$

where $i = 1, 2, 3$; $j = 1, 2$; and $k = 1, 2, ..., m$. In this case A_i represents the effect of the ith level of factor A ; B_j represents the effect of the jth level of factor$B$; $(AB)_{ij}$ represents the interaction effect of the ith level of factor A and the jth level of factor B, μ is the "mean" effect; and ε_{ijk} is the random error term.

In general, if factor A can assume a levels while B can assume b levels, the appropriate sums of squares to test for the significance of the effects are

$$SS_{TOT} = \sum_{i=1}^{a}\sum_{j=1}^{b}\sum_{k=1}^{m} Y_{ijk}^2$$

$$SS_{MEAN} = mab\overline{Y}_{\bullet\bullet\bullet}^2$$

$$SS_A = \sum_{i=1}^{a} mb\overline{Y}_{i\bullet\bullet}^2 - mab\overline{Y}_{\bullet\bullet\bullet}^2$$

$$S_B = \sum_{j=1}^{b} ma\overline{Y}_{\bullet j\bullet}^2 - mab\overline{Y}_{\bullet\bullet\bullet}^2$$

$$SS_{AB} = \sum_{i=1}^{a}\sum_{j=1}^{b} Y_{ij\bullet}^2 - \sum_{i=1}^{a} mb\overline{Y}_{i\bullet\bullet}^2 - \sum_{j=1}^{b} ma\overline{Y}_{\bullet j\bullet}^2 + mab\overline{Y}_{\bullet\bullet\bullet}^2$$

$$SS_E = SS_{TOT} - SS_{MEAN} - SS_A - SS_B - SS_{AB}$$

where just as before, a quantity such as $\overline{Y}_{\bullet j \bullet}$ means to average over the dotted subscripts.

Now to test for the various effects, we need to determine an estimate for the variances and form the F–ratios as before.

To test A

$$F = \frac{SS_A / (a-1)}{SS_E / ab(m-1)}, \qquad \text{reject if } F > F_{1-\alpha,\ (a-1,\ ab(m-1))}$$

To test B

$$F = \frac{SS_B / (b-1)}{SS_E / ab(m-1)}, \qquad \text{reject if } F > F_{1-\alpha,\ (b-1,\ ab(m-1))}$$

To test AB

$$F = \frac{SS_{AB} / (a-1)(b-1)}{SS_E / ab(m-1)}, \qquad \text{reject if } F > F_{1-\alpha,\ ((a-1)(b-1),\ ab(m-1))}$$

These results are readily extended to more than two factors. As the number of factors grow, so do the interaction effects that can be tested. Fortunately, three–and four–way interactions have very little meaning and are rarely tested.

Example 10.5 Consider again a queueing system in which arrivals occur according to a Poisson process and service times are exponentially distributed. Suppose that the manager of the system is investigating the feasibility of hiring a second server and simultaneously wants to compare the relative efficiency of two queueing disciplines, FIFO and SIRO. The quantity of interest is the number of customers in the system at the end of one hour of simulated time.

This experiment could be readily handled using a factorial design. Call the factor "number of servers" S, with levels 1 and 2, and the factor "queueing discipline" Q, with levels FIFO and SIRO. There are four treatment combinations to consider.

Suppose that two replications of the experiment were obtained and the data in the following table were recorded.

	Factor Q	
Factor S	FIFO	SIRO
1	62 71	48 47
2	32 22	28 28

Then

$$\begin{aligned} SS_{TOT} &= 16{,}531.0 \\ SS_{MEAN} &= 14{,}365.125 \\ SS_{Q} &= 153.125 \\ SS_{S} &= 1711.25 \\ SS_{SQ} &= 210.125 \\ SS_{ERR} &= 91.501 \end{aligned}$$

To test the various effects, we use the following.

To test Q

$$F = \frac{(153.125/1)}{(91.501/4)} = 6.69$$

To test S

$$F = \frac{(1711.125/1)}{(91.501/4)} = 74.8$$

To test SQ

$$F = \frac{(210.125/1)}{(91.501/4)} = 9.18$$

Then if $\alpha = 0.05$, $F_{0.05,(1,4)} = 7.71$, and we would conclude that the number of servers has an effect on the response variable, as does the interaction of the number of servers and the queueing discipline. There are insufficient data to prove that the queueing discipline alone has an effect.

Many simulation experiments involve a relatively large number of factors. Experiments of this type require a simulation run for each unique treatment combination to be considered one replication of the experiment. For example, suppose we are conducting a simulation experiment to investigate the effect of three factors, A, B, and C, on a response variable. Suppose further that factor A can assume three levels, factor B two levels, and factor C four levels. Then a total of 2 x 3 x 4 = 24 simulation runs are required to be considered one replication. The effects to be tested include the main factors A, B, and C; the two–way interactions AB, AC, and BC ; and the three–way interaction ABC. As the number of factors and levels per factor increase, the number of simulation runs required to be considered a complete replication can become intolerably large. This has led to the development and use of fractional replications.

The feature that allows the satisfactory use of fractional replications in factorial experiments is that in many applications, the higher–order (greater than two) interaction terms have little or no meaning to the analyst. These higher–order interaction terms can be omitted from the model, and their contribution can be lumped into the random error term. This confounding of the interaction effects with the random error term decreases the number of simulation runs necessary to test for the desired effects. The analyst must decide before the experiment which of the interaction effects to combine with the error term and design the experiment accordingly. The details behind the design of a fractional replication of a factorial experiment are too lengthy to include here. Hicks gives an excellent treatment using fractional replications (10.4).

10.3.1 Fractional Factorial Design

Thus, statistical design of experiments embodies principles for designing experiments to maximize the information per observation, and to permit valid inferences about the effects, on responses of interest, of variations in factors under the experimenter's control.

One of the key ideas underlying this principle is that of varying several factors simultaneously, rather than one at a time. By varying factors simultaneously, the simulator can capture information pertaining to their joint effects, or interactions, as well as their individual, or main effects. In addition, different effects can be estimated from the same set of observations.

As we have seen in the previous simple example, that of a **balanced experiment** with two factors (A, B), each to be varied at two levels, the total number of observations, or experimental runs, was $N = 4n$. The one factor at a time approach was to do n observations at each level of each factor.

A		*B*	
1	2	1	2
n	n	n	n

The effect of varying factor A can be measured by the difference in the means of the observations at each level of A. The appropriate t–test for assessing the statistical significance of this difference uses a sample standard deviation based on $2n - 2$ degrees of freedom, computed by pooling the sample standard deviations at each level. In essence, the standard deviation among observations recorded under identical experimental conditions, provides an estimate of the variation in model errors.

The main effect of varying factor B can be measured in exactly the same way, but because each factor is varied separately, the experiment does not provide any information about possible interactions between the factors. That is, we have no way of knowing whether the effect of varying factor A is different at different levels of factor B.

Let us now consider the **factorial approach**.

		A 1	*A* 2
B	1	n	n
	2	n	n

Here, for the same expenditure of observations, we now have $2n$ observations at each level of each factor, and can estimate the interaction between the factors.

We could, using the factorial design, cut the size of the experiment in half, leaving us with n observations at each level of each factor, as in the one time approach. We could then have approximately the same precision in estimating main effects as in the one at the time experiment, plus the ability to estimate the interaction of the two factors.

For experiments involving large numbers of factors, the number of factor level combinations increases geometrically, but higher order interactions can be estimated. In many applications, certain high order interactions may be assumed to be negligible, and further economies can then be realized by using designs which are **fractions of full factorials**. Fractions of full factorials are designed in such a way as to sacrifice information on specified interactions by eliminating particular factor–level combinations. The main idea behind **fractional factorials** is illustrated Figure 10.2.

Figure 10.2 depicts a 1/2 fraction of a 2^3 design (3 factors, each at 2 levels). The coordinates of each vertex of the cube represents the levels of factors A, B, and C. Note that the design corresponding to the four vertices denoted by the circles provides estimates of all three main effects.

For example, the B main effect is measured by the difference between the means of the pairs of points on the top and the bottom faces of the cube. The other main effects are measured by similar comparisons on the other pairs of the faces of the cube. In higher dimensions, smaller fractions of balanced points on the hypercube can be used.

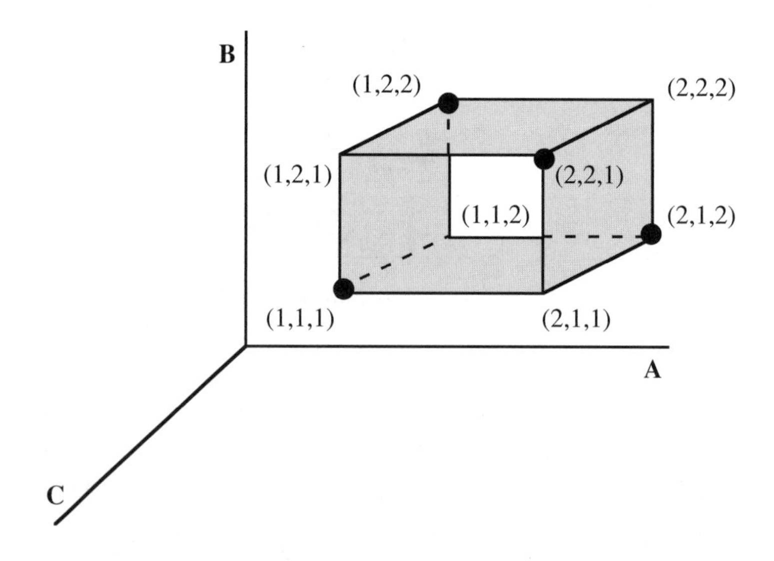

Figure 10.2. Fractional factorials.

The analysis of data from a designed experiment is based on the assumption of a linear model with normally distributed error terms. For example, for three factor design, the model would be

$$Y_{ijkl} = \mu + \alpha_i + \beta_j + \alpha_k + (\alpha\beta)_{ij} + (\alpha\gamma)_{ik} + (\beta\gamma)_{jk} + \varepsilon_{ijkl}$$

where ε_{ijkl} are independent errors from a normal distribution with mean zero and unknown variance σ^2.

This model states that each observation is composed of an overall mean μ, plus main effects due to the three factors, plus interaction terms and a random error term. The index l is used to represent possible replications of observations made under identical experimental conditions.

Usually, an analysis of variance is carried out on the data from a designed experiment. Essentially, the total sum of squares of all observations is partitioned into component sums of squares corresponding to the terms in the linear model. Each of these are then divided by their appropriate degrees of freedom to obtain sample variances. The sample variances corresponding to the main effects and their interactions are then compared to the error variance by means of, for example, F–tests in order to access the statistical significance of each of the observed effects.

Because the validity of inferences drawn from such analysis is dependent on the adequacy of the assumed model, various analyses and plots of **residuals** (differences between the observations and fitted values obtained from the model) may be employed to indicate whether there appear to be serious discrepancies.

10.4 Network Simulation Model Performance Analysis

10.4.1 Model Description

The 10–node network topology shown in Figure 10.3 was used to analyze integrated packet/circuit network performances. The generalized simulation model that was used in this analysis was developed at Texas A&M University (9.5–9.10).

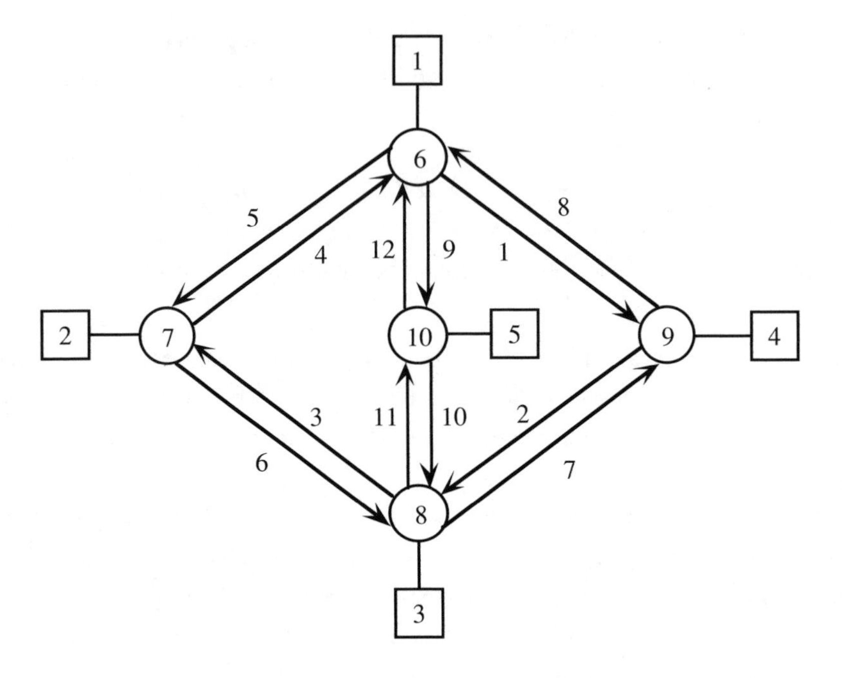

Figure 10.3. 10–node network topology.

The circuit switch (circular) nodes correspond to the major computing centers of Tymshare's TYMNET, a circuit–switched network, while the trunk lines are full–duplex (FDX) carriers, each of which is modeled by two independent, half–duplex (HDX) channels.

10.4.2 Goals and Scope of the Analysis

The specific goals of the analysis were as follows:

1. Design an experiment whereby the performance data can be efficiently and economically obtained.
2. Determine the effective ranges of network parameters for which realistic and acceptable network performance results.
3. Investigate the sensitivity of performance measures to changes in the network input parameters. Specifically, determine how network

performance is affected by changes in the network traffic load, trunk line or link capacity, and network size.

The scope of the analysis was restricted to those parameters that were closely related to the network topology and the workload imposed on that topology. With regard to performance sensitivity to workload and link capacity, the following four parameters were investigated:

CS: Circuit Switch Arrival Rate (voice calls / min)
PS: Packet Switch Arrival Rate (packets / sec)
SERV: Voice Call Service Rate (sec)
SLOTS: Number of Time Slots per Link (a capacity indicator)

The sensitivity to network size was also investigated by varying the number of nodes and links in a network. In all cases, the performance measures observed were mean packet delay (MPD), fraction of voice calls blocked (BLK), and average link utilization (ALU).

10.4.3 Experimental Design

A preliminary sizing analysis indicated that the effective range of input parameters were as follows:

CS: 0–8 (calls / minute at each node)
PS: 0–600 (packets / second at each node)
SERV: 60–300 (seconds / call)
SLOTS: 28–52 (a link capacity indicator)

The fixed parameter settings were such that each slot represents a capacity of about 33 Kbits / sec.

The experimental design selected for this analysis was a second order (quadratic), rotatable, **central composite design** (9.11, 9.12). Such a design for k (number of parameters) = 3 is illustrated in Figure 10.4. This design was chosen because it reduces the number of experimentation points considerably from what would otherwise be required if the classical 3^k factorial design were used. The "**central composite**" feature of the design replaces a 3^k factorial design with a 2^k factorial system augmented by a set of axial points together with one or more center points. A "**rotatable**" design is one in which the prediction variance is a function only of the distance from the center of the design and not on the direction.

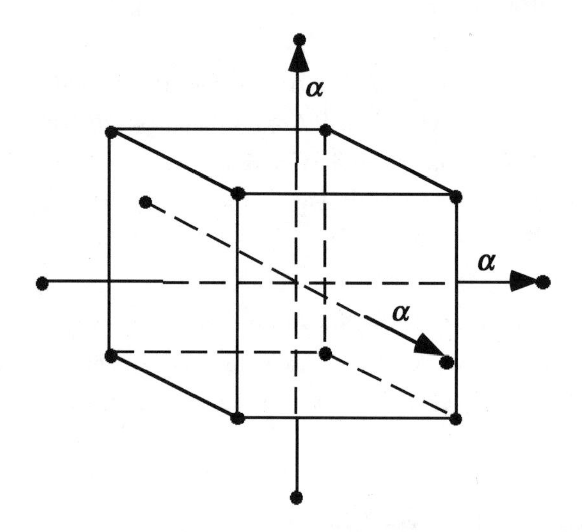

Figure 10.4. Central composite design ($k = 3$).

Myers (9.11) shows that the number of experimental points required for $k = 4$ is 31. That is, there are 16 factorial points, 8 axial points, and 7 replications at the center point. In order, for the design to be rotatable, Myers shows that α must be chosen as $(2^k)^{1/4}$. In this case, α , the distance from the center point to an axial point, is 2. The seven replications at the center point will allow an estimate of the experimental (pure) error to be made; thus, a check for model adequacy is possible 9.13. Myers recommends seven replications at the center point simply because it results in a **near–uniform precision design**. That is, it is a design for which the precision on the predicted value $\hat{y}$, given by

$$\frac{N \operatorname{var}(\hat{y})}{\sigma^2}$$

where,

N = total number of experimental points, and
σ^2 = the error variance,

is nearly the same at a distance of 1 from the center point as it is at the center point.

In this design, each parameter was evaluated at five different levels. The five levels for each of the four parameters were as follows:

CS:	0	2	4	6	8
PS:	0	150	300	450	600
SERV:	60	120	180	240	300
SLOTS:	28	34	40	46	52

The center point was defined as CS / PS / SERV / SLOTS = 4 / 300 / 180 / 40.

Analysis of the results of these 31 runs indicated that the original range of data was too large. Ten of these experimental data points resulted in network performance that would be totally unacceptable, e.g., an MPD of more than 10 seconds. These "**outliers**" were eliminated and 14 additional points in the moderate to heavy loading of the network were added. Two observations were taken at 8 of these 14 additional points for the purpose of estimating the error variance at points other that the center of the design.

10.4.4 Regression Analysis

The **Statistical Analysis System (SAS)** was used to perform a **multiple regression** analysis for each of the three performance measures. The regression variables were the four input parameters. SAS procedures REG and RSREG were used to analyze the data. The assumption of a quadratic response surface allowed for the estimation of 15 model parameters, including the intercept.

The model selection procedure was based on the following approach. RSREG was used first to check the full (quadratic) model for specification error (lack of fit test) and to determine significance levels for the linear, quadratic, and crossproduct terms. The models for MPD, BLK, and ALU exhibited a lack of fit that was significant at the .14, .05, and .82 levels, respectively. It was found, however, that these significance levels were heavily influenced by the observations at the extremes of the heavy–loading region. For example, by eliminating 13 of the observations in the heavy–loading range of data, the lack of fit for BLK could be raised to the .70 significance level. Hence, although the .05 level for the BLK model bordered on statistical significance, the lack of fit was not deemed sufficient to justify a more complex model.

10.5 Summary

In this chapter we have surveyed some of the more common experimental designs that seem to have application to simulation studies, both in the validation and the execution phases of the process. Many alternate designs have been omitted, since an in–depth coverage of all designs that have been proposed would require a complete and rather voluminous textbook by itself. Certain statistical experimental designs were mentioned to indicate their utility in simulation studies rather than to provide the analyst with the level of expertise to design an experiment.

10.6 Exercises

10.1 A simulation experiment to test the effect on a response variable by a single factor was conducted using a completely randomized design. Two replications at each of three levels of the factor were obtained, and the following data recorded.

Level	*Response*	
0	19.5	20.6
1	28.1	31.6
2	37.4	41.8

Test the significance of the effect of the treatments.

10.2 Two factors are believed to have an effect on a given response variable. The two factors can each assume two levels. Discuss the design of a simulation experiment that could be used to test the effects. Include in your discussion the model used to test the data, the effects to be tested, and the assignment of treatment combinations to the simulation runs. What advantage would be gained through replication of the experiment?

10.3 Describe a system for which the randomized complete block design might be useful. Why could one not use the factorial arrangement to analyze the system in lieu of the block arrangement?

10.4 Paging is a common technique for implementing virtual memory. In paging the physical memory of a computer system is subdivided into fixed-size page frames. A program submitted to the system is likewise subdivided into pages of like size. The illusion of unlimited memory is then created by requiring only thoose pages of a program that are ikely to be referenced soon to actually reside in memory. Other pages are retained on auxiliary storage and loaded only when referenced. A measure of the efficiency of a paging system is the page fault rate – the relative frequency with which references are made to a page not in memory. Investigate this idea of paging, and determine some factors that may affect the page fault rate.

10.5 Using the factors determined in Exercise 10.4, describe a simulation experiment by which you could test those effects.

10.7 References

10.1 Farrell, W., McCall, C. H., and Russell, E. C., "Optimization Techniques for Computerized Simulation Models". *Technical Report 1200-4-75*. Los Angeles: CACI, June 1975.

10.2 Ferrari, D. *Computer Systems Performance Evaluation.* Englewood Cliffs, NJ: Prentice-Hall, 1978.

10.3 Fisher, R. A. *The Design of Experiments*, Edinburgh: Oliver and Boyd, 1935.

10.4 Hicks, C. R., *Fundamental Concepts in the Design of Experiments*, New York: Holt, Rinehart and Winston, 1973.

10.5 Kiemele, M. and Pooch, "The Design of an Integrated Packet/Circuit Switched Network Simulator", in Vol. 5 (Ruschitzka, M., ed.) *Parallel and Large Scale Computers: Performance, Architecture, Applications*, North Holland Publishers, 1983.

10.6 Kiemele, M. and Pooch, U. W., "Topological Optimization of an Integrated Circuit/Packet–Switched Computer Network," *WSC 1984 Winter Simulation Conference*, Dallas, TX, 1984.

10.7 Kiemele, M. and Pooch, U. W., "Design of an Integrated Packet/Circuit Switched Network," *10th IMACS*, Montreal, Canada, August 1982.

10.8 Kiemele, M. and Pooch, U. W., "A Simulation Model for Evaluating Integrated Packet/Circuit Switched Networks," *WSC 1983 Winter Simulation Conference*, Arlington, VA, December 1983.

10.9 Kiemele, M. and Pooch, U. W., "A Sizing and Timing Analysis of an Integrated Computer Network Simulator," *WSC 1985 Winter Simulation Conference*, San Francisco, CA, December 1985.

10.10 Kiemele, M. and Pooch, U. W., "A Sensitivity Analysis of an Integrated Computer Network Simulator," *WSC 1985 Winter Simulation Conference*, San Francisco, CA, December 1985.

10.11 Myers, R. H., *Response Surface Methodology*, Boston, MA: Allyn and Bacon, Inc., 1971.

10.12 Naylor, T. H. (ed.), *The Design of Computer Simulation Experiments*, Durham, NC: Duke University Press, 1969.

10.13 Ostle, B. and Mensing, R. W., *Statistics in Research (3d edition)*, Ames, IA: The Iowa State University Press, 1975.

11

ESTIMATION OF MODEL PARAMETERS

The problem of response surface search may be characterized as that of determining an optimum, however this may be defined, on a multidimensional surface. The surface itself may or may not exist in its entirety in the appropriate representational space. It may be dynamically created, and searched at the same time.

11.1 Optimization of Response Surfaces

A **response variable** is a quantification of the aspect of the system's performance of interest in a simulation experiment and can be thought of as the output of the simulation experiment . Alternatively, it can be visualized as a function of the input parameters (factors). It is of interest in some simulation experiments to determine which input parameter settings optimize (either maximize or minimize) the response variable. We will survey some of the more common techniques that have been employed in this optimization process. Farrell, McCall, and Russell (11.1) list a total of 11 such optimization techniques and compare them with regard to convergence, nearness to optimum, and running time. This work should be consulted for more details, because we will only survey some of the more common techniques.

If the response variable Y is expressed as a function of n factors, call them X_1, X_2, ..., X_n, represented as $Y = f(X_1, X_2, ..., X_n)$, then the plot of the response variable over the range of the input parameters can be thought of as a surface (hyperplane) in an (n + 1)–dimensional space. Such a surface is called a **response surface**. The optimization problem then consists of finding the highest (lowest) point on this surface. This idea of a response surface can be readily interpreted for $n = 2$ [$Y = f(X_1, X_2)$], illustrated in Figure 11.1. Of course, visualization in higher than three dimensions becomes more difficult, but the idea is the same. We start from some initial point on the surface and through some algorithmic procedure move to the maximum (minimum) point on the surface. The objective is to reach the optimum in the minimum number of steps. Search methods are typically designed to use information already obtained about the shape of the response function to decide both the direction and size of steps to take next.

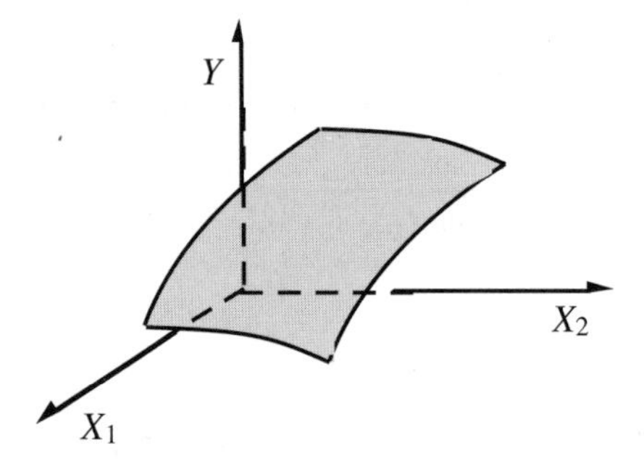

Figure 11.1. A three–dimensional response surface.

Although search procedures have to be efficient over a wide range of response surfaces, no current procedure can effectively overcome "**non–unimodality**", or a surface with more than one local maximum or minimum. The only way to overcome this would be to evaluate all the local maxima and then choose the optimum.

One technique that is used when non–unimodality is known to exist is called the **"Las Vegas" technique**. This search procedure is used to estimate the distribution of the local optima. That is to say, we plot the response measure for each local search against its corresponding search number. Those local searches which produce a response greater than any previous response are then identified and a curve is fit through the data (see Figure 11.2). This curve is then used to project the **"estimated incremental" response** that will be achieved by one more search. The search continues until the value of the estimated improvement in the search is less than the cost of completing one additional search.

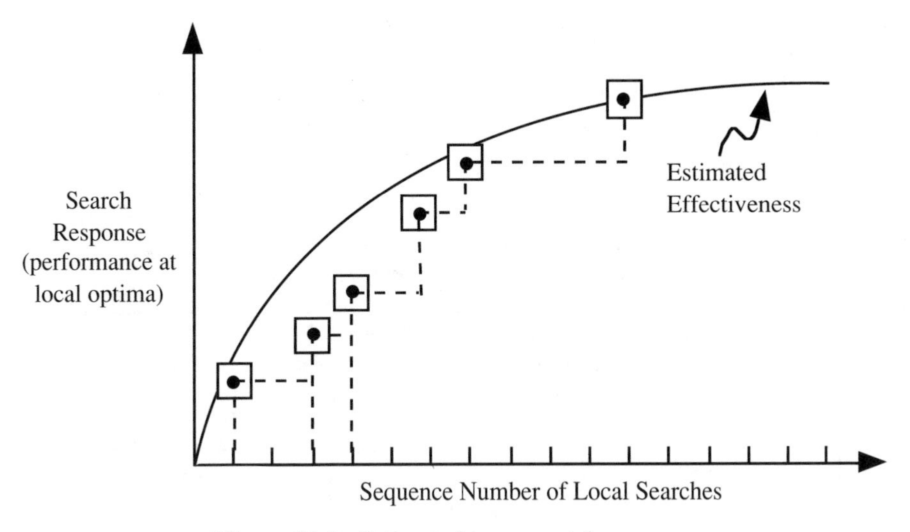

Figure 11.2. Estimated incremental response.

It should be noted that a well–designed experiment produces a sufficient number of replications of conditions so that the average response can be treated as a deterministic number for search comparisons. Whenever this is not the case, and because replications are expensive, it becomes necessary to effectively utilize the number of simulation runs.

Although each simulation is at a different setting of the controllable variables, smoothing techniques such as exponential smoothing can be used to reduce the required number of replications. As an illustration of how these techniques can be used as an alternative to replicating output, consider Figure 11.3, and a model representation given by the following equation

$$V = f(x_i, y_j, a_k)$$

where V is the measure of performance, x_i the controllable variables, y_j the uncontrollable variables, and a_k parameters. Assuming that the model structure is correct, the objective is to find the minimum of the difference between the model output and its historical data as a function of a_k. An example of such a measure of closeness is that of the **least square**, which can be written as

$$\min_{a_k} \sum [V_t^h - f(x_i^h(t), y_j^h(t), a_k)]^2$$

where f represents the simulation model and h the historical data.

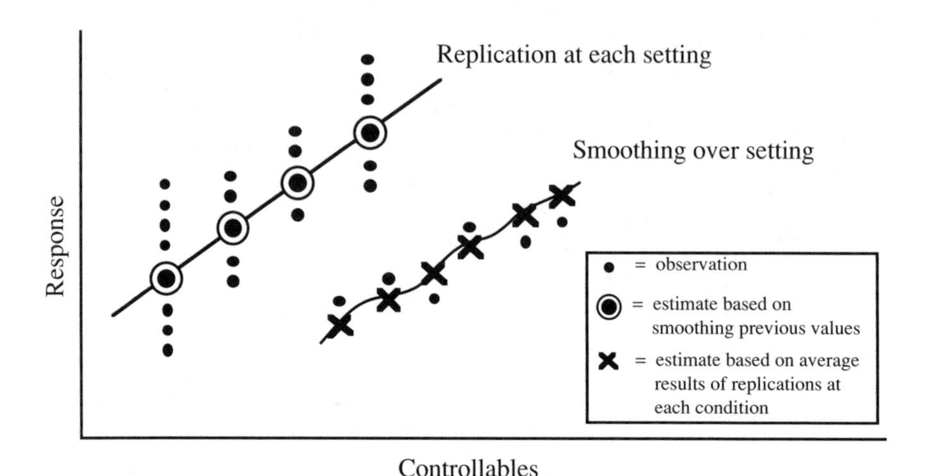

Figure 11.3. Exponential smoothing.

While the search for the best parameters is equivalent to that of parameter identification, it should be obvious that inverting the model for parameter estimation makes validation more difficult. This is because the data used for parameter estimation cannot be reused for validation, and historical data is, in general, not readily accessible or available to both estimate parameters and test output adequately.

11.2 Heuristic Search

The **heuristic search technique** is probably the most common of the techniques used in optimizing response surfaces. It also is the least sophisticated mathematically and can be thought of as a "seat of the pants" approach. The starting and stopping points are determined by the analyst based on previous experience with the system being simulated. The input parameters (factors) are set at levels that appear reasonable to the analyst, a simulation run is made with the factors set at those levels, and the value of the response variable is noted. If it appears to be maximum (minimum) to the analyst, the experiment is stopped.

Otherwise the parameter settings are changed, and another run is made. This process continues until the analyst is satisfied that the output has been optimized. Suffice it to say that if the analyst is not intimately familiar with the process being simulated, this procedure can turn into a blind search and can expend an inordinate amount of time and computer resources without producing corresponding results.

11.3 Complete Enumeration

The **complete enumeration technique**, although not applicable to all cases, does yield the optimal value of the response variable. All the factors must assume a finite number of values for this technique to be applicable. For such a system a complete factorial experiment is run and the value of the response variable at each of the treatment combinations is noted. Normally a number of replications of the basic experiment are made, and the output values averaged over these replications at each of the treatment combinations. The analyst can attach some degree of confidence to the determined optimal point when using this procedure. Although this technique yields the optimal point, it has a serious drawback. If the number of factors or levels per factor is very large, the number of simulation runs required to find the optimal point can be exceedingly large. For example, suppose that an experiment is conducted with three factors; one can assume three levels, the second can assume four levels, and the third can assume five levels. Suppose further that five replications are desired to provide the proper degree of confidence. Then a total of $5\ (3 \times 4 \times 5) = 300$ runs of the simulator would be required to isolate the optimal point. If, on the other hand, there are ten factors, each of which can assume only two levels, a total of $5\ (2^{10}) = 5\ (1024) = 5120$ simulation runs would be necessary for the five replications. Thus this technique should be used only when the number of unique treatment combinations is relatively small.

11.4 Random Search

The **random search technique** is similar to the complete enumeration technique except that a simulation run is not made at every unique treatment combination. Instead a set of treatment combinations is selected at random, and simulation runs are made at these points. The treatment combination of this set that yields the maximum (minimum) value of the response variable is taken to be the **optimal point**. This procedure reduces the number of simulation runs required to yield an "optimal" result; however, there is no guarantee that the point found is actually the optimal point. Of course, the more points selected, the more likely one is to achieve the true optimal point. Thus there is a trade–off between the number of runs one is willing to make and the likelihood that the value found is the true optimum. Note that the requirement that each factor assume only a finite number of values, inherent in the complete enumeration scheme, is not a requirement in the random search scheme. Just as with the complete enumeration scheme, replications can be made on those treatment combinations selected to increase the confidence in the optimal point. The trade–off is that is that it may be better to devote the simulation runs used in replicating to examining more points. The question which is better strategy, replicating a few points or looking at a single observation on more points, has not been answered.

11.5 Steepest Ascent (Descent)

The **steepest ascent (descent) method** uses a fundamental result from calculus – that the gradient (vector of first partial derivatives) points in the direction of the maximum increase of a function – to determine how the initial settings of the parameters should be changed to yield an optimal value of the response variable. The direction of movement is made proportional to the estimated sensitivity of the performance to each variable.

Assume that performance is "linearly related" (although sometimes quadratic functions are used) to the change in the controllable variables for small changes. Assume that a linear form is a good approximation, and that the linear relationship between the *n*–controllable variables and the system's performance is given by

$$y = \beta_0 + \beta_1 x_1 + \beta_2 x_2 + \ldots + \beta_n x_n$$

where β_0 is the intercept when all controllable variables are zero, and β_i is the slope of the function in the ith direction.

The basis of the **linear steepest ascent** is that each controllable variable is changed in proportion to the magnitude of its slope. The set of slopes $\beta_1, \beta_2, \ldots, \beta_n$ at a particular point is called the **gradient** of the surface, and thus the name "**gradient search**".

To determine sequentially the system's response when each controllable variable is changed by a small δ (delta) is analogous to determining the gradient at a point. For a surface containing n–controllable variables, this requires n points around the point of interest. That is,

$$\beta_i = (y_{i\delta} - y_0) / \delta$$

where $y_{i\delta}$ is the performance when x_i is moved by δ and all other x's remain the same, and y_0 is the performance at the point of interest. A one–dimensional search is then undertaken along the ray defined by the slopes β_i.

The new direction is proportional to the slopes in the original dimension, thus one of these directions can be used arbitrarily to determine the step size in the one–dimensional search. For instance, if $\hat{y}\beta_1$ is chosen to be the step size for variable x_1, $\hat{y}\beta_2$ for x_2, etc., then the equation for the change in value along the steepest ascent direction is given by

$$\Delta y = \hat{y}(\beta_1 + \beta_2 + \ldots + \beta_n)$$

The search is completed when all the β_i's are negative and the gradient yields a decrease in performance.

To visualize this method, consider a response surface in three dimensions, Y = $f(X_1, X_2)$, as illustrated in Figure 11.4.

Suppose that some initial starting values for the two parameters have been determined. By running the simulator with parameter settings in the "neighborhood" of initial values X_1^0 and X_2^0, it is possible to obtain an estimate of the shape of the response surface in that vicinity using the method of least squares. For example, assume that in the simple two–factor model, the vicinity of the initial point on the response surface can be postulated to be $Y = B_0 + B_1X_1 + B_2X_2$. To estimate the values of the coefficients B_0, B_1, and B_2, one needs at least three points. Thus the simulator would run using at least three points in the vicinity of (X_1^0, X_2^0). Using the output values from these points and applying least squares, one can obtain an estimate of the shape of the response surface, $\hat{Y} = \beta_0 + \beta_1 X_1 + \beta_2 X_2$. The gradient to this surface at the point (X_1^0, X_2^0) is grad $\hat{Y} = (\partial\hat{Y}/\partial X_1, \partial\hat{Y}/\partial X_2) = (\beta_1, \beta_2)$. The direction in which the values for X_1 and X_2 should be changed is along a line given by

$$(X_2^1 - X_2^0) = \frac{\beta_1}{\beta_2}(X_1^1 - X_1^0)$$

The magnitude of the change is up to the analyst; the gradient gives only the direction. A similar procedure is followed for the vicinity of the point (X_1^1, X_2^1) to determine the direction by which the parameters should be changed for the next step. The process is then continued until no further change is indicated.

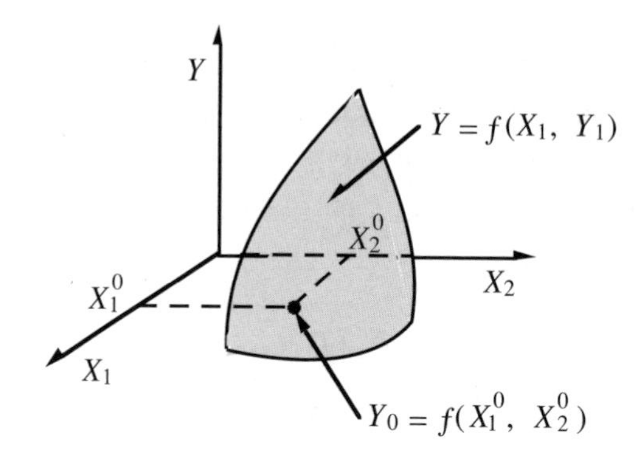

Figure 11.4. The method of steepest ascent (descent).

This description of the steepest ascent (descent) method is admittedly very brief, and many important factors have been omitted. For example, the goodness of fit of the model should be tested at each step, and a higher–order model used if the fit is not good enough. A procedure for the selection of starting values and the selection of points in the vicinity of the initial estimates has been studied. Other factors of interest for this method are the design of the experiment to collect the data in the vicinity of each of the points and the magnitude of the change required in the parameters at each step. These factors, although largely ignored in this survey of the method, must be considered before it is used. See Hicks (11.4) and Farrell, McCall, and Russell (11.1) for details.

Note that the method of steepest ascent (descent) is fairly inefficient on certain surfaces, and is guaranteed to find the maximum (minimum) not of the entire response surface but a local maximum (minimum). Whether the two points coincide depends on the application and the incrementing values that are used. This may be a matter of some concern and should be kept in mind.

11.6 Coordinate or Single–Variable Search

The **single–variable search technique** is similar to the steepest ascent technique in that some initial point on the response surface is chosen, and movement toward the optimal point is made through interactive adjustments of the parameter levels. In the steepest ascent method the parameters are adjusted simultaneously in the direction of the gradient to an estimate of the response surface at the initial point; in this method, on the other hand, a single parameter is adjusted at each step until it is no longer possible to vary any parameter individually and improve the response. For example, consider the simple two–factor model $Y = f(X_1, X_2)$ with associated response surface (see Figure 11.5). With the coordinate search method an initial point P_0 at $(X_1^0,\ X_2^0)$ would be chosen (based on the best guess to optimal operating conditions), and a simulation run made with these parameter settings. Two additional runs would be made with values of $(X_1^0 + \varepsilon, X_2^0)$ and $(X_1^0 - \varepsilon, X_2^0)$. The values of the response variable would be compared. If neither $f(X_1^0 + \varepsilon, X_2^0)$ or $f(X_1^0 - \varepsilon, X_2^0)$ is greater than $f(X_1^0,\ X_2^0)$, the process would be stopped in the X_1 direction and a similar procedure followed for X_2. Otherwise the parameter setting for X_1 would be adjusted in the direction of increasing response until no further increase is possible by varying only X_1. This would yield a point P_1 at(X_1^1, X_2^0). A similar procedure is then followed for X_2. That is, it is varied in the direction of increasing response until no further response is possible.

Although this method can easily be extended to more than a two–dimensional search, this technique, according to Farrell, McCall, and Russell (11.1), is useful only when the response surface is unimodal, and even then it is not always successful. In fact, the method's weaknesses are the excessive number of simulation runs beyond those using the diagonal searches involving multi–factor changes, and poor performance on surfaces where ridges exist. Since the shape of a given response surface is not known beforehand, it does not appear to have much applicability. In fact, for the same effort expended, one could use the gradient search scheme.

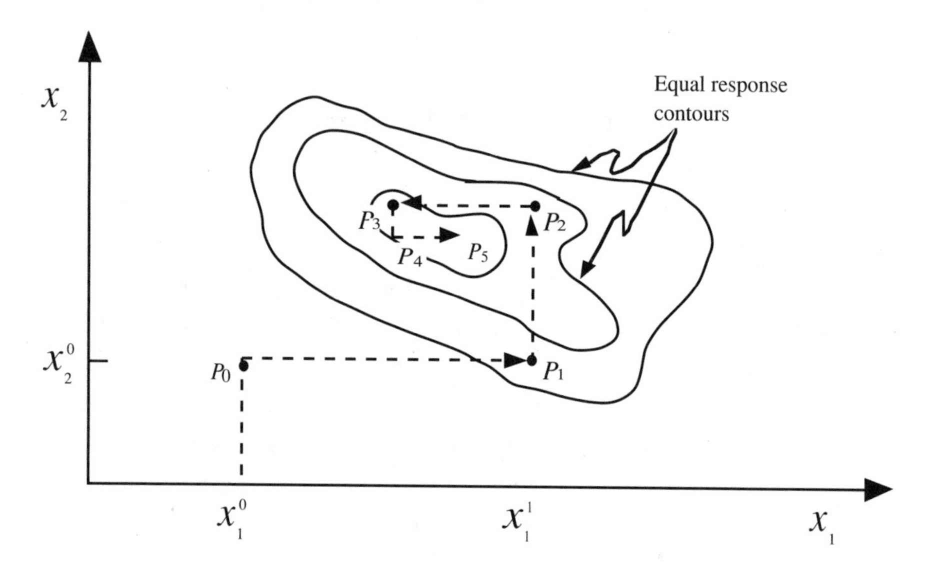

Figure 11.5. Two variable response surface.

11.7 Parallel Tangent Search

This search procedure augments the steepest ascent technique by also considering planes in the n–dimensional space to reduce the region wherein the optimum way occur. The optimum of an n–dimensional elliptical surface may therefore be found in $2n - 1$ searches. This search procedure is illustrated in Figure 11.6 and can be described as follows:

1. Let P_i represent points obtained from the gradient steepest ascent.
2. Let γ_i represent parallel planes in the n–dimensional space constructed as described below.
3. Pick an arbitrary point P_0 on a plane γ_0.

4. Search is now conducted on any line not tangent to plane γ_0, resulting in a maximum at P_2.

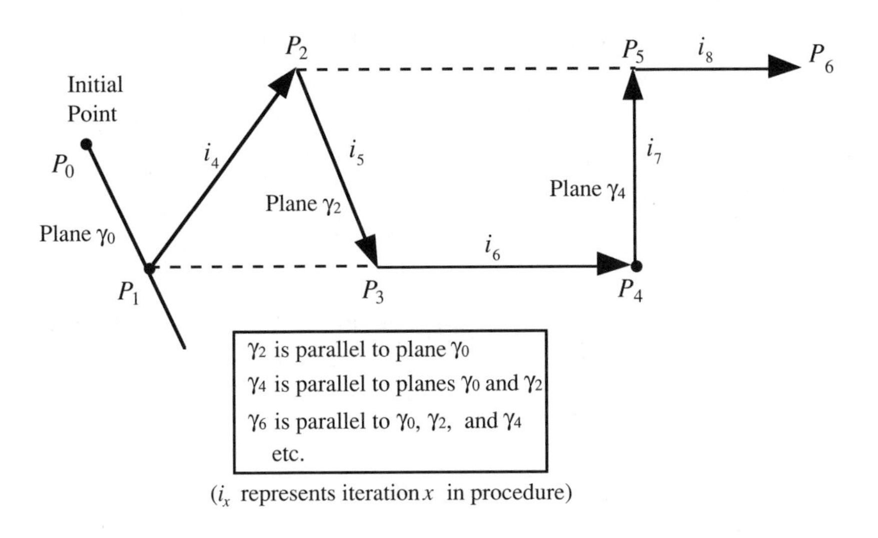

Figure 11.6. Parallel tangent search example.

5. From P_2 the gradient steepest ascent is used to determine the direction of search within plane γ_2 that is parallel to γ_0, resulting in a maximum point at P_3.
6. The next search continues along the line connecting points P_1 and P_3, resulting in a maximum at P_4.
7. Next a gradient search is made in plane γ_4 that is parallel to γ_0 and γ_2, resulting in point P_5.
8. The connection between points P_2 and P_5 is used as a direction of search resulting in a maximum point at P_6.
9. The search continues in a similar fashion building **planes parallel to preceding planes**, restricting the successive directions of future searches, resulting in the optimum in $(2n - 1)$ search iterations.
10. Should the surface not be quadratic, we can either start the parallel tangent search again from the point of the $(2n - 1)$ iteration, or continue the above process by constructing each plane parallel to the most recent $n - 1$.

11.8 Conjugate Direction Search

This search requires no derivative estimation yet finds the optimum of an n–dimensional quadratic surface after at most n–iterations. The procedure redefines the n dimensions so that the single variable search can be used successively. Single variable procedures can be used whenever dimensions can be treated independently. That is, whenever

$$f(x_1, x_2, x_3, \ldots, x_n) = g_1(x_1) + g_2(x_2) + \ldots + g_n(x_n)$$

the optimization along each dimension leads to the optimization of the entire surface.

> DEFINITION 11.1
> *Two directions are defined to be* **conjugate** *whenever the crossproduct terms are all zero.*

The **conjugate direction method** tries to find a set of n dimensions that describe the surface, such that each direction is conjugate to all others.

> THEOREM 11.1
> *If x_0 is a single variable stepwise optimum in a space containing α, and x_1 is also an optimum in the same space, then the direction defined by the points x_0 and x_1 is conjugate to α.*

Next, using the above theorem, the method attempts to find two search optima and replaces the nth dimension of the quadratic surface by the direction specified by the two optima. Successively replacing the original dimensions yields a new set of n dimensions which, if the original surface is quadratic, are all conjugate to each other and appropriate for n–single variable searches.

While this search procedure appears to be very simple on the surface, we should point out that the selection of appropriate step sizes is most critical. The step size selection is more critical for this search method because during axis rotation the step size does not remain invariant in all dimensions. As the rotation takes place, the best step size changes and becomes exceedingly difficult to estimate.

11.9 Summary

In the last chapter we surveyed some of the approaches used in optimizing response surfaces. Various alternate techniques were viewed. The work by Farrell, McCall, and Russell (11.1) contains an excellent description of these techniques as well as additional techniques. The work by Hicks (11.4) contains an illustrative example of the method of steepest descent. Many other excellent treatments of response surface optimization exist in the literature.

11.10 Exercises

11.1 Discuss the five optimization techniques described in Section 11.5 in terms of

a. Relative efficiency (number of simulation runs required)
b. Optimally of the result obtained
c. Ease of use

11.2 Suppose you were involved in a simulation project that required optimization of a response variable. Which technique would you use? List some of the factors that led to your decision.

11.3 Consider the response surface given by $f(x, y) = \sin(x + y)$; $0 \leq x \leq 5, 0 \leq y \leq 5$. Use the random search technique with 10 sample values of (x, y) drawn at random from the 36 integral pair values to find the maximum of $f(x, y)$.

11.4 Use the complete enumeration technique (x and y can take only integer values) to find the maximum of the response surface given in Exercise

11.3. Does the value agree with that obtained in Exercise 11.3 Should it agree?

11.5 Use the method of steepest ascent to find the maximum of the response surface given in Exercise 11.3. Start with $x = 2$, $y = 2$.

11.6 Use the coordinate search technique to find the maximum of the response surface given in Exercise 11.3. Start with $x = 2$, $y = 2$.

11.7 Compare the techniques used in Exercises 11.3 – 11.6 in terms of computation efficiency and results obtained.

11.11 References

11.1 Farrell, W., McCall, C. H., and Russell, E. C., "Optimization Techniques for Computerized Simulation Models." Technical Report 1200–4–75. Los Angeles: CACI, June 1975.

11.2 Ferrari, D., *Computer Systems Performance Evaluation.* Englewood Cliffs, NJ: Prentice–Hall, 1978.

11.3 Fisher, R. A., *The Design of Experiments.* Edinburgh: Oliver and Boyd, 1935.

11.4 Hicks, C. R., *Fundamental Concepts in the Design of Experiments.* New York:Holt: Rinehart and Winston, 1973.

12

OUTPUT ANALYSIS

12.1 Analysis of Simulation Results

Once the simulator has been designed, tested, and debugged, the actual simulation study can be carried out, the results can be interpreted, and inferences about the operation of the true system can be drawn. Numerous pitfalls await the unsuspecting analyst in the interpretation of simulation run data. In this chapter we survey some of these problems and review some of the techniques that can be used to overcome them.

A simulation study is normally conducted to investigate various aspects of a system's operation for certain operating conditions. The aspects of the system that are under study are normally referred to as the **performance measures** or **response variables**. The aspects of the system's operation that are controllable and are normally varied by the analyst to observe the effect on the response variables are referred to as the **control variables** or **input parameters**.

Example 12.1 Consider a single–channel queueing system like the one illustrated in Figure 12.1. In this model customers arrive for service according to some interarrival distribution, possessing some required interval of service that is assigned from some service–time distribution. If the service facility is idle, the customer moves immediately into service and departs after receiving the required service. Otherwise the customer joins a queue and waits to be selected for service. Note that the $M/M/1/\infty$/FIFO queueing system is an example of this type of model. A performance measure for this system could be the average number of customers in the queue. Control variables could include the interarrival distribution, service–time distribution, queueing discipline, and system capacity.

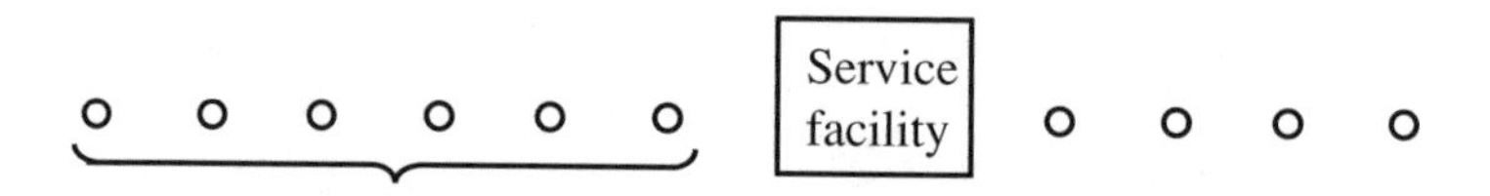

Figure 12.1. A single–channel queueing system

A problem commonly encountered in simulating systems similar to the one described in Example 12.1 is that the performance measures observed during the initial part of a simulation run are not typical of the system's true operation. This **transient period** normally occurs because the starting conditions chosen for the simulation model are not characteristics of the system. It normally takes some time for the effect of the unrealistic starting conditions to become insignificant and for the simulation model to stabilize, or reach **steady state**. The analysis of the simulation results under steady–state conditions is desirable, since it is normally under these conditions that the true system's operation is simulated. The concepts of transient and steady state phases in a simulation run was previously discussed for model validation in Chapter 9. Let us recall some of the more salient characteristics.

If the problem is to obtain performance measures for the model under steady–state conditions, then the analyst must eliminate the biasing effect of statistics generated during the start–up phase of the simulation. Consider, for example, the effect on the queue length of the different queueing disciplines in the system described in Example 12.1. If the starting conditions are not characteristic of the true operating conditions of the system, the statistics that are generated (average queue length) will be biased because they include this transient phase. Ideally, then, statistics should be collected only after the system has reached steady state. Such a selective collection of statistics is virtually impossible to program. Thus statistics are normally collected for an entire simulation, and then an attempt is made to eliminate the bias of the transient phase.

Two problems with this approach are immediately apparent. First, how can one recognize that steady state has been achieved? Second, how can the biasing effect of the transient phase be eliminated? Several techniques, discussed in Chapter 9, can be used to determine when the system has reached steady state.

Once the system has reached steady state, the analyst must decide how to reduce or eliminate the bias in the performance measures introduced during the transient phase. Once again we have previously discussed, in Chapter 9, several methods that can be used to either eliminate or adjust for any transient phase during simulation model validation. These same concepts apply to simulation output analysis.

Another problem encountered in the analysis of simulation output is deciding which methodology produces a valid statistical sample. A **valid statistical sample** is, of course, required before one can make valid inferences about the true value of a performance measure. Determining the appropriate methodology is generally restricted to the case in which there is uncertainty in some phases of the model. (The model is stochastic. In stochastic models the events generated are only a subset of the total event set.) If the model is both finite and discrete, the total event set could eventually be generated if the duration of the simulation run is long enough. This is not the case if the model is continuous. Even in the discrete case, few analysts can afford to let the simulator run as long as would be required to generate the entire event set. Thus the objective in most cases is to limit the total computer time used (minimize the cost of the simulation runs) and at the same time achieve a reliable performance measure. Statistical sampling techniques are needed to obtain some estimate of the reliability of the derived performance measures.

As an illustration of the procedures involved, suppose that the desired performance measure is the mean of some parameter. From the central limit theorem we know that the sampling distribution of the mean is approximately normal if sufficient observations are taken. Thus there is a probability of approximately .997 that the sample mean lies within three standard errors of the true population (or process) mean. The standard error of a sample is $\sigma/n^{1/2}$, where σ is the standard deviation of the true process and n is the number of observations. Thus if we know the size of the error that is acceptable, as well as the standard deviation of the process, we can calculate the number of observations needed to provide a valid estimate of the mean. Once the required size of the sample is determined, the analyst is confronted with the task of obtaining the sample.

Emshoff and Sisson (12.1) discuss several approaches for generating a valid sample. The first method is applicable when the underlying process is thought or known to be normally distributed. In this case as single run is considered to be a sample, and the individual observations of the performance measure are considered to be members of that sample. Next the standard deviation of the sample is taken as an estimate of the standard deviation of the true process and a confidence interval estimate of the true parameter is calculated. For example, if the performance measure is the mean of some quantity, a confidence interval estimate for the true process mean can be calculated from

$$I = \overline{X} \pm (t_{\alpha,n-1})(s / n^{1/2})$$

Where $\overline{X}$ is the sample mean, s is the sample standard deviation, and is the appropriate value of the t–statistic.

If the underlying process is not known to be normal, a confidence interval estimate can still be obtained by resorting to the central limit theorem, which says essentially that the sampling distribution of the mean is normally distributed is sufficient samples are obtained. Then if a sample means is generated, the performance measure may be estimated from this sample using the preceding statistical method.

12.2 Estimation and Confidence Limits

In general, the output of a simulation run may be viewed as a set of estimators of the stochastic variables. The simulationist must be concerned about the precision of these estimates, and a very natural and useful way to express this is by way of the so–called **confidence interval**. For example, Figure 12.2 is an illustration of an interval estimate of a parameter. Here if the probability of a parameter being included in the particular interval is $p = .95$, we can state instead that we are 95% confident that the interval includes the parameter value, and this interval is referred to as the 95% confidence interval for the parameter.

Since we rarely measure every single member of the population to determine a parameter value, we must resort to interval estimates of those parameters. This is done by considering the relationship between a sample statistic and the corresponding population parameter. In the most general case this can be written as

Statistic = Parameter ± error

That is to say, the observed sample statistic deviates from the fixed parameter value by some error. If we are dealing with the mean, this can be restated as

$\bar{x} = \mu \pm$ error

which says that the true population mean μ is equal to the sampled mean $\bar{x}$ ± some error. While we will never know the size of the error for any single sample mean, we would like to attach a probability to errors of a given size, if we knew the probability distribution of those errors. In other words, what is the mean, the standard deviation, and the shape of the distribution of the errors.

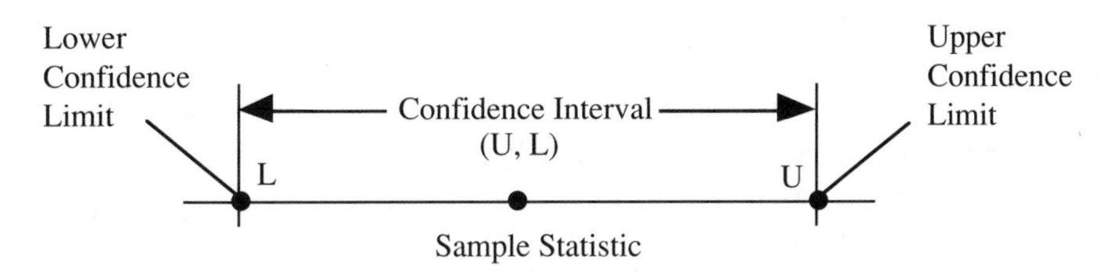

Figure 12.2. Interval estimate of a parameter.

Let us assume that we have taken repeated samples from the population, each time calculating a sample mean $\bar{x}$. When repeated a very large number of times, we have the equivalent of the sampling distribution of the mean, whose distribution variability (as we have previously seen) is the standard error of the mean, $\sigma_{\bar{x}}$. We also know that $\sigma_{\bar{x}}$ is the standard deviation of the distribution of errors. Thus, the variation of the unknown estimation errors is the same as the variation of the sample means. That is why it is called the **standard error of the mean.**

Now that we know that the standard deviation of sampling errors equals the standard error of the mean, all we are left to determine is the shape of the distribution of the sampling errors. Again from Chapter 3 we know that the shape of the distribution of sampling errors is identical to the shape of the sampling distribution of the mean $\bar{x}$. Furthermore, the sampling distribution of the mean will be normal in form if the parent population is normal or if $N \geq 30$ (large) regardless of the shape of the parent population.

We also saw in Chapter 4 that in a normal distribution, 95% of the observations fall within ±1.96 standard deviations from the mean. Since the standard deviation of sampling errors is equal to the standard error of the mean $\sigma_{\bar{x}}$,

$$\mu = \bar{x} \pm 1.96\ \sigma_{\bar{x}}$$

with 95% certainty. This may be rewritten as

$$(\bar{x} - 1.96\ \sigma_{\bar{x}}) < \mu < (\bar{x} + 1.96\ \sigma_{\bar{x}})$$

the 95% confidence interval for the population mean μ, when $\bar{x}$ is based on a **random sample** from a normal population. What this means is that with long simulation runs and steady state, 95% of the confidence intervals will contain the true population mean μ.

The equation for a 99% confidence interval would correspond to

$$(\bar{x} - 2.58\ \sigma_{\bar{x}}) < \mu < (\bar{x} + 2.58\ \sigma_{\bar{x}})$$

Since the **standard error of the mean** is given by $\sigma_{\bar{x}} = \sigma / n^{1/2}$, the width of the confidence interval will vary inversely as the square root of the sample size. However it must be remembered that not only estimates of the mean are required, but that the sample must also be obtained in a random fashion.

When the true standard deviation of the population σ is not known, the **estimate** or **sampled standard deviation** s must be used. Thus we would have

$$(\bar{x} - 1.96\ s_{\bar{x}}) < \mu < (\bar{x} + 1.96\ s_{\bar{x}})$$

and

$$s_{\bar{x}} = s / n^{1/2}\ .$$

Suppose we are interested in estimating the mean $\hat{\mu}$ of some process, and for the moment assume that we know $\hat{\sigma}$. We realize that as we increase the sample size *N*, we can generate an increasingly more precise estimate of $\hat{\mu}$, which we will call $\bar{x}$. We can express the precision of $\bar{x}$ as follows

$$\text{Prob}(\hat{\mu} - \Delta \le \bar{x} \le \hat{\mu} + \Delta) = \text{Prob}\left(-\frac{\Delta}{\sigma_{\bar{x}}} \le z \le \frac{\Delta}{\sigma_{\bar{x}}} \right)$$
$$= 1 - \alpha$$

where α is the confidence level.

This of course, assumes that $\bar{x}$ is normally distributed with mean $\hat{\mu}$ and $\sigma_{\bar{x}}$ = $\sigma / n^{1/2}$ (from the central limit theorem), and

$$z = (\bar{x} - \hat{x})/\sigma_{\bar{x}}$$

Using symmetry, $\text{Prob}\left(z \le \frac{\Delta}{\sigma_{\bar{x}}} \right) = 1 - \frac{\alpha}{2}$, and letting $z_{1-\alpha}$ be defined such that

$$\text{Prob}\left(z \le z_{1-\alpha}\right) = 1 - \frac{\alpha}{2}$$

then

$$\sigma_{\bar{x}} = \frac{\Delta}{z_{1-\alpha/2}}$$

and the required sample size for a given value of $1 - \alpha$ is given by

$$N = (\sigma z_{1-\alpha/2})^2 / \Delta^2$$

For example, say we wish to estimate the process with $\sigma = 200$ to within $\pm$ 100, with confidence level $\alpha = 0.05$. Then

$$N = \frac{40000\,(3.84)}{10000} = 15.36$$

We can then develop a confidence interval (L, U), where $L = \bar{x} - z_{1-\alpha}\left(\frac{\sigma}{n^{1/2}}\right)$ and $U = \bar{x} + z_{1-\alpha}\left(\frac{\sigma}{n^{1/2}}\right)$, respectively.

This analysis so far has one minor problem. If we knew σ, we probably would not need to simulate much of the system. Thus we cannot use this approach to determine N, but we might use it to make some educated guess by assuming some value for σ. A second, and perhaps better approach would be to iterate, that is to simulate for N_1 observations and then computing σ. We now can calculate a confidence interval δ_1. If this confidence interval is satisfactory we quit, else we simulate to obtain some more observations and repeat the above procedure. It should also be noted that for accuracy the z statistics should be replaced by the t statistics, but for $N > 100$ the z statistic is a good approximation.

Table 12.1 provides appropriate confidence interval formulae for the most common simulation outputs.

Table 12.1 100(1 – α)% Confidence Intervals

Parameter Estimated	Qualification	*Confidence Intervals* *L*	*U*
Mean (μ)	Variance (σ^2) known	$\bar{x} - x_{1-\alpha/2}\frac{\sigma}{\sqrt{n}}$	$\bar{x} + x_{1-\alpha/2}\frac{\sigma}{\sqrt{n}}$
Mean (μ)	Variance (σ^2) unknown	$\bar{x} - t_{1-\alpha/2}(n-1)\frac{s}{\sqrt{n}}$	$\bar{x} + t_{1-\alpha/2}(n-1)\frac{s}{\sqrt{n}}$
Variance (σ^2)		$(n-1)s^2/\chi^2_{1-\alpha/2}(n-1)$	$(n-1)s^2/\chi^2_{\alpha/2}(n-1)$
Standard Deviation (σ)		$\sqrt{(n-1)s^2/\chi^2_{1-\alpha/2}(n-1)}$	$\sqrt{(n-1)s^2/\chi^2_{\alpha/2}(n-1)}$
Proportion (p)	k/n is the estimate of p where n is the sample size and m is the number of events of interest occur–ring in the n observations	$\frac{k}{k+(n-k+1)F_{1-\alpha/2}(r_1,r_2)}$ $r_1 = 2(n-k+1)$ $r_2 = 2k$	$\frac{(k+1)F_{1-\alpha/2}(r_1,r_2)}{n-k+(k+1)F_{1-\alpha/2}(r_1,r_2)}$ $r_1 = 2(k+1)$ $r_2 = 2(n-k)$

Example 12.2 Consider a simulation that is to be used in following the average system cost which is to be estimated to within ± \$15.00 with 95% confidence. For iteration increments of 100, the output might look as follows:

Simulation Iterations	$\bar{x}$	L	U	$\Delta = U - L$
100	197.63	156.01	239.24	83.23
200	202.40	174.50	230.30	55.80

300	220.45	195.50	245.06	49.22
400	219.69	198.50	240.87	42.37
500	219.05	200.11	238.00	37.89
600	216.08	198.02	234.13	36.11
700	219.97	202.99	236.96	33.97
800	222.01	206.27	237.75	31.48
900	222.26	207.14	237.39	30.25
1000	220.60	206.63	234.57	27.94

We can see from this example how the confidence interval may be used to control the length of the simulation run. Table 12.2 provides an example of an algorithm that uses confidence intervals as such an optimization control mechanism.

Table 12.2 Optimization with Confidence Intervals

Step 1	Identify and select a performance measure, ϕ.
Step 2	Select a level of confidence, $1 - \alpha$.
Step 3	Define allowable error, Δ.
Step 4	Using the simulation model, obtain samples i of size $N = k$. Compute ϕ_ι and (L_i, U_i) for each sample group.
Step 5	Determine any samples i for which $U_i < L_j$ (max. value), and $L_i < U_j$ (min. value) Those samples i, whose intervals meet the above criteria are eliminated.
Step 6	Terminate when only one sample i is left or the $\left\| \max U_i - \min L_j \right\| < \Delta$.

In summary, if a simulation analyst's interest is in estimating the mean of a process, then a confidence interval will provide such a measure where we expect to find the mean μ with approximate probability of $1 - \alpha$. As the random sample size increases, this interval diminishes for fixed α, and a more precise determination of μ can be made.

12.3 Initial Conditions and Inputs

Initial conditions bias results obtained from short simulation runs. Not so apparent is the fact that quite a long simulation run may be required to offset these initial biases (transients), until steady state results become viable. For example, if a user wishes to obtain an average queue length estimate, starting the system empty will bias the estimate for quite some time. Thus, running a simulation long enough to offset the transient bias is only second best. Other options include discarding some of the initial portions of the statistics, or choosing more typical start–up (initial) conditions. The reader is referred to Chapter 9 to deal with the question of transient and steady state behavior.

Let us address the question of better **starting conditions**. Choosing a reasonable set of starting conditions should not be difficult for an existing system. However, for a conceptual system it might be very difficult. This is particularly troublesome when considering the number and range of the parameters for which starting conditions may vary. The most reasonable choice is to select some average condition for all parameters that are completely unknown and to consider some nonparametric statistical method for "**goodness–of–fit**" testing to identify desirable distributions and associated parameters.

The overall procedure, given in Table 12.3, is to fit simple distributions to data. That is, one favors distributions which may be efficiently produced and which may be easily understood and related to the process being simulated.

Table 12.3 Goodness–of–Fit Methodology

1	Obtain or synthesize a representative set of data.
2	Summarize the data, using the relative frequency or histogram method.

Table 12.3 Goodness–of–Fit Methodology (continued)

3	Compare the data with various density functions and form a hypothesis as to the likely distribution.
4	Estimate the parameters of the distribution using the data.
5	Apply a goodness–of–fit test and produce the test statistics.
6	Conclude whether the fit is acceptable. If it is not, repeat steps 3 – 5 until satisfied.

Before proceeding to examples of the total process, let us recall the use of the chi–square and the Kolmogorov–Smirnov (K&S) tests for goodness–of–fit testing.

Chi–square test:

First the data is classified into a set of mutually exclusive cells, and the chi–square computed from

$$\chi^2 = \sum_{i=1}^{k} \frac{(f_{ok} - f_{ek})}{f_{ek}}$$

where k is the number of cells, f_{ok} the observed frequency in cell k, and f_{ek} the expected frequency in cell k. The following should be noted:

1. f_{ok} and f_{ek} are actual and not relatively frequency counts.
2. f_{ok} must be greater than or equal to five for each cell.
3. The critical values are given in tables for a given confidence (α) and degrees of freedom (δ),

   $$\delta = k - 1 - m$$

   where k is the number of cells and m is the number of distribution parameters. A table of critical values is found in most simulation and statistics books, as well as in most mathematics/science handbooks.

Kolmogorov–Smirnov test:

To apply this test form the empirical cumulative distribution from the observed data and the difference between the observed cumulative function and the hypothesized theoretical cumulative distribution. Next, determine the maximum difference between the two and compare its absolute value to the critical value given by

$$D_{\text{critical}} = \frac{k_\alpha}{\sqrt{n}}$$

where k_α is a function of α and the degrees of freedom (sample size = degrees of freedom). Once again, tables can be found in mathematics/science handbooks.

It would also be useful to have graphical representations of various distributions with corresponding possible parameter values available. These will form a valuable aid to forming a hypothesis as to the distribution of the data.

Once a hypothesis for the distributional form has been selected, the analyst must then estimate the parameters to be used in completing the hypothesis. The methods that can be used depend largely on the hypothesized distribution, and on approaches that combine intuition, knowledge, and mathematical tools. These approaches generally revolve around using the mean and the variance of the data to estimate the parameters, especially when there are only two or fewer. For a greater number of parameters, higher moments may be computed for the data and equated to theoretical moments, resulting in solved parameters.

The method of moments has been criticized by some simulationists, preferring a more sophisticated approach of maximum likelihood estimation (see Chapter 5). For many cases the two approaches yield the same results.

When calculating the sample mean and variance the usual formulae for observations are applicable, namely

$$\bar{x} = \frac{1}{N}\sum_{i=1}^{N} x_i \quad ,$$

and

$$S^2 = \left(\frac{1}{N-1}\right)\sum_{i=1}^{N}(x_i - \overline{x})^2 = \left(\frac{1}{N-1}\right)\left(\sum_{i=1}^{N} x_i^2 - N\,\overline{x}\right).$$

However, in the process of the actual data collection, and certainly for the development of histograms, the data will be grouped by cells, and the following formulae for the sample mean, variance, and sample moments, about zero, will be more appropriate.

$$\overline{x} = \frac{1}{N}\sum_{i=1}^{k} m_i x_i$$

$$S^2 = \frac{1}{N}\left(\sum_{i=1}^{k} m_i^2 x_i \; - \; N\,\overline{x}^2\right)$$

$$M_j = \frac{1}{N}\left(\sum_{i=1}^{k} m_i^j x_i\right)$$

where k is the number of cells or groups, m_i the midpoint of the ith cell or group, x_i the number of observations in the ith cell or group, and N the total number of observations.

Example 12.3 Let us model a computer system where resource blockages have been a problem. Data on these resource blockages are as follows:

Number of blockages	Number of days
0	0
1	6
2	16
3	30
4	31
5	14

6	5
7	1

If we desired to include the number of blockages per day into our general computer system simulation model, what would be the most appropriate distribution and what would be the recommended parameter values?

Solution: The tabulation of the data results in

Number of blockages	Number of days (f)	fx	fx^2
0	0	0	0
1	6	6	6
2	16	32	64
3	30	90	270
4	31	124	496
5	14	70	350
6	5	30	180
7	1	7	49
Totals	103	359	1415

with sampled mean $\bar{x} = 359/103 = 3.48$

and variance $$S^2 = \left(1415 - \frac{(359)^2}{103}\right) \Big/ 102 = 1.605 \; .$$

See Figure 12.3 for a graphical representation of the observed frequency data.

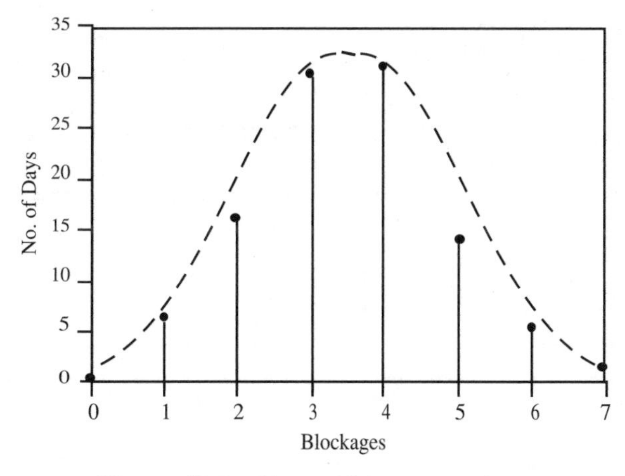

Figure 12.3. Observed frequency data.

Let us consider either a binomial or a Poisson distribution. For a binomial we have

$$\mu = np$$
$$\sigma^2 = np(1-p)$$

Thus, $np = 3.48$ and $np(1-p) = 1.605$ resulting in $p = 0.538$ and $n = 6.47$. But n must be an integer, thus we would have $n = 6$, $p = 0.580$ or $n = 7$ $p = 0.497$.

Next, let us use or goodness–of–fit test to check on the validity of the distributions. For the chi–square we have

Cell	Theoretical		Observed	$(f_{ok} - f_{ek})^2 / f_{ek}$	
	$N = 6$	$N = 7$		$N = 6$	$N = 7$
0–1	5.253	6.633	6	0.1062	0.0604
2	16.171	17.201	16	0.0018	0.0836
3	29.777	28.325	30	0.0016	0.0991
4	30.838	28.047	31	0.0008	0.3109
5	16.985	16.542	14	0.5245	0.3906
6–7	3.976	6.252	6	1.0306	0.0102
Totals	103	103	103	1.673	0.956

If we choose $\alpha = 0.05$, then from the tables, $\chi^2(3, 0.95) = 7.81$. Thus both $N = 6$ or $N = 7$ are easily accepted. This is also true for $\alpha = 0.1$, where $\chi^2(3, 0.90) = 7.78$.

Repeating the same data analysis but for the K&S test, we reach a similar conclusion.

No. of blockages	$F(x)$ observed	$F(x)$ $N = 7$	| Diff |	$F(x)$ $N = 6$	| Diff |
1	0.0000	0.0081	0.0081	0.0055	0.0055
2	0.0583	0.0644	0.0061	0.0510	0.0073
3	0.2135	0.2314	0.0178	0.2080	0.0056
4	0.5049	0.5064	0.0015	0.4971	0.0077
5	0.8058	0.7787	0.0276	0.7965	0.0093
6	0.9417	0.9393	0.0026	0.9614	0.0003
7	0.9903	0.9924	0.0021	1.0000	0.0097
Totals	1.0000	1.0000	0	1.0000	0

$$D_{.05} = \frac{1.36}{\sqrt{n}} = \frac{1.36}{10.15} = 0.134$$

Another technique of obvious utility in simulation model development is that of both **regression** and **correlation** analysis. We are faced with the task of developing relationships among various elements, variables, etc. When the relationship is deterministic, at least to the same approximation, the obvious first step is to acquire and display the pertinent data by means of a scatter diagram (see Figure 12.4).

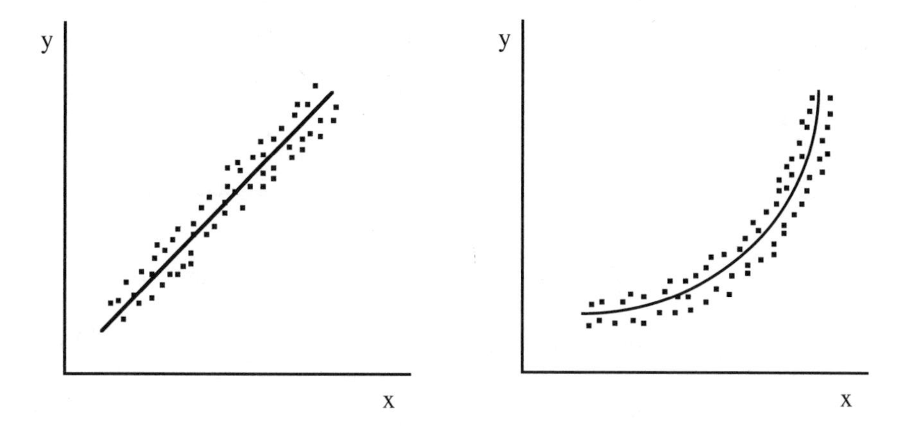

Figure 12.4. Scatter diagrams.

We obviously may make use of a variety of scales to attempt to fit the randomly sampled data. Some common functions, including rectangular, semi–log, log–log, etc., are illustrated in Figure 12.5. The difficulty in performing this fit depends on the function chosen and the number of variables.

One approach that is widely accepted is to minimize the **sum of the squared deviations** of the data points from the fitted equation. Furthermore, when several possible fits are compared, it may be difficult to decide on the most appropriated, especially when the underlying assumptions of causation are not justified. To help the analyst in this dilemma, we need a well–defined procedure for choosing a best fitting regression line.

$$y' = a + bx$$

where y' represents the **predicted** value of the criterion variable for a given value of the predictor variable x, and a and b are the y–intercept and slope of the line and must be determined from the data.

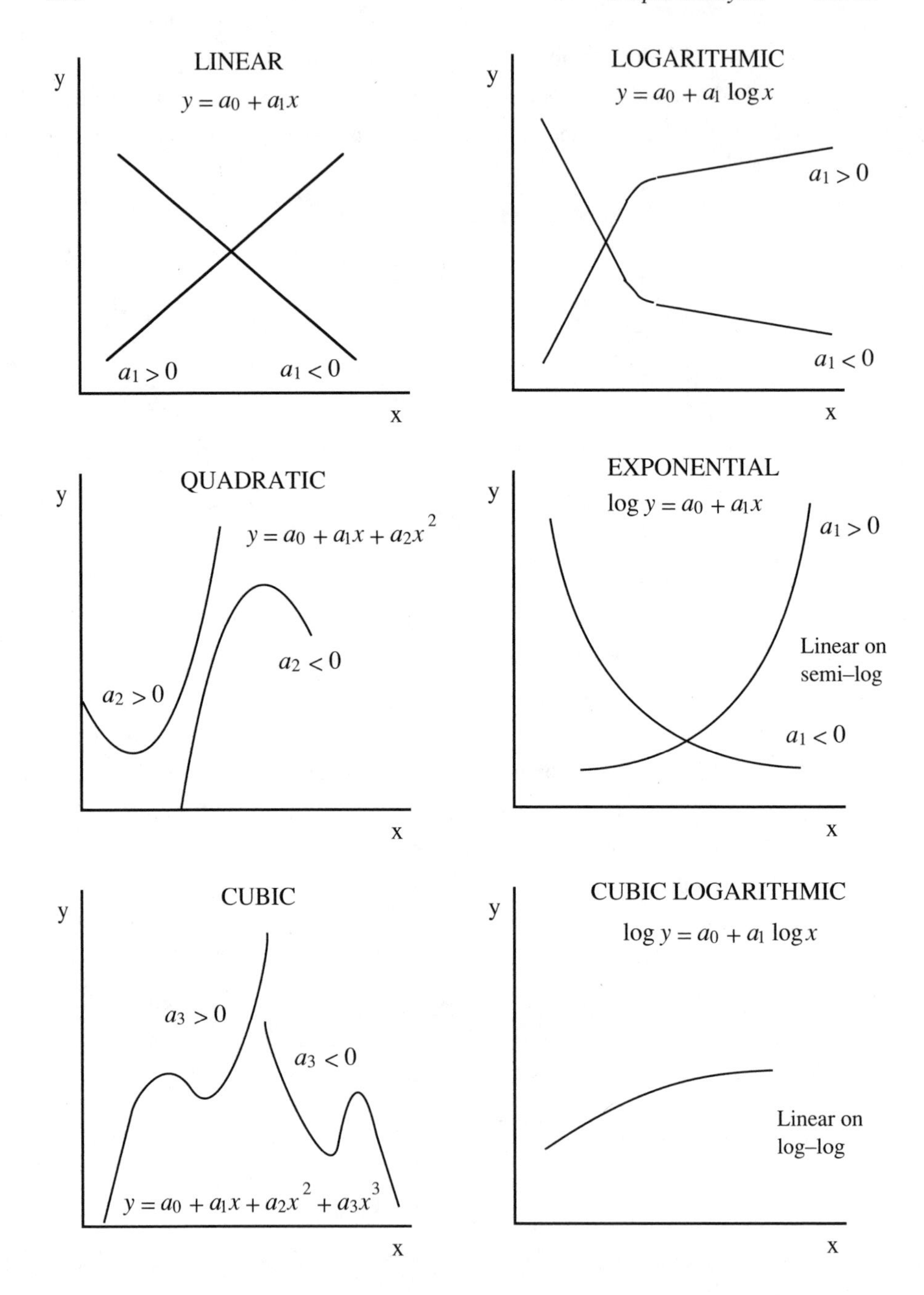

Figure 12.5. Some common functions.

Using the least squares criterion, we choose the line with the smallest sum of the squared deviations of the data points from the line. That is to say, we need to identify the values of a and b which will minimize $\sum_{i=1}^{n}(y_i - y_i')^2$. The slope b and y–intercept a of the best fitting line, based on the least squares criterion, can be shown to be

$$b = \frac{\sum_{i=1}^{n}(x_i - \bar{x})(y_i - \bar{y})}{\sum_{i=1}^{n}(x_i - \bar{x})^2}$$

and

$$a = \bar{y} - b\bar{x}$$

The slope b can also be expressed as

$$b = r(s_y / s_x)$$

where r is the **correlation coefficient** between two variables x and y, while s_x and s_y are their respective standard deviations.

A number of assumptions concerning the population of the sampled data must be met, for the resulting regression equation to be properly interpreted. These assumptions are as follows:

1. For each value of x, there is a probability distribution of independent values of y. From each of these y distributions, one or more values is sampled at random.
2. The variances of the y distributions are all equal to one another.
3. The means of the y fall on the regression line

$$\mu_y = \alpha + \beta x$$

where μ_y is the mean of the y distribution for a given value of x, β is the slope and α is the y intercept.

Thus, for any given value of x, the values of y vary randomly about the regression line (see Figure 12.6). Any individual observation of y_i will deviate from the population regression line by $\pm\, e_i$ depending upon whether the observation falls above or below the true regression line.

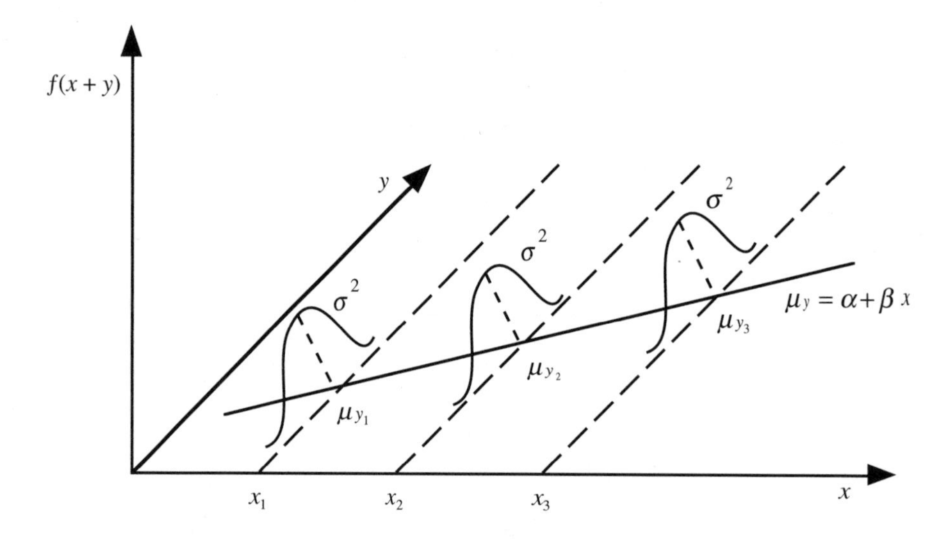

Figure 12.6. Regression model of independent y populations with equal σ^2 and μ.

Based on the above, the individual observations of y_i are

$$y_i = \alpha + \beta x + e_i$$

a part due to the true regression line and a part due to the natural variation of the y values about the regression line. Thus we are assured that the least squares method will yield a sample regression line $y' = a + bx$, which is an unbiased estimate of the true, but unknown population regression line $\mu_y = \alpha + \beta x$. However, the a and b estimates of α and β are subject to sampling error just like any other sampling statistics. Also the random variation of the y_i's about the regression line, the e_i's are a great source of error. These variation can be estimated using a **standard error of estimate** given

$$S_{yx} = \sqrt{\frac{\sum_{i=1}^{n}(y_i - y_i')^2}{n-2}}$$

Although S_{yx} alone is not sufficient to precisely estimate the expected magnitudes of the prediction errors, as n becomes large, it can provide approximate confidence intervals, since errors in estimation of α and β become small relative to S_{yx}. Thus, for $n \geq 100$, 95% of the prediction errors will be within $\pm 1.96 S_{yx}$ of the sample regression line, as long as the y populations for each value x are normally distributed.

We may conclude by saying that regression defines a proposed relationship, while correlation addresses how good that relationship is. The correlation coefficient is defined between ±1 and a sample of the degree of correlation is evident in the scatter diagrams of Figure 12.7.

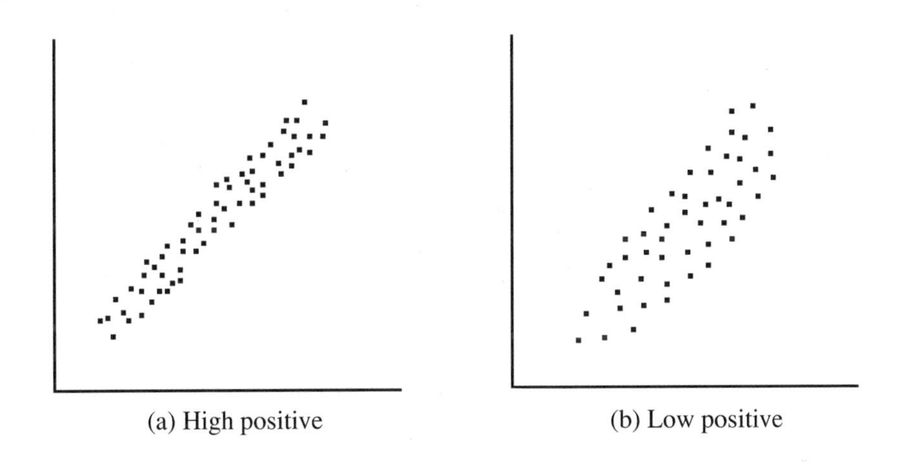

Figure 12.7. The relationship of correlation coefficients to the scatter diagram.

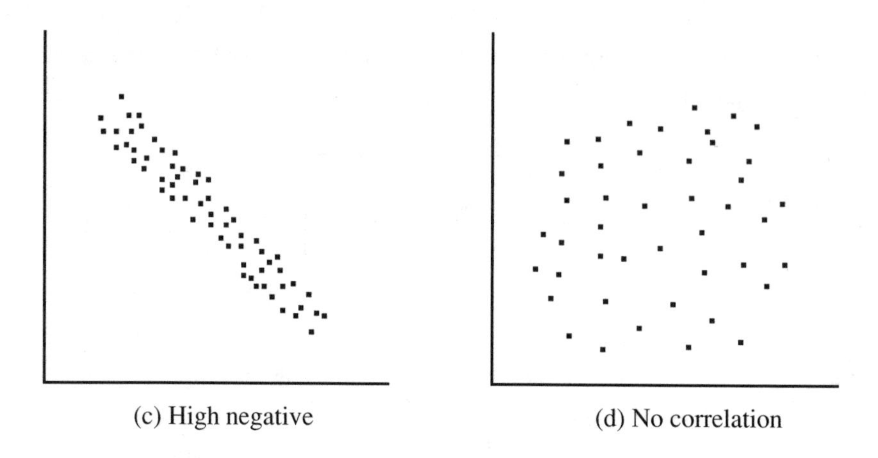

Figure 12.7. The relationship of correlation coefficients to the scatter diagram (continued).

12.4 Simulation Model Run Length

Let us determine the **length of a simulation run**. That is for one specified combination of parameter values, how long should the simulation run (iterate) before we are willing to accept the results as representative. We may choose as criteria, performance, efficiency, economy, and/or effectiveness of the simulated/modeled system.

Suppose α_i is the ith criteria, and has been evaluated j times (i.e., through j iterations)

$$\alpha_{ij} \ , \quad j = 1, 2, \ldots, J$$

Let us assume that this criterion measure is a random variable, with probability density function, $F(\alpha_i)$, mean or expected value $E(\alpha_i)$, and variance $\sigma^2(\alpha_i)$. Since a distribution function cannot be completely determined by iterations of the simulation, and is not known *a priori* its true mean and variance is also not obtainable. However from randomly sampling we can obtain the sample mean

$$\overline{\alpha}_i(J) = \frac{1}{J}\sum_{j=1}^{J} \alpha_{ij} \ ,$$

and the sample variance

$$S^2(\alpha_{ij}) = \frac{1}{J-1}\sum_{j=1}^{J}[\alpha_{ij} - \overline{\alpha}_i(J)]^2 \quad .$$

And since α_{ij} is a random variable, so is the sample mean $\overline{\alpha}_i(J)$.

Using

$$\boxed{\sigma^2[\overline{\alpha}_i(J)] = E\left\{[\overline{\alpha}_i(J) - E(\alpha_i)]^2\right\}}$$

$$\sigma^2[(\alpha_{ij})] = \frac{1}{J}\,\sigma^2(\alpha_i)$$

and that

$$\lim_{J\to\infty} \sigma^2[\overline{\alpha}_i(J)] = 0$$

we see that $\overline{\alpha}_i(J) \to E(\alpha_i)$.

Thus for large J,

$$\boxed{\overline{\alpha}_i(J) = E(\alpha_i)}$$

Using **Tchebycheff's inequality**, for $\Delta > 0$

$$\boxed{\text{Prob}\left[\,\left|\,\overline{\alpha}_i(J) - E(\alpha_i)\right| > \Delta\,\right] < \frac{\sigma^2(\alpha_i)}{J\Delta^2}}$$

The above inequality states that the expected value (mean) of α_i is captured the interval 2Δ, centered at $\overline{\alpha}_i(J)$, and can be made as close to one by choosing J sufficiently large.

Motivation:

Because $\text{Prob}\left[\,\overline{\alpha}_i(J)-\Delta < E(\alpha_i) < \overline{\alpha}_i(J)+\Delta\,\right] > 1 - \dfrac{\sigma^2(\alpha_i)}{J\Delta^2}$ and since $\sigma^2(\alpha_i)$ is not known, we use the estimate $S^2(\alpha_{iJ})$. Then choosing the interval Δ, and increasing J until

$$1 - \frac{S^2(\alpha_i)}{J\Delta^2} \to 1$$

the desired result is obtained.

Example 12.4 For example, assume a value of $\Delta > 0$, and the J_1 simulation iterations have been completed, and we want

$$\text{Prob}\left[\,\overline{\alpha}_i(J)-\Delta < E(\alpha_i) < \overline{\alpha}_i(J)+\Delta\,\right] > 1-\varepsilon$$

where $\varepsilon \doteq 0.05$.

Thus, for some value of J we have

$$\frac{S^2(\alpha_{iJ})}{J\Delta^2} \le \varepsilon$$

Set $\varepsilon = S^2(\alpha_{iJ})/(J_1\,\Delta^2)$.

If $\varepsilon_1 > \varepsilon$ then estimate the required number of additional iterations by choosing

$$J^2 = \left\langle \frac{S^2\left(\alpha_{iJ_1}\right)}{\varepsilon\,\Delta^2} \right\rangle + 1$$

and obtaining $(J_2 - J_1)$.

Recompute $\varepsilon_1 = S^2(\alpha_{iJ_2})/(J_2\Delta^2)$, check $\varepsilon_2 \le \varepsilon$ and keep iterating until

$$S^2(\alpha_{iJ_x})/(J_x\Delta^2) \le \varepsilon$$

is satisfied.

12.5 Variance Reduction

In the analysis of the simulation model we have assumed random sampling and then averaged the result of this random sampling to estimate the output variables. The variance in our estimate was a function of the sample size. More specifically the $\sigma_{\bar{x}}$ of the sample mean was proportional to $n^{-1/2}$. Thus, if we wish to reduce the sample standard deviation by one half, we must quadruple the random sample size!

Unlike the early days of simulation when large sample sizes were completely out of the question, high speed modern computers frequently provide the opportunity to obtain as large a sample size as desired. Nonetheless, for cases when this is not possible, or whenever as much information as can be gotten from a given sample size is necessary, variance reduction techniques are appropriate. While many simulation practitioners are frightened away because of overly elegant mathematical treatments, many of the variance reduction techniques are quite simple conceptually and may be readily applied with little difficulty.

Let us consider a small example, due to Hillier and Lieberman (12.2), which will convey the concepts and power of two of the variance reduction approaches. Let us assume we have an exponential distribution with a mean of one and

$$f(x) = e^{-x} \quad ,$$

and

$$F(x) = 1 - e^{-x} \ .$$

Furthermore, let us assume that this is the character of the random sampled output, and that we do not know that $\mu = 1$. The problem is to estimate the mean of the randomly sampled output. Table 12.4 provides this sampled data, with sample size of length of 10.

Table 12.4 Example Data for Variance Reduction Technique

Iteration i	Random number r_i	Random observation $x_i = -\ln(1-r_i)$
1	0.495	0.684
2	0.335	0.408
3	0.791	1.568
4	0.469	0.633
5	0.279	0.328
6	0.698	1.199
7	0.013	0.014
8	0.761	1.433
9	0.290	0.343
10	0.693	1.183
	Totals:	7.793
	Estimate of mean:	0.779

When this data is analyzed, we notice that there are no random numbers r_i from 0.014 – 0.038, and no random numbers $r_i > 0.791$. An analyst might conclude that this is the reason for the poor result of obtaining an estimated population mean of 0.779, or perhaps conclude that a sample of ten data points is insufficient. In fact, one is able to obtain much better results.

The very nature of random sampling may result in a less than uniform distribution of random numbers for small samples. Certain portions of the distribution may thus become more crucial in determining the mean. For example, the tail of the exponential distribution.

Let us consider a procedure called **stratified sampling**. This method is used to treat problems as described above, by partitioning the distribution into regions or strata. Each strata is then individually sampled, usually with more emphasis placed on the critical areas of the distribution (such as the tail of the exponential distribution). To illustrate this procedure let us reconsider the previous example. If three strata (0–1, 1–3, 3–∞) are defined, Table 12.5 results.

Table 12.5 Stratification of Data from Table 12.4

Stratum	Portion of Distribution	Stratum Random Number	Sample Size	Sampling Weight
1	$0 \le F(x) \le 0.64$	$r_i' = 0 + 0.64 r_i$	4	$w_i = \frac{4/10}{0.64} = \frac{5}{8}$
2	$0.64 < F(x) \le 0.96$	$r_i' = 0.64 + 0.32 r_i$	4	$w_i = \frac{4/10}{0.32} = \frac{5}{4}$
3	$0.96 < F(x) \le 1$	$r_i' = 0.96 + 0.04 r_i$	2	$w_i = \frac{2/10}{0.04} = 5$

Continuing this scheme and using the same stream of 10 random numbers as before, an improved estimate of the mean can be developed (see Table 12.6).

Table 12.6 Example of Stratified Sampling

Stratum	i	Random number r_i	Stratum random number r_i'	Stratum random observation $x_i' = -\ln(1 - r_i')$	Sampling weight w_i	$\frac{x_i'}{w_i}$
1	1	0.495	0.317	0.381	5/8	0.610
	2	0.335	0.215	0.242	5/8	0.387
	3	0.791	0.507	0.707	5/8	1.131
	4	0.469	0.300	0.357	5/8	0.571
2	5	0.279	0.729	1.306	5/4	1.045
	6	0.698	0.864	1.995	5/4	1.596
	7	0.013	0.644	1.033	5/4	0.826
	8	0.761	0.884	2.154	5/4	1.723
3	9	0.290	0.9716	3.561	5	0.712
	10	0.693	0.9877	4.398	5	0.880
						Total = 9.485
						Estimate of mean = 0.948

While the stratified sampling greatly improves the mean estimate of the randomly sampled data, frequently by an order of magnitude, there are additional tradeoffs to be considered. First, the user must choose the strata. Ideally the user can choose the strata such that the variances are all equal. The simplest approach might be to make all strata of equal size. This however is much less desirable in general. The next problem is to decide how many random observations are to be obtained for each strata. One guide, due to Tocher, might be to make the number of observations in the *i*th strata proportional to the probability of a random observation falling *i* times into this strata.

The second variance reduction technique makes use of the **antithetic variate** concept, first encountered in Chapter 9. The underlying concept is to use pairs of variates which compensate for the variation in each other. That is, assume y is estimated by both x_1 and x_2, and that x_1 and x_2 are negatively correlated. The estimate of y given by

$$\hat{y} = (x_1 + x_2)/2$$

will have a variance

$$\boxed{\hat{\sigma}_{\hat{y}}^2 = \frac{1}{4}\left(\sigma_{x_1}^2 + \sigma_{x_2}^2\right) + \frac{1}{2}\beta\left(x_1, x_2\right)}$$

where β is the covariance between x_1 and x_2. Obviously, there will be a very large variance reduction if x_1 and x_2 are highly negatively correlated.

We have chosen to use the method of complementary random numbers from among many possible implementations of this concept. That is, we simply use both r_i and $1 - r_i$ as random numbers. The results of using the procedure on the same randomly sampled data from Table 12.4, is given in Table 12.7.

A third variance reduction procedure uses the idea of **importance sampling** to distort the true random probability distribution in such a way as to more heavily sample from the important areas. A correction is then made by using an appropriate weighting factor for each randomly sampled observation. For example, suppose that it is desirable to observe the behavior in a certain region which has a probability of occurrence of only 0.01. Then for a sample size of $N = 100$, we may get one or sometimes no observation of the randomly sampled observations. On the other hand, if we distorted the actual random process so that the probability of occurrence is 0.05, and then corrected the observations, we would obtain more observations yet with only one fifth the experiments. The correction factor is obtained from

$$w(x) = p(x)/p^*(x)$$

where $p(x)$ is the probability of the random observations, and $p^*(x)$ the distorted probability of the random observations.

Table 12.7 Example for Method of Complementary Random Numbers

i	Random number r_i	Random observation $x_i = -\ln(1 - r_i)$	Complementary random number $r_i' = 1 - r_i$	Random observation $x_i' = -\ln(1 - r_i')$
1	0.495	0.684	0.505	0.702
2	0.335	0.408	0.665	1.092
3	0.791	1.568	0.209	0.234
4	0.469	0.633	0.531	0.756
5	0.279	0.328	0.721	1.275
6	0.698	1.199	0.302	0.359
7	0.013	0.014	0.987	4.305
8	0.761	1.433	0.239	0.272
9	0.290	0.343	0.710	1.236
10	0.693	1.183	0.307	0.366
	Totals:	7.793		10.597
		Estimate of mean = 1/2 (0.799 + 1.060) = 0.920		

Example 12.5 Suppose the population of computer jobs in a system is 80% compute bound and 20% I/O bound jobs. We wish to estimate the amount of time spent per job in the waiting queue, given that compute bound jobs spend about 50 msec. with a range of 25 – 100 msec., and I/O bound jobs spend from 0 – 500 msec. with a great variation. A sample size of $N = 15$ is to be used. Normal sampling would probably result in 12 compute bound jobs and

3 I/O bound jobs. Instead let us distort the sampling heavily toward the I/O bound jobs (they, after all, have the high variance), and choose a sample of 10 I/O bound jobs and only 5 compute bound jobs.

Let us assume the sample came out as follows:

Compute bound jobs: 45, 50, 55, 40, 90
I/O bound jobs: 80, 50, 120, 80, 200, 180, 90, 500, 320, 75

with sampled means $\bar{\mu}_{CB} = 56$ and $\bar{\mu}_{I/O} = 169.50$. Weighted properly, the corrected mean of $\bar{\mu} = 79.49$ is obtained from

$$\frac{1}{15}\left[\frac{0.80}{0.33}\left(\sum_{i=1}^{5} CB_i\right)+\frac{0.20}{0.66}\left(\sum_{i=1}^{10} IO_i\right)\right] = 79.49$$

12.6 Summary

The purpose of analyzing the simulation model output is to make one or more inferences about one or more parameter of the stochastic processes that produced the output. In this chapter we have attempted to provide techniques useful to this output analysis and the associated inference procedures. Included were parameter estimation and confidence limits, run length determination, variance reduction, and model inputs.

12.7 References

12.1 Emshoff, J. R. and Sisson, R. L., *Design and Use of Computer Simulation Models*, New York: MacMillan Company, 1970.

12.2 Hillier, F. S. and Lieberman, G. J., *Introduction to Operations Research*, San Francisco, CA: Holden–Day, Inc., 1980.

13

LANGUAGES FOR DISCRETE SYSTEM SIMULATION

Once the aspects of the system to be simulated have been decided and the model formulated in terms of a logical flowchart, the model must be translated into a form suitable for the computer. That is, the model must be coded using some language. The coding process is probably one of the best understood of the steps in model building, yet difficult decisions sometimes accompany the choice of a suitable programming language. The range of programming languages that have been used in discrete system simulation covers the entire spectrum, from the low–level, machine–oriented assembly languages to the specialized simulation–oriented languages such as GPSS. A number of factors can influence the choice of a language, including the programmer's familiarity with the language; the ease with which the language is learned and used if the programmer is not already familiar with it; the languages supported at the installation where the simulation is to be done; the complexity of the model; and the need for a comprehensive analysis and display of the results of the simulation run.

In this chapter we survey some of the characteristics that any simulation language should have. We also discuss briefly how a simulation model may be implemented in FORTRAN and in three of the more common specialized simulation languages: GPSS, SIMSCRIPT, and GASP. The purpose of this chapter is not to provide the analyst with the required level of expertise to code a detailed simulation model. Rather it is to survey the characteristics of each language and to aid the analyst in choosing a suitable language. After choosing a language, the analyst should consult the programming manuals and texts devoted to that language to obtain all the details. The choice of a language should not be made lightly, however, since this choice can have a significant impact on the time required to develop the simulation model and on the expense of running the simulation.

13.1 Language Characteristics

Many simulation models perform similar functions. Some of these functions are

1. Generating random variates
2. Managing simulation time
3. Handling routines to simulate event executions
4. Managing queues
5. Collecting data
6. Summarizing and analyzing data
7. Formulating and printing output

Many systems are characterized by some **stochastic** behavior, whether it is interarrival times and service requirements in a queueing system or the time between equipment failures in a production system. To simulate such probabilistic events usually requires the use of the simulated sampling techniques outlined in Chapter 7. Then any language to be used in implementing a simulation model should either provide or allow easy development of a facility to generate and transform standard uniform random variates.

Whether the **periodic scan** (fixed–time increment) or the **event scan** (variable–time increment) method of **time management** is employed, some means of representing time is required. This is important not only for controlling or "driving" the simulator but also for collecting, summarizing, and analyzing the data. A simulation programming language must allow for easy representation and manipulation of simulation time.

When a scheduled event is to be executed, the simulation program normally effects the required changes in system state by invoking a program module designed specifically for that purpose. Depending on the complexity of the required changes, the execution could be simulated with one or two statement routines inserted at the appropriate points in the model or with more complex, lengthy routines implemented as sub–programs.

Queue management is common to many models because many systems involve competition for limited resources. The representation and manipulation of waiting lines can be accomplished in many ways. A convenient way of representing queues is by a list because the primary operations in queue management are the addition and deletion of members. These two operations are easily performed using the list representation and the manipulation of pointers. Lists can be singly linked (forward) or doubly linked (forward and backward). The chain is generally ordered on some entity attribute or else on the basis of when the entity entered the file. The simple list processing approach implies an exhaustive scanning of the pointers to locate a position in the list. Pointers are then updated as required – entities are not physically moved.

Recently, considerable research and development has centered around new efficient list processing methods that include binary trees, indexed lists, multi–level indexed lists, and partitioned lists. The implementation of these techniques in a simulation model often results in an order of magnitude reduction in computation time. Thus a language with efficient list–processing capability offers a significant advantage in simulation.

Many simulation models are implemented to assess the effect on the system of varying certain conditions or parameters. This measurement and comparison requires some facility for the **collection of data**. There are two distinct philosophies for collecting data within a simulation, the classical approach and the total approach. In the **classical approach** we define precisely what information and statistics are to be collected **prior** to the simulation and use effective methods for collecting and calculating those within the simulation. In the **total approach** we create a database during the course of the simulation which consists of all the data and information that can be collected. We then use an inquiry mechanism to extract any desired information from this database **after** the simulation. This requirement is not a strong criterion, however, because the most common data collected are count data, and these are easily represented and collected by integer variables. Nearly all languages that have gained any degree of recognition have this capability.

In most cases the simple collection and representation of raw data are not enough. In addition, some summarization and analysis of the data are normally required. This analysis could be relatively simple requiring only the calculation of summary measures such as means or standard deviations.

In collecting data to describe a random variable there are two class distinctions statistics based on observations (e.g., waiting time), and statistics based on time persistent values (e.g., number in the system).

In order to calculate the mean and variance of the first class of data, we can save during the course of a simulation the following:

1. Number of observations, N;
2. Sum of the values of the observation, $\sum_{i=1}^{N} x_i$;
3. Sum of squares of the values of the observations, $\sum_{i=1}^{N} x_i^2$;
4. Maximum and minimum values observed, $x_{\max}$, $x_{\min}$.

Then we can calculate the **sample mean and variance**

$$\bar{x} = \frac{1}{N}\sum_{i=1}^{N} x_i$$

$$S^2 = \frac{1}{(N-1)}\left(\sum_{i=1}^{N} x_i^2 - N\bar{x}^2\right)$$

The second class of data collection requires more skill. In essence what is needed to be saved is the following:

1. The integral of the persistent value through time, $\int_0^T xdt$;
2. The integral of the persistent value squared through time, $\int_0^T x^2dt$;
3. The true period, T.

In practice this is simple if the value is constant during the interval between 0 and T. For items 1 and 2, we simply multiply the value and the value squared by the time period, respectively. If on the other hand, the value of x is fluctuating over the interval, then we usually employ some type of trapezoidal approximation. Note that the value x generally describes the state of the system. For example, x may be the number of jobs in a computer system waiting for CPU time. Our data collection points will most often be points in time when the state of the system is changing (i.e., an event), and we will most often collect the data before a change in the system state has occurred. The following equations are then used to find the sample mean and variance:

$$\bar{x} = \frac{1}{T}\int_0^T x(t)\,dt$$

$$S^2 = \frac{1}{T}\int_0^T x^2(t)\,dt - \bar{x}^2$$

This simple form of analysis can be easily done in nearly all languages, although some languages carry out these calculations as standard functions, while others require that the programmer provide these routines. Other simulation projects may require a more extensive analysis of the data, perhaps regression analysis, analysis of variance, time series analysis, or other sophisticated tools of statistical analysis. A language that provides this level of analysis as part of its normal operation offers an advantage in simulation. The programmer can spend much time designing and coding the statistical analysis modules if they are not automatically provided in the programming language.

Some simple simulation models may require only simple output. Others may require extensive **report writing**. In any case the model must be able to display the results of a simulation run. A language that allows flexibility in formatting and presenting the data offers a significant advantage, particularly if the model is complex and if the results are to be presented to management personnel.

In addition to assessing how well each candidate language supports these seven common functions, another major consideration in selecting the best language is the assistance provided during the debugging phase of program development. In some versions of assembly language, for example, one must analyze a system dump, which is a snapshot of all or portions of the computer's memory, to verify that the results obtained are the correct ones or to trace an error in the code's logic or syntax. This is a tedious, seemingly interminable operation. Some languages, particularly the special–purpose simulation languages, provide extensive **debugging assistance**. The impact that the selection of a language with debugging assistance can have on a simulation project has been pointed out by Emshoff and Sisson (13.3). They report that it is not uncommon to reduce the time spent on coding and debugging a simulation model by a factor of ten through the use of a language with facilities that enhance the debugging process.

Another criterion for choosing a suitable programming language is the **programmer's knowledge** of the language. A programmer will probably choose a familiar language to implement a simulation model even if some other language provides better support of the common functions or better debugging assistance. This tendency is natural, and in most cases the decision is wise. Becoming familiar enough with a language to write efficient programs is time–consuming. The time needed to learn the new language may well be greater than the time needed for coding additional routines to support the common functions or debugging in the old language. A related consideration is the **ease of use** and **ease of learning**. Given two languages equally suited for the task, one would normally select the language that is easiest to learn and easiest to use.

A final consideration in selecting a simulation language is the inherent **flexibility** of the language. Having selected and learned a simulation language, the analyst is likely to use it for more than one model. It must therefore have a capability for expansion and adaptation, which would allow it to be applied to a wide range of models.

In summary then, the languages available on a given computer system should be rated according to their

1. Support of the basic functions
2. Debugging assistance provided
3. Familiarity to the programmer

4. Ease of learning and use
5. Flexibility

The relative weights given to these factors depend somewhat on the problem. The aim in all cases, however, is to minimize the time spent in the coding and debugging phases.

13.2 Use of Multipurpose Languages

Many programmers tend to select multipurpose languages such as FORTRAN, ALGOL, PL/I, Pascal, and C for use in simulation. One of the chief reasons is the widespread availability of the languages. Even a very small computer installation probably has a FORTRAN, ALGOL, PL/I, Pascal or C compiler. Along with COBOL, FORTRAN is probably the most common programming language in use and is probably the first language to which a programmer is exposed. It is understandable than that this language would be selected for implementing a simulation model. In this section we survey some of the advantages and disadvantages of using multipurpose languages such as FORTRAN, ALGOL, PL/I, Pascal, or C in simulation projects. Although there are significant differences in the capabilities of these languages, we consider them as a group in assessing their applicability to simulation projects. To assess these languages we ask: How well do these languages support the basic functions of simulation? What assistance is provided in debugging? How flexible are the languages?

There is no express capability in any of these languages to generate random variates. Many installations have among their library routines a function that generates standard uniform variates. Programmers are sometimes reluctant to use these standard functions; they may not understand the techniques used and thus may shy away from the generation of numbers using the black–box approach, preferring to write their own routines. Even if they use the standard functions to generate the uniform variates, they still must code the routines to transform the standard uniform variates to a normal, exponential, or Poisson distribution. Most of these routines are fairly trivial to code and do not significantly increase the time required to develop the simulation model.

Management of simulation time is generally easily done in the procedure–oriented multipurpose languages. A counter variable, say CLOCK, can be initialized at the beginning of the simulation run. If the periodic scan approach is used, some fixed increment is added to CLOCK each time the simulation clock is to be advanced. If the event scan approach is used, the pending events must be scanned to determine the size of the increment before it is added to CLOCK. In any case the programmer must define and initialize the clock and code routines to update the clock, but management of simulation time is not a big problem in these languages.

FORTRAN, ALGOL, PL/I, Pascal, and C all have well–defined sub–program capabilities for simulating event executions. Generally they define a subroutine or procedure for each type of event to be executed. For example, in the simulation of a single–channel queueing system, such events are the arrival of a customer to the system, the completion of service to a customer (the customer's departure from the system), and the action taken when the system is full or closed. For each type of event a routine is coded that effects the changes in the system state in a way that reflects the event occurrence in the real system and takes care of housekeeping chores such as updating statistics. Since the definition and invocation of subprograms is straightforward in these languages, no major problems should be incurred in this function.

Languages such as PL/I, Pascal, or C offer some advantages over FORTRAN in queue management. List processing in FORTRAN is weak and is usually implemented by arrays. This approach can cause problems, since the maximum size and dimension of the arrays must be determined and declared beforehand. Hence it is not really possible to simulate the operation of say an $M/M/1/\infty$/FIFO queueing system. Manipulation of pointers in FORTRAN is also inefficient, since pointers are normally included as a part of a multidimensional array. Accessing a particular pointer can then become time–consuming. The actual penalty that results from FORTRAN's list–processing capability depends on the model. The amount of list processing required in the model has to be assessed and balanced against FORTRAN's advantages such as widespread availability and programmer familiarity. PL/I eases the problem of inefficient list processing at the cost of losing some advantage in availability and familiarity.

The collection of count data in FORTRAN, ALGOL, PL/I, Pascal, or C is normally done with integer variables. The variables are initialized at the beginning of the simulation and incremented when an event occurs. Frequency data can be accumulated through the definition of classes and the use of arrays.

The summarization and analysis routines required in most simulation programs can greatly lengthen the programming time. Each routine must be coded by the programmer. Some statistical analysis routines are complex, and care must be taken to interface them properly with the model. Models that do not require a great deal of analysis pose less difficulty.

When using multipurpose languages, the programmer must consider the formatting and printing of results. Unlike the specialized languages, multipurpose languages have no automatic output. Input and output routines that are part of the implementation of multipurpose languages provide for flexible formatting, under programmer control. Many installations provide plot routines that may be invoked to provide a visual presentation of the output. However, some effort is required on the part of the programmer to define, interface, and initialize the parameters needed by the routines.

Debugging aids in the multipurpose languages are somewhat limited. They all identify syntactic errors, such as the use of undefined variables and errors made during input. Errors in logic, however, must be detected by the programmer. To do this, the programmer must have some idea of what results to expect from the model. The model is then run and its output compared with the expected results. Debugging in these languages is in many respects a trial–and–error process. The programmer finds one bug and eliminates it, only to expose another.

In summary then, multipurpose languages can be used successfully in the programming of simulation models. In fact, there may be more simulation models in FORTRAN than in any other language. The main drawback to using these languages is that the entire model must be coded by the programmer. The languages provide few simulation–oriented functions and little debugging assistance other than pointing out syntactic errors. The advantages of these languages, however, are numerous. First, most programmers have at least been exposed to FORTRAN, Pascal, or C. Thus user familiarity is not likely to be a problem. Second, FORTRAN or C are among the most common languages now available. Thus the installation is likely to support FORTRAN or C. Third, models developed in standard FORTRAN or C are likely to be highly portable because different implementations of the language are similar. A model developed at one location can probably be run at another location with only minor changes. Models developed in these general–purpose languages are likely to cost less to run. The programmer does not have to share the cost of a large software system, of which only a small part may be used. Another advantage, which may be hard to quantify for a particular project, is that programmers using one of the general–purpose languages to develop a simulation model are likely to be more conversant with the details of the model than if the model was coded in one of the especially designated simulation languages. Since a programmer is required to address each aspect of the model, the programmer obviously will be familiar with the design details. In selecting a language for a simulation model, one must weigh the advantages of the general–purpose languages against the almost guaranteed longer program development and debugging time.

Example 13.1 Consider the following system as an example of how a multipurpose programming language, such as FORTRAN, can be used.

Arrivals occur to a single–channel queueing system according to the Poisson distribution with arrival rate λ. The arrivals, each requiring an exponentially distributed service time, immediately try to enter the service facility. If the facility is busy, they join a queue. If the service time required is assigned when the arrival event occurs, there is no need to distinguish between the arrivals as far as the service facility's operation is concerned. If the periodic scan approach is used, the operation of this system can be simulated using the model depicted in the flowchart given in Figure 13.1.

Suppose that the statistics of interest are the average time an arrival spends in the queue, the average queue length, the average number in the system, and the total time the system is idle. The model could be implemented using the following FORTRAN program (see Figure 13.2). In most cases the actions are well documented; others should be apparent after careful scrutiny.

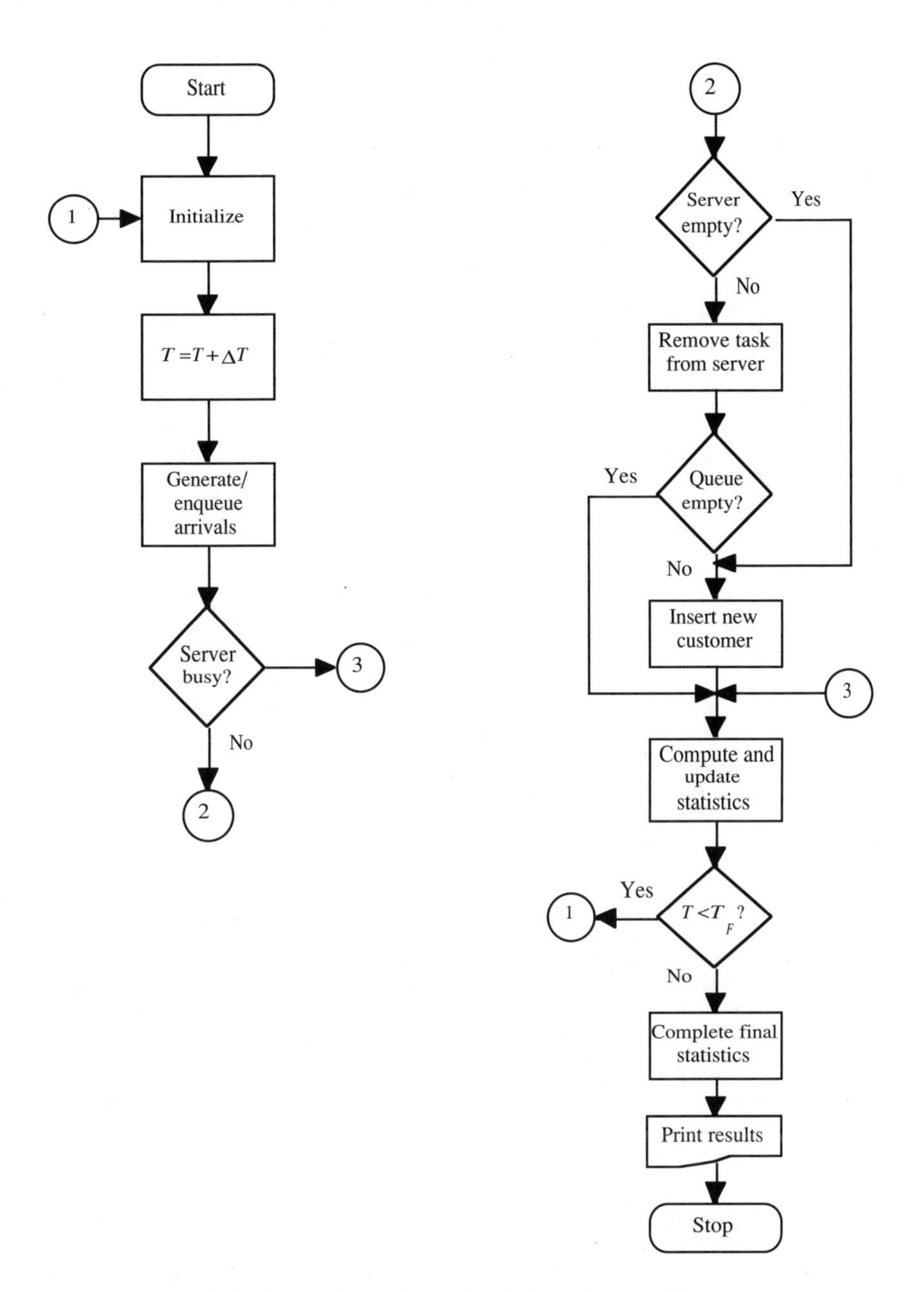

Figure 13.1. Flowchart for the simulation of a queueing system.

```
C
C       SIMULATION OF A SINGLE CHANNEL QUEUE
C
C       THIS PROGRAM SIMULATES THE OPERATION OF A SINGLE CHANNEL
C       QUEUEING SYSTEM WITH POISSON ARRIVALS AND EXPONENTIAL
C       SERVICE TIMES.
C
C       THE INPUT DATA IS
C           DELT THE TIME INCREMENT
C           SLEN THE LENGTH OF THE SIMULATION
C           LAMBDA THE ARRIVAL RATE
C           MU THE SERVICE RATE
C           ISEED AN INTEGER SEED TO PRIME THE RANDOM NUMBER
C               GENERATOR
C
        INTEGER QUEUE
        LOGICAL SFULL
        REAL LAMDA,MU
C
C       INITIALIZATION
C
        T = 0.
        TIDLE = 0.
        AVEQ = 0.
        AVES = 0.
        NSER = 0
        WTIME = 0.
        ARRTIM = 0.
        TSER = 0.
        QUEUE = 0.
        SFULL = .FALSE.
        READ(5,500)DELT,SLEN,LAMDA,MU,ISEED
500     FORMAT(4F10.5,I5)
        LAMDA = 1./LAMDA
        MU = 1./MU
C
C       SIMULATION ROUTINE
C
C       GENERATE FIRST ARRIVAL
        CALL URANDX(ISEED,IRAND,URAND)
        ISEED = IRAND
        ARRTIM =    LAMDA * ALOG(URAND)
C       CHECK IF ARRIVAL HAS OCCURRED
1       IF(ARRTIM.LE.0.)CALL ARRVL(ARRTIM,QUEUE,ISEED,LAMDA)
C       CHECK IF SERVICE TIME HAS EXPIRED
        IF(TSER.LE.0.)CALL SERCOM(TSER,QUEUE,ISEED,NSER,MU,SFULL)
C       COMPUTE UPDATES
        IF(QUEUE .NE.0)GO TO 40
        IF(SFULL)GO TO 40
        TIDLE = TIDLE + DELT
```

Figure 13.2. Program listing for the simulation of a queueing system.

```
40      AVEQ = AVEQ + QUEUE
        IF(SFULL)AVES = AVES + 1
        AVES = AVES + QUEUE
        WTIME = WTIME + QUEUE * DELT
C       TEST TO SEE IF PERIOD HAS EXPIRED
        T = T + DELT
        ARRTIM = ARRTIM   DELT
        TSER = TSER   DELT
        IF(T .LT. SLEN)GO TO 1
C
C       COMPUTE AND PRINT FINAL STATISTICS
        AVEQ = AVEQ/(SLEN/DELT)
        AVES = AVES/(SLEN/DELT)
        IF(SFULL)I = 1
        WTIME = WTIME/(NSER + I + QUEUE)
        WRITE(6,610)
610     FORMAT(1X)
        WRITE(6,603)DELT
603     FORMAT(1X, 'FOR A DELTA T OF',5X,F10.5,5X,'MIN THE RESULTS WERE')
        WRITE(6,600)NSER,QUEUE,I
600     FORMAT(1X,'THERE WERE',5X,I3,5X,'ITEMS SERVICED,' ,5X,I3,5X,'ITEMS
        +LEFT IN THE QUEUE AND',5X,I1,5X,'ITEM(S) LEFT IN SERVICE.')
        WRITE (6,601)AVEQ,AVES,WTIME
601     FORMAT(1X,'AVERAGE QUEUE LENGTH WAS',5X,F5.2,5X,'AVERAGE NUMBER
        +IN THE SYSTEM WAS',5X,F5.2,5X,'AVERAGE WAITING TIME WAS'5X,F5.2,
        +5X,'MIN.')
        WRITE (6,602)TIDLE
602     FORMAT(1X,'THE SYSTEM WAS IDLE FOR A TOTAL OF',5X, F5.2,5X,'MIN.')
        STOP
        END
C
        SUBROUTINE URANDX(JSEED,IRAND,URAND)
C
C       THIS ROUTINE GENERATES A PSEUDORANDOM STANDARD UNIFORM VARIATE.
C       THE PARAMETERS FOR THIS ROUTINE ARE
C           JSEED AN INPUT PARAMETER EQUAL TO THE INTEGER SEED READ IN.
C           IRAND AN OUTPUT PARAMETER REPRESENTING A PSEUDO RANDDOM VARIATE
C               USED AS THE SEED FOR THE NEXT ITERATION.
C           URAND AN OUTPUT PARAMETER REPRESENTING A PSEUDO RANDOM STANDARD
C               UNIFORM VARIATE.
C
        IRAND = JSEED * 65539
        IF(IRAND)5,6,6

5       IRAND = IRAND + 2147483647 + 1
6       URAND = IRAND
        URAND = URAND * .4656613E 9
        RETURN
        END

        SUBROUTINE ARRVL(ARRTIM,QUEUE,ISEED,LAMDA)
        INTEGER QUEUE
        REAL LAMDA
C
```

Figure 13.2. Program listing for the simulation of a queueing system (continued).

```
C       ROUTINE TO SIMULATE AN ARRIVAL TO THE SYSTEM
C
        QUEUE = QUEUE + 1
        CALL URANDX(ISEED,IRAND,URAND)
        ISEED = IRAND
        ARRTIM =   LAMDA * ALOG(URAND)
        RETURN
        END

        SUBROUTINE SERCOM(TSER,QUEUE,ISEED,NSER,MU,SFULL)
        REAL MU
        INTEGER QUEUE
        LOGICAL SFULL
C
C       ROUTINE TO SIMULATE A SERVICE COMPLETION
C
        IF(SFULL)NSER = NSER + 1
        IF(QUEUE .NE. 0)GO TO 10
        SFULL = .FALSE.
        RETURN
10      QUEUE = QUEUE   1
        CALL URANDX(ISEED,IRAND,URAND)
        ISEED = IRAND
        TSER = MU * ALOG(URAND)
        SFULL = .TRUE.
        RETURN
        END
```

Figure 13.2. Program listing for the simulation of a queueing system (continued).

13.3 Simulation Languages

In addition to developing models in a multipurpose language, specialized simulation languages have been developed. These languages were especially designed for developing and executing simulation applications. Characteristics that distinguish these languages from multipurpose languages include:

1. Lower time requirement to represent a simulation operation;
2. Built–in error checking facilities;
3. Built–in (automatic) world view of the simulation structure;
4. Automatic data collection;
5. Built–in process generation schemes; and
6. Built–in time management.

While for most simulation applications the use of a simulation language may be desirable, the reader should be made aware of the following disadvantages:

1. Requirement to learn a special language;
2. Requirement for unusual (more standard) compilers;
3. Restriction for simulation model construction based on the constructs of the language, and not the application;
4. Processing inefficiencies; and
5. Loss of portability.

Although simulation languages attempt to provide basically equivalent capabilities, there are many ways in which the various languages differ. Examples of some of the differences are given in Table 13.1. Features which may be used to evaluate various simulation languages are given in Table 13.2.

Table 13.1 Examples of Differences Among Various Simulation Languages

- Mode and nature of data entry
- Procedures for generating random numbers and random deviates
- Base code of language
- Time management
- Initialization requirements
- Methods of data collection and analysis
- Form and extent of output
- Ease of use / Difficulty to learn
- Level and degree of documentation

The advantage in using a simulation language can best be explained by considering the process of translating an abstract or real–world problem into a simulation model. In general, the simulationist has the objective of the study and the scope of the simulation model well defined. Furthermore, the simulationist will usually be proficient in at least one programming language.

The first step in the modeling creation process would be a clear definition of the problem. This may be in either a symbolic, descriptive, or conceptual (i.e., prototype models) form. As soon as the problem to be studied is clearly defined and categorized, the simulationist is ready to proceed with the actual construction of the simulation program. In the process of the model construction, the program will, no doubt, have to be checked, debugged, and modified, probably many times.

Table 13.2 Simulation Language Evaluation Features

Portability	• Language/compiler availability
Flexibility	• Degree to which the language supports various modeling concepts.
Programming Considerations	• Ease of programming • Self–documentation features • Availability of simulation methodology constructs • Automatic statistics collection features • Compiling requirements • List processing capabilities • Dynamic storage management • Standard report facilities
Debugging Capabilities	• Ease of debugging • Reliability and availability of support systems
Run–time Considerations	• Speed (both run–time speed as well as computation speed)
Ease of Learning	• Base language • Documentation • On–line tutorials

If the simulationist cannot program, but must rely on an outside source, serious communication problems might arise. If the simulationist happens to know a common programming language, such as FORTRAN, and does his/her own programming, the communication problems will be eliminated. However, this does not eliminate the efforts required to prepare, develop, test, and debug the simulation program after the basic model, its components, and their interactions are well–understood.

Simulationists had to face these problems repeatedly in the early years of software development. It was the repetition of those common processes which subsequently led to the development of specialized simulation languages. As we pointed out before, the major advantage in using a simulation language is in the savings of time and effort required to structure and develop the total simulation model. Although there will undoubtedly be a process of trial and error, modification and debugging, the overall total time is drastically reduced.

13.3.1 Special–purpose Languages: GPSS

Special–purpose simulation languages were developed (beginning in the late 1950's) because many simulation projects needed similar functions across various applications. Although several such languages were created, few have gained any degree of acceptance. This section introduces three of the more commonly used languages, GPSS, SIMSCRIPT, and GASP.

The **General Purpose Simulation System (GPSS)** language was first published in 1961 by Gordon (13.4). The language has evolved over the years to the point where there are now two versions: GPSS/360 and **GPSS V**. The language was designed for the express purpose of simulating the operation of discrete systems. The system that is to be simulated is represented by a set of blocks connected by lines. Each **block** represents some activity, and each **line** represents a path to the next activity. The programmer is usually responsible for the contents of the block in programming languages that use block diagrams, but GPSS has predefined blocks to which the programmer is restricted. Each block symbol is unique, thus providing a ready interpretation of the block diagrams. For a more complete description of GPSS, see Gordon (13.5) or Bobillier, Kahan and Probst (13.1).

The entities that pass through the system are called **transactions**. Some examples of transactions are customers arriving at a service station, messages passing through a communications center, and jobs arriving for processing at a computer system. The **attributes** of these entities, such as the required service time, are represented as **parameters**.

Time is advanced in **fixed units** as "transactions" flow through a specified sequence of block commands. **Transaction** might possess certain "attributes" which can be used to make logical decisions at chosen block commands. Each block type might have names, symbols, or numbers associated with it and each block consumes a specified amount of time to process the transaction. Block types can handle one item (facilities) or multiple items simultaneously (store).

Transactions enter the system with the GENERATE block, which is represented by the following symbol.

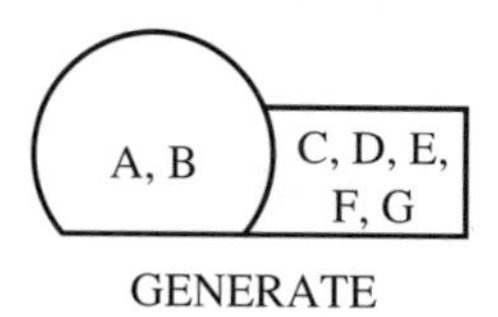

GENERATE

The operands for the GENERATE block are A, the mean interarrival time; B, the spread or mean modifier; C, the time of the generation of the first arrival; D, the total number of arrivals to be generated; E, the priority level of the transaction; F, the number of parameters to be attached to the transaction; G, the parameter type (F, fullword; H, halfword). If some of these parameter fields are omitted, default values are assumed. See the particular implementation for appropriate default values.

Transactions leave the system by way of the TERMINATE block, which is depicted by the following symbol.

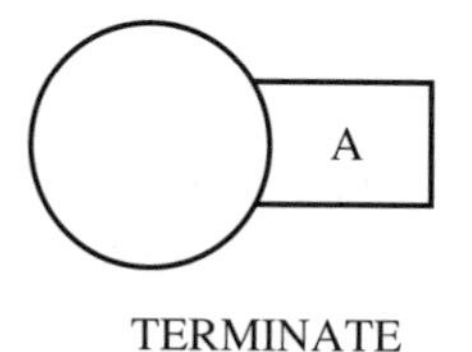

TERMINATE

The operand A for this block indicates the number by which the termination counter is incremented. The number may be zero, but at least one TERMINATE block in the model must have a nonzero operand (value).

GPSS automatically keeps track of where each transaction is in the system and when it is to be moved. Transactions are classified according to priorities (0–127), and the transaction with the highest priority is moved first. If more than one transaction has the same priority, transactions are moved in a first–come, first–

serve manner or in the order in which they were generated. A transaction's progress through the system may be held up for two reasons. First, it may enter an ADVANCE block, depicted by the following symbol.

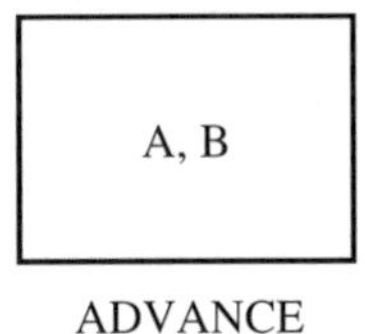

ADVANCE

The ADVANCE block represents some activity that involves an expenditure of time, such as when a customer is receiving service at a service station. For the ADVANCE block operand A is the mean and operand B is the mean modifier such that the time that the transaction is delayed is equal to A ± B. This time is selected so that any time within the interval has equal probability. When a transaction encounters an ADVANCE block, its progress is blocked and the system will advance another transaction. A transaction's progress may also be halted if it attempts to enter a block already occupied by another transaction. In this case the system holds the transaction at the preceding block until the requested block is free.

A transaction always goes to the next sequential block unless a TRANSFER block is encountered. This block is depicted by the following symbol.

TRANSFER

The selection factor S specifies a decision rule used to determine which of two specified paths is to be taken. There are several selection rules. The simplest rule is to set S to a three–digit fraction that specifies the percentage of time that the second of the two specified paths is to be taken. Of course, the first of the two paths is taken the remainder of the time. Normally the path selection is made at random, so that only the long run frequency (or steady–state behavior) approaches this specified percentage.

With these concepts, it may be useful to assess GPSS in terms of the criteria of Section 13.2. We will then give a simple example showing the application of GPSS to the modeling of a system.

GPSS provides eight random number generators, RN1–RN8; each provides a source of standard uniform pseudorandom numbers. To allow for the generation of nonuniform random numbers, the concept of a function is provided. The function accepts the output of one of the random numbers as its input and provides as output a random variate from the appropriate distribution. The method used is the **inverse transformation method**, which was discussed in Section 7.4. The programmer must still define the appropriate function but does not have to program a random number generation routine. An example of a function used in this way is included as a part of Example 13.2.

The management scheme used by GPSS more closely resembles the event scan approach than the time increment mechanism. The simulation clock is maintained by the GPSS control program. The progress of transactions through the system is monitored by using two lists; the current events chain and the future events chain. One attribute of every transaction is the block departure time (BDT), or the simulation time at which the transaction is scheduled to depart its current block. All transactions whose current BDT is less than or equal to the current clock time reside on the current events chain; all transactions whose current BDT is greater than the current simulation time reside on the future events chain. Transactions with a BDT earlier than the current clock time can still be on the current event chain because a transaction in the next block could have precluded the first transaction from departing the current block. The two event chains are ordered within each priority class by ascending BDT. The simulation control program scans the current events chain for a transaction move. Once it has selected a transaction, it moves that transaction through the system as far as it can, that is, until the transaction encounters an ADVANCE block or becomes blocked. At this time the control program selects another transaction from the current events chain and repeats the process. Transactions are moved from the future events chain to the current events chain whenever their BDT is reached by the simulation clock. Transactions are moved from the current events chain to the future events chain whenever an ADVANCE block is encountered.

Since GPSS uses a limited number (e.g., 48) of different blocks, no more than that number of events can be simulated. The GPSS control program takes whatever actions are necessary whenever a transaction enters any of the blocks. Thus the programmer does not have to handle any event simulation routines; these are handled automatically. The programmer does have to ensure that the proper blocks are included in the proper sequence to effect the required actions of the model.

Many systems have entities that compete for limited resources. It is this competition for limited resources that results in the formation of queues. In GPSS there are two types of resources for which transactions compete – facilities and storages. A **facility** can be used by only one transaction at a time, whereas a **storage** resource can be used by multiple transactions as long as its maximum capacity is not exceeded. Several controlling blocks are associated with facilities and storages. These four blocks are given in Figure 13.3.

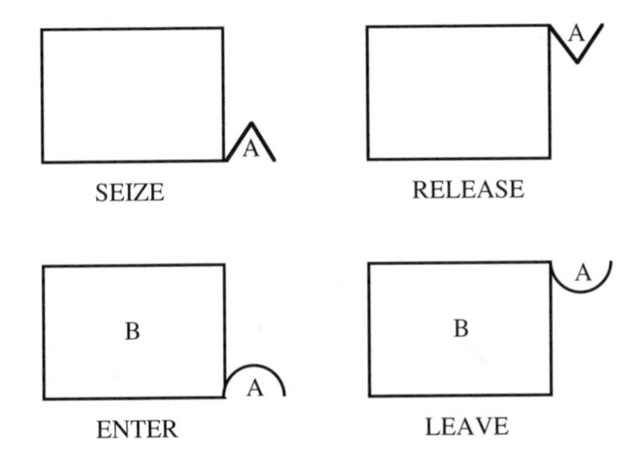

Figure 13.3. Block symbols for managing storages and facilities.

In each block the operand A specifies the number (programmer–assigned) of the facility or the storage involved. The SEIZE and RELEASE blocks allow a transaction to engage or disengage a facility if it is available, while the ENTER and LEAVE blocks serve the identical purpose for a storage. The B operand of the ENTER block specifies the amount of storage capacity that the transaction is requesting. If it is omitted, a default value of one unit is assumed. If transactions attempt to enter a SEIZE block for a facility already in use or an ENTER block for a storage with insufficient residual capacity to satisfy the request, it is prohibited from entering the block. Blocked transactions remain on the current events chain in ascending order of BDT within each priority class. These queues are then serviced on a FIFO basis.

A queueing system could be coded and run in GPSS using only the blocks outlined so far. If this is done, however, no data would be collected or statistics prepared on the system's performance. Four additional control blocks are needed to gather data. These block diagrams are illustrated in Figure 13.4.

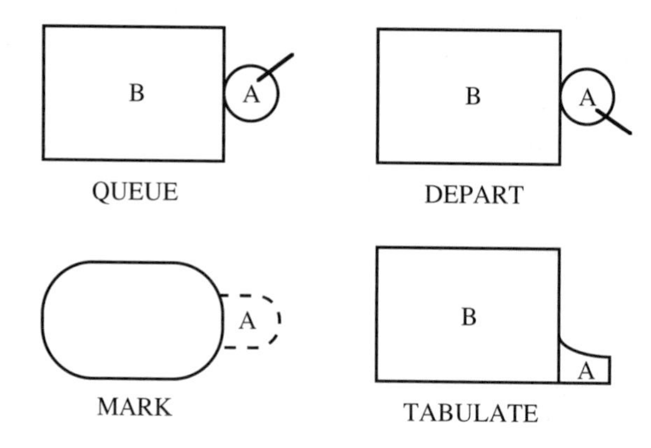

Figure 13.4. Block diagrams for data gathering.

Queues formed by transactions attempting to enter a block are assigned a number or name by the programmer. The QUEUE block increases the size of the queue named by operand A by the amount specified by operand B, with a default of one. The DEPART block decreases the size of the queue in a similar manner. The MARK and TABULATE blocks are used to compute the transit time for a transaction between two points of the model. The time that a transaction passes through the MARK block is noted. When the same transaction traverses the TABULATE block, the transit time between the two blocks is computed and stored in a table named by operand A of the TABULATE block. The TABULATE block can be used without the MARK block. In this case the time computed is the total time elapsed since the transaction was generated.

The QUEUE, DEPART, MARK, and TABULATE blocks are used for gathering statistics. The statistics that are automatically generated by GPSS include the utilization of storages and facilities, lengths of queues (both maximum and average), and the frequency distribution of transit times. The GPSS data–gathering and summarization capabilities are extensive and in most cases provide all the information required.

Output routines in GPSS are precoded and formatted. Thus the results of a GPSS run are output in a report form at the conclusion of the simulation. The programmer therefore need not be concerned with the coding of output routines.

GPSS handles nearly all the details of the basic simulation functions. This is not surprising, since the language was designed for this purpose. These built–in features can drastically reduce the amount of time that must be devoted to coding and debugging a simulation model. This benefit is not gained without cost; most simulation models written in GPSS execute more slowly and hence are more expensive than models written in high–level general–purpose languages such as FORTRAN. Furthermore GPSS is not as widely available as FORTRAN.

GPSS provides much assistance in debugging a simulation model. These aids are not restricted to the assembly phase but extend to execution errors as well. The diagnostic messages are usually descriptive and helpful in eliminating routine errors.

One of the best features of GPSS is its flexibility. Logical changes in a system's operation can normally be effected by changing or replacing one or two statements. This feature is in sharp contrast to a language such as FORTRAN, in which entire routines may have to be rewritten.

Example 13.2 Consider the simulation of a single–channel queueing system. Arrivals occur to the system according to a Poisson distribution. Service–time requirements are exponentially distributed. This model can be expressed in block–diagram form as illustrated in Figure 13.5.

Coding from this block diagram is quite straightforward, as given in Figure 13.6. Most of the actions should be clear from the previous discussion of the blocks. The function denoted XPDIS is a tabular representation of the inverse of the cumulative distribution function. It is used in conjunction with the system's random number generation routines to produce the interarrival and service times for transactions in the model. In line 1 the GENERATE block produces the arrival time for the next arrival by multiplying the mean interarrival rate, in this case 15, by the output of function XPDIS. When the scheduled arrival occurs, that transaction enters QUEUE 1 through the block in line 2. It then attempts to seize facility 1 (the server). Having seized the server, it departs the queue through the DEPART block on line 4. It then enters the ADVANCE block of line 5, where it is delayed an exponentially distributed service time. Control of the server is released in line 6 and departs the system (line 7). The START block controls the length of the simulation. In this case the simulation will proceed until 100 customers have been served. The statistics automatically produced by this run are given in Figure 13.7.

Finally, we must consider the ability of a simulation written in GPSS to represent the real world. The latest version of the language permits part of the simulation model and its data to be stored on random access devices. Therefore, any desired level of simulation detail may be achieved. Efficient use of core storage is implemented through the structuring of data in byte, half word, and full word arrays. Logical modeling can be well represented and implemented using Boolean equations. The mathematical capability of GPSS is adequate for problems that do not require complex equations such as those requiring double precision. When complex mathematical needs do arise, they must be handled outside of GPSS using the HELP block. When there is a large need for this form of mathematical assistance, the HELP block can be set up to call FORTRAN or other programming language routines.

List processing is readily handled with both the SET and CHAIN concepts. These permit complex data to be processed and handled on a FIFO (LIFO) basis, or by any other arbitrary ranking. Any item anywhere in the system may be treated and modified.

The maximum size of the simulation model is determined from primarily a tradeoff between available computer storage and elapsed execution time. When really large scale models need to be constructed, it is necessary to first consider how the model will fit into core storage, what will overflow onto auxiliary memory (e.g., DASD device), and the effect this will have on the simulation model's running time.

Hybrid systems can only be accommodated through the HELP block or the use of special ties between the systems. This leaves open the basic question of whether GPSS or the hybrid system has control of the system. Since there is no guarantee that either will finish its tasks in an allotted time slot, the results could be catastrophic.

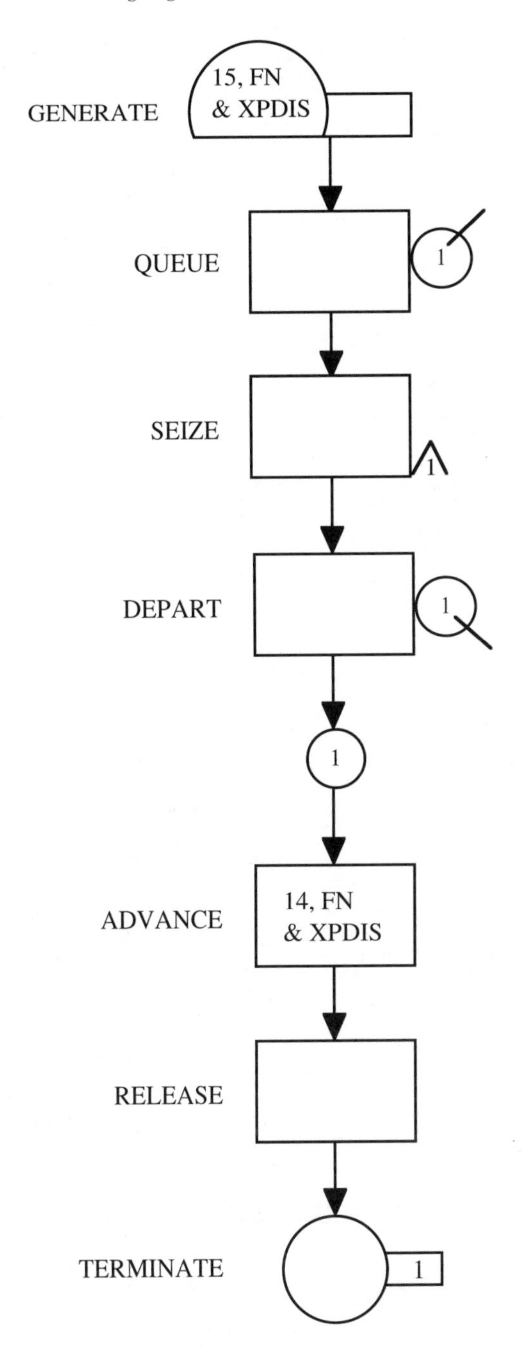

Figure 13.5. GPSS block diagram for the simulation of a queueing system.

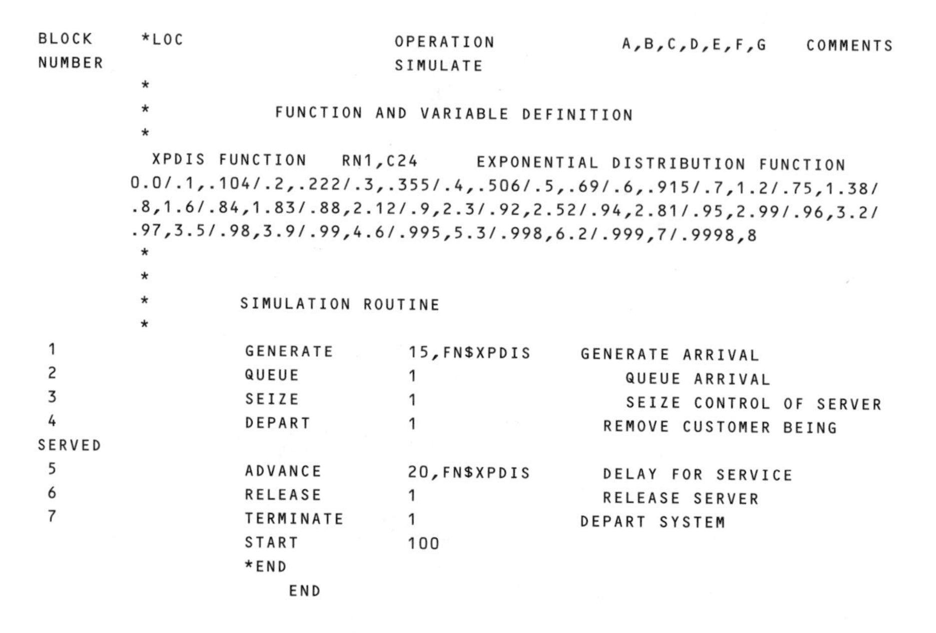

```
BLOCK     *LOC                     OPERATION          A,B,C,D,E,F,G    COMMENTS
NUMBER                             SIMULATE
          *
          *          FUNCTION AND VARIABLE DEFINITION
          *
           XPDIS FUNCTION    RN1,C24      EXPONENTIAL DISTRIBUTION FUNCTION
         0.0/.1,.104/.2,.222/.3,.355/.4,.506/.5,.69/.6,.915/.7,1.2/.75,1.38/
         .8,1.6/.84,1.83/.88,2.12/.9,2.3/.92,2.52/.94,2.81/.95,2.99/.96,3.2/
         .97,3.5/.98,3.9/.99,4.6/.995,5.3/.998,6.2/.999,7/.9998,8
          *
          *
          *        SIMULATION ROUTINE
          *
 1                 GENERATE       15,FN$XPDIS     GENERATE ARRIVAL
 2                 QUEUE          1                   QUEUE ARRIVAL
 3                 SEIZE          1                   SEIZE CONTROL OF SERVER
 4                 DEPART         1                 REMOVE CUSTOMER BEING
SERVED
 5                 ADVANCE        20,FN$XPDIS       DELAY FOR SERVICE
 6                 RELEASE        1                 RELEASE SERVER
 7                 TERMINATE      1               DEPART SYSTEM
                   START          100
                   *END
                       END
```

Figure 13.6. GPSS program listing for the simulation of a queueing system.

```
FACILITY     AVERAGE          NUMBER         AVERAGE         SEIZING       PREEMPTING
           UTILIZATION        ENTRIES        TIME/TRAN       TRANS. NO.     TRANS.
NO.
   1          .999              100           19.289

QUEUE     MAXIMUM       AVERAGE        TOTAL        ZERO       PERCENT     AVERAGE
          CONTENTS      CONTENTS       ENTRIES     ENTRIES      ZEROS      TIME/TRAN
  1         46          24.798          146           1          .6         330.191

&AVERAGE        TABLE       CURRENT
TIME/TRANS      NUMBER      CONTENTS
 332.468                      46

$AVERAGE TIME/TRANS=AVERAGE TIME/TRANS EXCLUDING ZERO ENTRIES
```

Figure 13.7. Statistics gathered by the GPSS model.

13.3.2 Special–purpose Languages: SIMSCRIPT II.5

SIMSCRIPT is probably the second most commonly used special–purpose programming language developed for simulation modeling. The original version of SIMSCRIPT was developed at the RAND Corporation in the early 1960s by Markowitz, Karr, and Hausner (13.6). Although a number of versions of

SIMSCRIPT have been produced, the features of the language described in this section are specifically those of SIMSCRIPT II.5. Unlike GPSS, SIMSCRIPT approaches a general–purpose language and can therefore be taught to the beginning programmer as a first language. The language possesses powerful simulation verbs, however, by which it is possible to simulate highly complex systems. The language is free format and similar to English in structure, which enhances both the learning of the language and the readability of the finished model.

SIMSCRIPT II.5, as described by CACI (13.2), is organized into five separate levels, supposedly structured to aid the beginning programmer in learning the language. Level 1, when considered alone, is a simple language designed to introduce the beginner to programming. Level 5 is a self–contained simulation language. This discussion concerns the features of level 5.

The terminology involved in SIMSCRIPT is consistent with that introduced earlier. The objects of the simulation model are called **entities**, which are characterized by a fixed collection of parameters called **attributes**. **Sets** are collections of individual entities having common properties. The state of the system at any given time is completely described by the current list of individual entities, their attributes, and set memberships. **Temporary entities** are created and destroyed during the course of the simulation, while **permanent entities** remain throughout. Changes in the system state at discrete points in system time are called **events**. When an event is scheduled, an **event notice** is created that causes the transfer to the appropriate event routine. This language distinguishes between exogenous and endogenous events. Those events which are generated internal to the simulator are endogenous events, and those events created outside of the simulation framework are exogenous events. Each event that is desired for the simulation model requires the construction of a separate small event subroutine.

A SIMSCRIPT program is generally composed of three parts: the preamble, the main program, and the appropriate event routines. The preamble contains a description of all data structures used in the model. Each entity must be named and described by listing its attributes. The computational mode is also specified in the preamble. The default mode is single–precision real arithmetic. Example 13.3 shows the types of information contained in the preamble. The main program consists of initialization of all entities, plus the instructions used to control the simulation. The instructions are like English in structure and hence self–explanatory most of the time. No attempt will be made in this brief introduction to include the many instructions available in SIMSCRIPT. The purpose of this section is only to provide a flavor of the language. There must be an event routine for each event named in the preamble. The event routine effects the appropriate changes in the system state called for by the occurrence of the event.

In assessing how SIMSCRIPT supports the basic simulation functions, note that the programmer must write more of the routines than was necessary with GPSS. In GPSS the programmer was limited to the predefined blocks. The action required by the use of each block was well defined entirely by the system. This is not the case with SIMSCRIPT.

SIMSCRIPT supports the generation of random numbers by supplying ten independent streams. The technique used in generating each stream is the **multiplicative congruential method** outlined in Section 7.2.2. Each of the ten generators uses the same multiplier, $a = 14^{29}$, and the same modulus, $m = 2^{31} - 1$. The difference in the generations is in the seed that is used. The seeds that are automatically supplied by the system are as follows:

SEED.V(1) =	524267	SEED.V(6) =	1157240309
SEED.V(2) =	683743814	SEED.V(7) =	15726055
SEED.V(3) =	964393174	SEED.V(8) =	48108509
SEED.V(4) =	1217426631	SEED.V(9) =	1797920909
SEED.V(5) =	618433579	SEED.V(10) =	477424540

Users can supply their own seeds if they wish. This may be desirable if one is performing replications of the same basic experiment. **Access** to the random number generators is by a function call. For example, if a uniformly distributed number is desired, the call would appear as

RANDI.F (A, B, N)

The first two parameters specify the range desired (for a standard uniform variate, $A = 0$ and $B = 1$) while the third parameter specifies the random number stream (N can range from 1 to 10). Generation of non–uniform deviates is also supported for most of the common distributions. For example, an exponentially distributed random variable can be obtained from a call such as

EXPONENTIAL.F (A, N)

where A is the mean of the distribution, and N specifies the random number stream.

SIMSCRIPT uses the event scan method of time management. When an event is scheduled by using the SCHEDULE statement, an **event notice** is created. The existing event notices are scanned, and the earliest event is selected for execution. The simulation clock is then advanced to that scheduled time, and the appropriate execution routine invoked. To reflect the instantaneous nature of an event occurrence, the simulation clock is not advanced during the time an event routine is executing. The current value of the simulation clock can be accessed by way of the system variable TIME.V. This variable measures the simulation time in days; thus if another unit of time is to be used, the value of TIME.V has to be scaled.

An event routine has to be defined for each possible event. The types of event notice to be created are named in the preamble. The event routines are given the same names and are simply subprograms to effect the given changes to the system state. Example 13.3 provides a sample of an event routine.

If a queueing system is to be modeled, the preamble must list that the system has a queue, as well as the queueing discipline that is to be used. These tasks are accomplished through the two statements

```
THE SYSTEM OWNS THE QUEUE
DEFINE QUEUE AS A FIFO SET
```

Most of the standard queueing disciplines are supported. As with GPSS, the user does not have to be concerned with the management of the queue; it is handled by the system.

The data collection and analysis features of SIMSCRIPT are quite flexible. Data is collected by the use of statements such as TALLY and ACCUMULATE. Statistics are computed using the COMPUTE verb. SIMSCRIPT provides a number of standard functions that support statistical analysis, including

AVERAGE
MEAN
SUM
VARIANCE
STD.DEV
SUM.OF.SQUARES
MEAN.SQUARE

The uses of some of these functions appear in Example 13.3.

The input–output features of SIMSCRIPT are also flexible. The language allows for free format of input, with data items separated by one or more blanks. Output is produced by the PRINT statement, which lists the variables to be printed along with the format. Thus the programmer controls the desired output.

Debugging aids in SIMSCRIPT are not as extensive as in GPSS, because more of the coding is left to the programmer. Syntactic errors are detected and reported by the compiler, just as in FORTRAN and other general–purpose languages. The English–like structure of SIMSCRIPT aids in debugging, since the language can be essentially self–documenting. The ease of debugging in SIMSCRIPT can probably be best described as somewhere between that of FORTRAN and that of GPSS.

SIMSCRIPT does provide some flexibility in model development. Major changes in a model can be made by replacing or modifying event routines. Minor changes can be affected by reading in different data or possibly modifying statements.

SIMSCRIPT was designed as a higher order programming system instead of being merely a simulation language. Thus, it may be used for applications other than simulation, as well as for a broad range of simulation applications. To use the language the simulationist must develop a programming competence in SIMSCRIPT (i.e., a closer understanding as for FORTRAN, rather than GPSS). Because there is no inherent simulation structure in the language, an extensive and detailed problem definition must be developed before any programming can begin. The structure for a specific simulation must be developed by the system designer. Model relationships are set up utilizing the entity–attribute–set relationships.

Statistics obtained either during or after the simulation model execution must also be programmed by the simulationist. While the language allows access to anything at anytime, the specific structure has to be added by the user. Flexibility is tied to the basic subroutine structure and capabilities of the programming approach. Individual subroutines can be compiled separately and added or substituted in the model.

Example 13.3 Consider the single–channel queueing system outlined in Example 13.1. This system can be simulated in SIMSCRIPT using the program given in Figure 13.8.

```
PREAMBLE
NORMALLY, MODE IS INTEGER
EVENT NOTICES INCLUDE ARRIVAL AND CLOSING EVERY SERVICE.END HAS A CUSTOMER
TEMPORARY ENTITIES
EVERY PERSON HAS AN ARR TIME, AND MAY BELONG TO THE QUEUE
DEFINE ARR.TIME AS A REAL VARIABLE
THE SYSTEM OWNS THE QUEUE
DEFINE QUEUE AS A FIFO SET
DEFINE IDLE, NRCUST AS VARIABLES
DEFINE LAMDA, MU AS REAL VARIABLES
DEFINE SYSTIME, QTIME AS DUMMY REAL VARIABLES
ACCUMULATE LQ AS THE AVG OF N.QUEUE
ACCUMULATE L AS THE AVG OF NRCUST
TALLY WQ AS THE AVG OF QTIME
TALLY W AS THE AVG OF SYSTIME
END

MAIN
PRINT 1 LINE THUS
SINGLE CHANNEL QUEUEING SYSTEM EXAMPLE 13.3
SKIP 3 OUTPUT LINES
PRINT 2 LINES THUS
ARRIVAL  SERVICE  LENGTH   LQ    L    WQ    W
RATE     RATE     OF SIM
READ LAMDA, MU, LEN.SIM
LET IDLE=1
SCHEDULE AN ARRIVAL NOW
SCHEDULE A CLOSING IN LEN.SIM HOURS
START SIMULATION

PRINT 1 LINE WITH LAMDA, MU, LEN.SIM, LQ, L.
WQ*HOURS.V*MINUTES.V, AND W*HOURS.V*MINUTES.V THUS *.**/HR   *.**/HR   *
       *.**   *.**    *.**MIN  *.**MIN
STOP
END

EVENT ARRIVAL SAVING THE EVENT NOTICE
CREATE PERSON
LET ARR.TIME(PERSON)=TIME.V
IF IDLE=0 FILE PERSON IN QUEUE
       GO NEXT.ARRIVAL
ELSE
```

Figure 13.8. SIMSCRIPT program listing for the simulation of a queueing system.

```
LET IDLE=0
SCEDULE A SERVICE.END(PERSON) IN EXPONENTIAL.F(1./MU,1)HOURS
'NEXT.ARRIVAL'
RESCEDULE THIS ARRIVAL IN EXPONENTIAL.F(1./LAMDA,2)HOURS
RETURN
END

EVENT SERVICE.END(PERSON)
LET SYSTIME=TIME.V ARRTIME
DESTROY PERSON
LET NRCUST=NRCUST 1
IF QUEUE IS EMPTY, LET IDLE=1
       RETURN
ELSE
REMOVE FIRST PERSON FROM QUEUE
LET QTIME=TIME.V ARRTIME
SCHEDULE A SERVICE.END(PERSON) IN EXPONENTIAL(1./MU,1)HOURS
RETURN
END

EVENT CLOSING
CANCEL THE ARRIVAL
DESTROY THE ARRIVAL
RETURN
END
```

Figure 13.8. SIMSCRIPT program listing for the simulation of a queueing system (continued).

This example should be self–explanatory if one keeps in mind the events that are necessary to simulate a queueing system.

Finally, we again consider the ability of a simulation written in SIMSCRIPT to represent the real world. The size of the simulation model that can be contained in core storage at one time can be maximized by using "ragged" or sparse data tables and assembly language code. These require a much clearer understanding of the simulation problem before any programming can be started.

Logical and complex situations can be well represented using the language's Boolean capability. Because SIMSCRIPT is a separate programming environment, its mathematical capability will be installation dependent. Subroutines and interfaces to other high–order languages have to be independently user developed. List processing capabilities are very strong, due to its structured data, and extensive ordered data structures, including FIFO and LIFO, can be easily established.

The maximum size of a simulation model is dependent on the available core storage, the efficient utilization of ragged tables (sparse data tables), the degree of program overlays, and the amount of data packing. Hybrid systems can be designed when either the continuous simulation is in SIMSCRIPT or links are established to other existing languages.

When compared to GPSS, SIMSCRIPT simulation models that are of trivial size require the same effort, those of medium size require more effort, and large models require considerably more effort. In all cases, SIMSCRIPT requires a very well–defined problem statement, well thought–out debugging procedures, and clearly enunciated system constraints.

13.3.3 Special–purpose Languages: GASP IV

GASP IV is the latest in a series of simulation programming languages carrying the GASP name. This version was developed by A. A. B. Pritsker and N. Hurst in the early 1970s. Rather than being an independent programming language in the vein of GPSS and SIMSCRIPT, GASP IV is a package of FORTRAN subroutines used to perform the basic simulation functions, and can be used to write discrete, continuous, or combined hybrid simulation programs. In fact, it was the first simulation language to completely integrate the concepts of discrete event simulation and continuous time varying event simulation under a common framework. As described by Pritsker (13.7) the routines of GASP IV are designed to accomplish the following functions: event control, state variable up–dating, information storage and retrieval, initialization, data collection, program monitoring and event reporting, statistical computation, report generation, and random deviate generation. The GASP IV package to support these functions contains a number of sub–programs (e.g., 34 or more) coded in ANSI standard FORTRAN. No attempt will be made here to describe all these subprograms, since the work of Pritsker (13.7) provides detailed descriptions and program listings. In addition to the GASP IV subprograms, there is an interface for a number of user–defined subprograms. These user–defined modules are used to simulate event occurrences, initialize state variables, provide additional error messages, and provide output in addition to the standard GASP IV output. These routines are represented in the GASP IV package as stubs, or dummy subprograms, eliminating the need for all the routines to be present when only a few are needed.

GASP IV generates standard uniform deviates by a function that employs the **multiplicative congruential scheme** described in Chapter 7. As mentioned by Pritsker (13.7), the effectiveness of the multiplicative scheme is affected by the machine on which GASP IV is implemented. For this reason Pritsker provides a generalized version of the function DRAND, as well as two specialized versions for CDC and IBM equipment. The GASP IV package includes, in addition to the function DRAND, functions to produce deviates from the uniform, triangular, normal, lognormal, Erlang, gamma, beta, and Poisson distributions. An exponentially distributed deviate can be obtained from the function that produces Erlangian distributed variates.

GASP IV, when used to model discrete systems, uses the event scan method of time management. The scan of the future events list, as well as the advancing of the simulation clock, is accomplished by a system subprogram called GASP. Because the language can also be used to model continuous as well as hybrid systems, a modified method of time advance is used. Events are described in terms of the mechanism by which they are scheduled. Those events occurring at a specified projected point in time are referred to as "**time events**" (i.e., analogous to events used in next–event simulations), while events that occur when the system reaches a particular state are called "**state events**." Unlike time–events they are not scheduled in the future, but occur when state variables satisfy prescribed conditions. State events can initiate time events, and time–events can initiate state events.

The user must supply routines to simulate the occurrence of a given event. A subroutine called EVNTS is defined to handle all events, with a computed GO TO statement used to invoke the appropriate routines.

Since GASP IV is composed of FORTRAN subprograms, the management of queues is just as handicapped in GASP IV as it is in FORTRAN. The lack of an efficient list–processing capability in FORTRAN makes queue handling somewhat inefficient. As with FORTRAN, the user must assess this disadvantage as it applies to the particular model.

GASP IV provides an extensive and flexible data collection computation, and reporting capability. Several subprograms support these functions. See Pritsker's work for the details of these routines (13.7). The user may define a subprogram, subroutine OTPUT, to provide output other than that automatically provided by GASP IV. The subroutine exists in stub form, so the user need only supply the particular logic.

GASP IV, composed of FORTRAN routines, enjoys many of the advantages of FORTRAN. GASP IV can be used at any installation equipped with a FORTRAN compiler. Programmers who are proficient in FORTRAN will need only to study the overall makeup of GASP IV and its naming conventions to be able to use the language.

In addition to the diagnostic capabilities provided by the FORTRAN compiler (compile–time errors), GASP IV possesses a subroutine ERROR to assist in debugging the simulation model. This subroutine is called whenever an illogical condition is detected. The routine identifies the type of error and provides a snapshot of the system state at the time that the error was detected. The user can define a subroutine UERR which can be written to provide any other information desired.

GASP IV, then, provides a great deal of assistance in the development of a simulation model. It retains many of the advantages of its parent language, FORTRAN, while at the same time relieving the programmer of the responsibility for coding many of the common routines. Because of the overall size of the GASP IV package, no example of its use is given here. See Pritsker (13.7) for more details of the language and numerous examples of its use.

13.4 Summary

In this chapter we have attempted to explain why new languages were designed for programming simulation models. We described functions common to many models; the generation of random variates, management of simulation time, simulation of event occurrences, queue management, data collection, summarization and analysis of data, and the formatting and printing of output data. In the later sections of this chapter we showed how three of the more popular simulation languages accomplished these functions. These three languages were selected because of their widespread use in this country. A fourth language,

SIMULA, widely used in Europe and based on ALGOL, has failed to gain widespread acceptance in this country outside of the academic community.

The choice of a suitable programming language is an important consideration in the development of a simulation model. The choice must be made based on the characteristics of the individual project. Considerations such as the languages supported at the installation, the programmer's level of proficiency, and the complexity of the model being developed all have an impact on this decision.

13.5 Exercises

13.1 Customers arrive at a barber shop according to a Poisson process at an average of five per hour. The shop is open from 8:00 a.m. to 5:00 p.m., and customers waiting at the time the shop closes are served. There is a single barber, who can give a haircut in an average of 15 minutes. This service time can be assumed to be exponentially distributed. The shop has chairs to seat ten customers in addition to the one being served. Customers who arrive and find the shop full leave. Develop a FORTRAN simulation model of this system and use it to simulate one day's activity. How many customers are lost during the course of the day? How many customers remain in the shop at closing time?

13.2 Develop a GPSS simulation model for the system described in Exercise 13.1.

13.3 Develop a SIMSCRIPT simulation model for the system described in Exercise 13.1.

13.4 Develop a GASP IV simulation model for the system described in Exercise 13.1.

13.5 Compare and contrast the models developed in Exercises 13.1–13.4.

13.6 Suppose the barber operating the shop described in Exercise 13.1 purchases seating for five additional customers. How does this affect the average number of customers waiting, the number of customers lost, and the number remaining at the time the shop closes?

13.7 Jobs arrive at a monogrammed computer facility according to a Poisson process of an average of 500 per hour. The CPU–time requirements are exponentially distributed with an average of 10 seconds. The shortest–job–first dispatching scheme is used. Develop a FORTRAN simulation model for this system, and collect statistics on the average number in the system and the average waiting time.

13.8 Develop a GPSS model for the system described in Exercise 13.7.

13.9 Develop a SIMSCRIPT model for the system described in Exercise 13.7.

13.10 Develop a GASP IV model for the system described in Exercise 13.7.

13.11 Suppose that the jobs arriving to the system are assigned one of three priorities, priority 1 being the highest, and that the total job mix is distributed as follows: 20% in priority 1, 25% in priority 2, and 55% in priority 3. As the CP finishes one job, it is assigned to the highest priority job that is waiting. The tie–breaking discipline within each priority class is shortest– job–first (SJF). Develop a GPSS model for this system and simulate one hour of operation.

13.12 Develop a SIMSCRIPT model for the system described in Exercise 13.11.

13.13 Change the program of Example 13.1 to allow multiple servers. Investigate the impact of multiple servers on factors such as the average queue size, average time in the system, and average time in the queue.

13.14 Using one of the languages surveyed in this chapter, design and implement a model to simulate the operation of a banking facility. Why did you choose that particular language?

13.15 If you were to design a simulation programming language, what features would you include? What general–purpose language would you base it on and why?

13.6 References

13.1 Bobiller, P. A., Kahan, B. C., and Probst, A. R., *Simulation with GPSS and GPSS V*, Englewood Cliffs, NJ: Prentice–Hall, 1976.

13.2 Consolidated Analysis Centers, Inc., *SIMSCRIPT II.5 Reference Handbook*, Santa Monica, CA: 1971.

13.3 Emshoff, J. R., and Sisson, R. L., *Design and Use of Computer Simulation Models*, New York: Macmillan, 1971.

13.4 Gordon, G., "A General Purpose Systems Simulation Program." in *Proc, EJCC*, Washington, D.C. New York: Macmillan, 1961.

13.5 Gordon, G., *System Simulation. 2d ed.* Englewood Cliffs, NJ: Prentice–Hall, 1978.

13.6 Markowitz, N. M., Karr, H. N., and Hausner, B., *SIMSCRIPT:A Simulation Programming Language*, Englewood Cliffs, NJ: Prentice–Hall, 1963.

13.7 Pritsker, A. A. B. *The GASP IV Simulation Language*. New York: John Wiley and Sons, 1974.

14

DISTRIBUTED SIMULATION

Simulations are among the most computationally intensive computer applications. Complex real–world problems often lead to complex models that generate high volumes of data. Thus, the need for distributed simulation systems is growing as the modeling of real world systems seem to require intractable mathematical processes.

14.1 The System Simulation Problem

Typically, simulations have been carried out as a repetition of a set of sequential steps. This process is extremely slow in the case of complex discrete event systems and is proving to be inadequate because of the magnitude of the problem.

One solution is to take advantage of distributed computer systems and high bandwidth channels by partitioning the simulation problem and executing parts in parallel. The best way to assess the advantages of the distributed system is in terms of "acceleration".

Acceleration can be defined as the time it takes a single processor to perform a simulation divided by the time it takes a multiprocessor system to perform that same simulation. This is the equivalent to the effective number of processors used for the simulation . The ideal speedup for N processors is N. The real acceleration is often difficult to obtain because the time required for a uniprocessor to run a large complex simulation is often impractical and not precisely determined. An alternative is to estimate how long it would take a uniprocessor to run a "distributed simulation". This can be approximated by measuring the total busy time of all the processors during the distributed simulation run. The resulting simulation acceleration is greater than the actual speedup obtainable by distributing the simulation (14.8).

Another measure can be the efficiency of the distributed simulation. **Efficiency** is defined as acceleration divided by the number of processors and is, thus, a measure of the effective utilization of the processors.

14.2 Decomposition of a Simulation

The decomposition of a simulation for effective use by multiple processors is key to the successful implementation of a distributed discrete event simulation system. There are five fundamental approaches to simulation decomposition. A sixth approach involves a hybrid combination of the previous five fundamental techniques. Each approach has its own strengths and weaknesses with regards to the exploitation of concurrency and parallelism and each poses unique challenges for synchronization.

The six possible approaches for simulation decomposition are as follows (14.8):

a. Parallelizing Compilers

This method uses a parallelizing compiler to find sequences of code in a sequential simulation that can be processed in parallel and on separate processors. Parallelizing compilers are transparent to the user and could be used to convert much of the sequential simulation software in existence today. However, such compilers ignore the inherent structure of the problems present in existing simulation models, and thus, only a small portion of the naturally occurring parallelism would be identified and rescheduled.

b. Distributed Experiments

An intuitive approach is to run independent replications of the same simulation on N processors and average the results. No coordination is required among processors except for the averaging. One could expect to see a virtual acceleration of N with N processors. The individual replication of the simulation on separate processors is most effective if the system quickly reaches steady state, and if the simulation runtimes are long.

This approach would also allow simultaneous simulations, each with different parameters. A hierarchical configuration, where low level processors use different parameters, may provide valuable data for optimization or factor screening. Although this method is effective, no single simulation run is accelerated. Therefore, any decisions about simulation model parameters that are to be used for parallel simulation runs must be made *a priori* before actual scheduling. This precludes interactive decision making, especially important for optimization , and promotes sequential decision making.

Given that distributed experiments require that all processors have enough memory to contain the entire simulation program, distributing experiments may not be practical in many cases. One advantage of multiprocessor simulations is to permit, based on time and memory constraints, much larger and more realistic simulations than have been possible on a uniprocessor.

c. Distributed Language Functions

This approach involves the assignment of simulation support tasks (e.g., random variable generation, event set processing, statistics collection, graphics generation, etc.) to various individual, but different, processors. The advantages of distributed language functions is that they avoid the deadlock problem and are transparent to the user. A disadvantage is that it does not exploit any inherent parallelism in the system being modeled. Related research has shown that significant simulation acceleration occurs when only a few processors are used and that the marginal speedup with additional processors drops off rapidly.

d. Distributed Events

The fourth approach is to distribute the scheduling of events from a global event list. Protocols, preserving consistency, are required because currently processed events may affect the next event on the list. This approach is particularly appropriate for shared memory systems because in such an environment the event list can be accessed by all processors.

Distributed events are suitable for a small number of processes or when the components of the system require a large amount of global information.

e. Distributed Model Events

The decomposition of the simulation model into loosely coupled components makes it possible for components to be assigned to a process, where several processes could be run on the same processor. This approach promotes the inherent parallelism in the model but requires careful synchronization.

Synchronization of the processes is usually controlled by message passing. The few primitives required for message–based simulations are constructs to (14.1):

a. **create** and **terminate** processes;

b. **send** messages to processes;

c. **wait** for messages and/or simulation time to elapse.

Such a system is usually modeled as a directed graph where the nodes represent processes and links represent possible interactions or message paths.

For a fixed topology, such as a queueing system, a good decomposition is to assign a process to each station and have the movement of customers be represented by message passing. Furthermore, the possible routes of customers are fixed. Alternatively, both stations and customers can be modeled by processes with messages used to change the customer's states.

For a dynamic topology, such as a battlefield scenario, processes can be used to represent the components that are interacting, such as tanks and aircraft. In this topology, components can move in any direction and can interact with all other components. Messages are used to define the interaction taking place. A more efficient variation of this method is to divide the physical space into regions or sectors and assign a process to a particular region. Messages are then used to describe the interactions among the components in different sectors or the movement of components from one sector to another.

Righter and Walrand (14.8) rate this approach as having the greatest potential in terms of exploiting inherent parallelism in the system. This procedure is especially appropriate for systems that require little global information and control.

f. Combined Approach

The ideal decomposition of a simulation model may involve a combination of the above approaches. Such a combined approach provides a means to tailor the decomposition effort to specific models.

14.3 Synchronization of Distributed Model Components

The distribution of model components requires explicit schemes for synchronization. **Synchronization** refers to the action necessary to ensure that simulated events are processed in their correct simulation time sequence. This sequence may be a partial ordering based on the dependencies of the events. It would allow for multiple sequences that could be considered correct and that individually would have no effect on the correctness of the simulation.

In a distributed system, time synchronization is a critical task. This task is complicated by numerous factors that include such activities as having separate processors advance at different times (e.g., events being simulated at different speeds, differences in hardware, etc.), with each processor maintaining a local clock. The resulting order of events is only a partial ordering and thus it may be difficult, if not impossible, to determine the precedence of the two events (14.4).

As stated previously, each processor has a local clock maintaining perceived local correct time. This physical clock is of little importance to the distributed simulation. In the distributed system, **global time** is based on the contents of time–stamped messages and frequently described in terms of an upper and lower bound rather than an absolute time (14.4). The larger the network becomes, the greater the difficulty in determining the global time.

Schemes for synchronization are influenced by the way the simulation model advances time. Our concern is with **simulation time**, which is an abstraction of real time. States of events in the real world with real times have corresponding simulated event states with simulated times.

Based on the simulation method, time is advanced in fixed increments (as in time–driven simulation) or moved from one event to the next (as in event–simulation). Another consideration is whether the simulation is synchronous (a global clock with all processes having the same simulated time) or asynchronous (each process having a local clock where the simulated time for different processes may be different).

a. Time–Driven Simulation

In **time–driven simulation**, the simulation time values are an increasing arithmetic sequence, $\Delta s = c$, where s denotes simulation time and c is a constant such that $c > 0$ (14.7). Simulated time advances in fixed increments called ticks. **Clock ticks** must be short to ensure accuracy. Longer simulation times can result when short clock ticks are used, because nothing may be happening during the short tick interval.

In a synchronous simulation, the simulation at a tick must be completed before any other process can begin. The simulation at a tick usually proceeds in two phases: a simulation or computation phase followed by a state update and communication phase.

In an asynchronous simulation, a process can begin simulating at the next tick as soon as its predecessors have finished at the last tick. The asynchronous system allows a greater degree of concurrency, but has an increased communications cost resulting from passing messages. This overhead can be avoided in a synchronous system, once the global clock is synchronized, because messages signaling an interaction or a state update are the only messages that can subsequently be transmitted.

On the surface it may seem that time–driven simulations are less efficient than event–driven simulations because there may be ticks during which no events occur. However, time–driven simulations may avoid the overhead required for synchronization in both synchronous and asynchronous event–driven simulations. Time–driven simulations are particularly appropriate for modeling continuous time systems (e.g., dynamic topologies such as a battlefield), and may be made more efficient by tailoring the interval of the ticks to the level of activity occurring in the system. For example, the tick interval may be lengthened when there is little activity.

b. Event–Driven Simulation

In **event–driven simulation**, the time sequence still increases monotonically; however, not in an arithmetic sequence (that is, $\Delta s \geq 0$ (14.7)). Event–driven simulations increment time based on the occurrence of an event that represents a change in state. Thus, a greater potential for acceleration exists in event–driven simulations than in time–driven simulations. As with time–driven simulations, the computation may occur either synchronously or asynchronously.

Synchronous event driven simulation is controlled by a global clock that may be centralized (with a dedicated process to act as a synchronizer) or distributed. The global clock is set to the minimum time of the next event for all processes.

Asynchronous event driven simulations spend less time waiting for other processes, thus have the potential for the highest performance. Events that have no effect on each other can be processed simultaneously even if they occur at different simulated times. This leads to non–determinism with respect to the activities of the processes yielding a particular simulation trace. **Time–stamped message passing** is the primary means for synchronization of the events.

Implementations vary from conservative to optimistic. **Conservative** implies that events are processed in a manner that never violates the correct chronology. On the other hand, **optimistic** implies that a process clock may run ahead of incoming activities, resulting in errors in chronology (**time warp**). In the time warp case, simulation time must be rolled back to correct for the errors.

Conservative approaches require that messages may only be transmitted in chronological order according to their time–stamp. A process P2 may update its local clock if there are unprocessed messages on each input link. The clock will be updated to the minimum time–stamp of the waiting messages. Since, messages are received chronologically along each link, it is not possible to receive a message with an earlier time. However, if process P2 does not have an unprocessed message from process P1 on an input link, the local clock cannot be updated because it is possible that process P1 would send a message with an earlier time from the local time. Therefore, process P2 is forced to wait for process P1 to send a message before process P2 can update its local clock. This can lead to deadlock, thus avoidance schemes will be discussed next.

An **optimistic approach** to asynchronous distributed simulation has been termed time warp. **Time warp** represents an **object–oriented view of simulation** in which the real system objects are modeled as logical objects and the interactions between the objects are represented by the transmission of **time–stamped event messages** (14.2). The time warp algorithm requires two time–stamps: a **send time–stamp** and a **receive time–stamp**. An object may simulate forward with other objects until it receives an event message with a time–stamp from its past. This inconsistency in time is corrected by the use of **rollback** or **antimessages**. The object is rolled back to the earlier time where the inconsistency occurred, with the goal of canceling any unwanted side effects resulting from this earlier event. If an event A causes event B, then the execution of A and B must be scheduled in real time so that A is completed before B starts (14.3).

The rollback mechanism requires that every object contain local state information. Every object in the logical system has its own clock called the **Local Virtual Time (LVT)**. A rollback occurs when the time of any received message is less than the LVT of the receiving object. The system is rolled back to the **Global Virtual Time (GVT)** rather than to the beginning of the simulation. GVT is defined to be the LVT of the object farthest behind at any snapshot point. The farthest behind object has the minimal LVT of all the objects with at least one unprocessed message in its input message queue (14.2).

14.4 Deadlock Resolution in Distributed Simulations

As with any system of shared resources, deadlock is a problem that must be resolved. Numerous deadlock algorithms are in existence and each has its own unique features and limitations. Table 14.1 shows a few algorithms that are described by Mukesh Singhal (14.9). However, it is beyond the scope of this chapter to cover these algorithms in any detail. An overview of a few of the more important algorithms is provided below.

One approach suggested by Misra (14.6) is the use of null messages as a mechanism to avoid deadlock. A **null message** (t, m) is used to announce the absence of messages. Reception of a null message is treated as any other message. The purpose is to guarantee that any future messages from the node sending the null message will have a time component greater than t. This technique is quite efficient for acyclic networks. Factors affecting the efficiency of this algorithm, to general networks, include the following (14.6):

a. Degree of Branching in the Network

The number of paths between the source and the sink is a (rough) measure of the amount of branching in the network. Null messages tend to get created at branches, and thus the greater the number of branches, the greater the reduction in efficiency. Proposed topology insensitive algorithms are accompanied by other difficulties such as the danger of cascading rollbacks, or a low degree of achievable parallelism (14.5).

Table 14.1 Performance Comparison of Various Distributed Deadlock Detection Algorithms

Algorithm	Number of Messages	Delay	Message Size
Goldman	$<m \bullet n$	$\tau + nT$	Variable (medium)
Isloor–Marsland	$r(N-1)$	0	Constant (small)
Menasce–Muntz	$m(n-1)$	nT	Variable (small)
Obermarck	$m(n-1)/2$	nT	Variable (medium)
Chandy et. al.	$<m \bullet n$	$\tau + n$ T	Constant (small)
Haas–Mohan	$m(n-1)$	$\tau + (n-1)T$	Variable (medium)
Sugihara et. al.	$<m \bullet n$	$(n-1)T$	Constant (small)
Sinha–Natarajan	best = $2(n-1)$ worst = $m(n-1)$	$2(n-1)T$	Constant (small)
Mitchell–Merritt	$m(n-1)$	$(n-1)T$	Constant (small)
Bracha–Toueg	$4m(n-1)$	$4dT$	Variable (medium)

N: number of sites
n: number of sites in deadlock cycle
m: number of processes involved in deadlock
T: intersite communication delay
τ: deadlock initiation delay
r: Transaction–Wait–For (TWF) graph update rate
d: diameter of TWF graph

b. Time–Out Mechanisms to Prevent Null Message Transmissions

A null message has no effect if it is followed immediately by a message t' such that $t' > t$. Therefore, it may be more efficient to wait some period before sending the null message. An active area of research is to determine such an optimum time–out period.

c. Amount of Buffering on Channels

The size of the buffer on a particular channel affects performance substantially. If the buffer size is reduced to zero, the sender may spend a large time waiting to send messages. Buffer management is enhanced by the application of the rule that any message put in a buffer after a null message annihilates the null message if it is still in the buffer. This is possible because the later message has a higher t component.

Another approach is to allow the deadlock to happen and to have a mechanism for detecting and recovering from an occurrence. One such approach is the implementation of a marker that continuously circulates in a network. The marker is a special type of message that begins at some logical process. The marker is then passed around the network. Each logical process has a one bit flag to show whether the process has sent or received a message since the marker's last visit. The process is said to be "white" if no such messages have been sent or received, otherwise the process is "black". Initially, all logical processes are black. The marker declares deadlock when it finds that the last N logical processes it has visited were all white, where N is the number of channels in the network (14.6).

The marker also carries the minimum of "next–event times" for the white logical processes it visits. Therefore, when the marker detects deadlock, it knows the next event time and the logical process at which this next event occurs. The process can then be restarted.

14.5 Summary

Distributed simulation is a relatively new area of investigation. Many algorithms have been proposed, but relatively little work has been done in terms of performance evaluation. Many issues must be addressed before unqualifyingly presenting distributed simulation as an alternative to uniprocessor simulation; however, the outlook appears promising.

According to Misra (14.6), the most important problem in distributed simulation is the empirical investigation of various heuristics on a wide variety of programs to establish (1) which heuristics work well for which problems, and on which machine architectures, (2) how to partition the physical system among a fixed set of processors, and (3) how to set simulation parameters such as time–outs and buffer sizes.

14.6 References

14.1 Bagrodia, R. L., Chandy, K. M., and Misra, Jayadev, "A Message–Based Approach to Discrete–Event Simulation". *IEEE Transactions on Software Engineering*, Vol. 13, No. 6, June 1987: 664–665.

14.2 Chandrasekaran, U. and Sheppard, S., "Discrete Event Distributed Simulation: A State of the Art Survey", *Computer Science Technical Report TAMUDCS: 87–005*, March 1987.

14.3 Jefferson, D. R., "Virtual Time", *ACM Transactions on Programming Languages and Systems*, Vol. 7, No. 3, July 1985, pp. 404–425.

14.4 Lamport, L., "Time, Clocks, and the Ordering of Events in a Distributed System", *Communications of the ACM*, Vol. 21, No. 7, July 1978, pp. 558–565.

14.5 Lubachevsky B. D., "Efficient Distributed Event–Driven Simulations of Multiple–Loop Networks, *Communications of the ACM*, Vol. 32, No. 1, January 1989, pp. 111–131.

14.6 Misra, J., "Distributed Discrete–Event Simulation", *Computing Surveys*, Vol. 18, No. 1, March 1986, pp. 39–65.

14.7 Peacock, J. K., Wong, J. W., and Manning, E. G., "Distributed Simulation Using a Network of Processors", *Computer Networks*, Vol. 3, 1979, pp. 44–56.

14.8 Righter, R. and Walrand, J. C., “Distributed Simulation of Discrete Event Systems”, *Proceedings of the IEEE*, Vol. 77, No. 1, 1989, pp. 99-113.

14.9 Singhal, Mukesh, "Deadlock Detection in Distributed Systems", *IEEE Computer*, Vol.22, No. 11, November 1989, pp. 37–47.

15

QUEUEING THEORY AND SIMULATION

Waiting lines, or queues, are encountered in nearly every facet of life. Queues range from waiting lines at the barber shop, supermarket, or filling station to a backlog of messages at a communication center or jobs at a computing center. The reason that waiting lines form is quite simple: there are simply not enough serving facilities (or servers) to satisfy all the customers simultaneously. The reason for an adequate number of servers is simple economics. Customers seem to arrive at random; thus to guarantee that there will be no waiting lines, the service station manager would have to hire as many servers as there are customers. This is not economically feasible, and hence a fixed number of servers are normally hired with the hope that the waiting lines do not become intolerably long. Should the customers become discouraged and leave before being served, the manager would want to hire more servers to avoid losing business.

Waiting lines or queues are so common in real life that it should not be surprising that one of the most common problems encountered in modeling or simulating the operation of a system involves queues. In fact, the requirements of queue handling led to the development of simulation languages.

A queueing system is a system in which customers arrive, wait if that service is not immediately available, receive the necessary service, and then depart. A simple queueing system is illustrated in Figure 15.1.

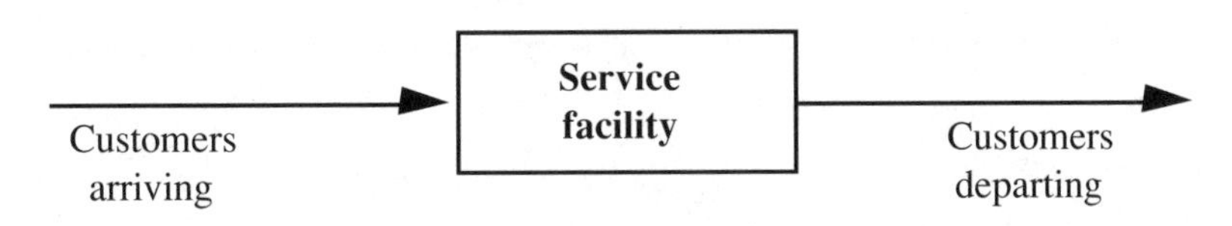

Figure 15.1. A queueing system.

There are a number of characteristics of any queueing system that need to be discussed.

1. The **arrival pattern** concerns itself with the distribution of arriving customers, whether customers are allowed to balk (leave without receiving service), and whether customers arrive singly or in batches. A simple arrival pattern is one that is deterministic or devoid of all uncertainty. A more general, and more common, pattern includes some uncertainty.

2. The **service process** considers such details as the distribution of service time requested, whether customers are served singly or in batches, and whether the level of service changes or remains constant as the queue forms. The server could change its service depending on the length of the waiting line that is forming or as the demand requires.

3. The **queue discipline** considers the technique by which customers are selected from the queue for service. Probably the most common queue discipline is FIFO (first-in, first-out) in which the customer at the head of the line is selected for service. Other disciplines include LIFO (last-in, first-out) and priority schemes in which customers other than those at the head of the queue (line) may be selected for service.

4. The **system capacity** in most systems is finite. For instance, the length of the queue in a barber shop may be limited by the number of chairs available. The maximum backlog of a message-switching center, on the other hand, may be limited by the amount of buffer (reserve) space available to hold these messages. The maximum capacity of a system has a pronounced effect on the operation of the system; hence when the operation of the system is being simulated or otherwise modeled, these capacities must be considered.

5. The **number of parallel servers** varies. In some systems several servers simultaneously serve customers. These servers can all select customers from a single queue or each can service its own queue. Regardless of the queue mechanism, the servers are normally considered to be operating independently of one another.

To summarize these characteristics of queueing systems, Kendall (15.2) developed a widely accepted notational convention. With this convention, a queueing system is described by a series of symbols separated by slashes.

$$A/B/C/D/E$$

In this notation A represents the interarrival time distribution, B the service time distribution, C the number of parallel servers, D the system capacity, and E the queue discipline. Some of the more common interarrival distributions are M (exponential), D (deterministic), E_k (Erlang type k), and G (general). The same letters are also used to denote similar service time distributions. Queueing disciplines are indicated by FIFO (first-in, first-out), LIFO (last-in, first-out), SIRO (service in random order), PRI (priority), and GD (general discipline). For example, $M/M/1/\infty$/FIFO indicates a single-server system with infinite system capacity, exponentially distributed interarrival times (Poisson-distributed arrivals), exponentially distributed service times, and a first-in, first-out queueing discipline.

Once a queueing model has been developed, the following are the items of interest concerning the model.

1. **Queue length**. Both the maximum and the average queue lengths are useful in characterizing the behavior of a system.
2. **Time in the system**. The expected length of time that a customer will spend in a system is of interest to the analyst as well as to the customer.
3. **Idle and busy time of the server**. Optimal utilization of the service facility is one of the aims of a system designer.

The rest of the chapter surveys techniques useful in the analysis of a queueing system. As the complexity of the queueing model grows, so does the complexity of the analysis; for this reason only relatively simple models are analyzed.

15.1 Review of the Poisson and Exponential Distributions

Many systems have been successfully modeled by a queueing model in which both the interarrival and the service distributions are exponentially disturbed. In Chapter 4 we showed that an exponential interarrival distribution implied that the arrival process is Poisson. Because the application of this model is widespread, the next two sections are devoted entirely to it. The properties of the Poisson and exponential distributions are reviewed in this section, and the equations necessary to analyze the model are developed in the next. The exponential and Poisson distributions are both single-parameter distributions with the following underlying assumptions (15.1).

1. The probability that a customer arrives within a small time interval Δt is $\lambda \Delta t + O(\Delta t)$, where λ is the arrival rate and $O(\Delta t)$ includes all higher-order terms in Δt such that $\lim_{\Delta t \to 0} [O(\Delta t) / \Delta t] = 0$.
2. The probability of two or more arrivals in Δt is $O(\Delta t)$ and hence can be neglected.
3. The number of arrivals in nonoverlapping time intervals is statistically independent.

It is easy to establish that if the arrival process is Poisson, then the interarrival distribution is exponential. Suppose that the arrival process to a queueing system is Poisson, with an arrival rate of λ. Let T be the random variable that measures the time between successive arrivals. Then

$$P(T > t) = P(\text{zero arrivals in time } t) = e^{-\lambda t}$$

from which

$$F(t) = P(T \le t) = 1 - P(T > t) = 1 - e^{-\lambda t}$$

This is precisely the exponential distribution function, proving the supposition. It is somewhat more difficult to prove the converse. The argument of this proof is given in Gross and Harris (15.1).

One of the main reasons the exponential distribution is so widely applicable is that it is memoryless. This property was mentioned in Chapter 4 and is reviewed here. Assume that the random variable X has an exponential distribution. Then for any $X_1, X_2 > 0$, $X_2 > X_1$, $P(\mathbf{X} \le X_2 | \mathbf{X} \ge X_1) = P(0 \le \mathbf{X} \le X_2 - X_1)$. In terms of a queueing model this property states that the probability of an arrival in a given time interval is not affected by the fact that no arrival has taken place in the preceding interval or intervals. This fact greatly simplifies the analysis of queueing systems with Poisson arrivals and exponential services.

The Poisson distribution also possesses a property that is valuable in the analysis of many queueing systems: the **aggregation** and **disaggregation** property.

Consider an arrival process that is the confluence of n independent Poisson arrival streams. Then the combined stream is Poisson with rate $\lambda = \lambda_1 + \lambda_2 + ...\lambda_n$. The converse also holds. That is, suppose a Poisson arrival stream with rate λ feeds n independent streams. Then the ith stream is also Poisson, and it has a rate equal to λP_i, where P_i is the probability that a given customer takes path i.

Recall from Chapter 4 that the Poisson distribution is used to model random events. This idea is formalized as follows. Suppose that the input stream to a queueing system follows a Poisson distribution. Suppose also that an arrival has occurred during the interval $(0, t)$. Then the exact instant of the arrival follows the uniform distribution. That is, the arrival occurs at random in the interval.

Having mentioned some of the more useful properties of the exponential Poisson distribution, we are ready to analyze the first queueing model, the $M/M/1/\infty$/FIFO system.

15.2 The *M*/*M*/1/∞/FIFO System

The $M/M/1/\infty$/FIFO system is a single-server system whose interarrival and service times are exponentially distributed with parameters $1/\lambda$ and $1/\mu$ respectively. There is no restriction on the system's capacity, and the queue discipline is a first-in, first-out discipline.

Of crucial interest to the analysis of any queueing system is the number of customers in the system. Let S_j denote the state of the system when there are j customers present, $j \geq 0$. Let $P_j(t)$ denote the probability of state S_j at some time t. Now the system is in state S_j at time $t + \Delta t$ if and only if one of the following mutually exclusive events occurs.

1. The system was in state S_{j-1} at time t, and one arrival but no departures occur during the interval $(t, t + \Delta t)$.
2. The system was in state S_j at time t, and no arrivals or departures occur during the interval $(t, t + \Delta t.)$.
3. The system was in state S_{j+1} at time t, and one departure but no arrivals occur during the interval $(t, t + \Delta t)$.

Now recall that the probability of a single arrival during the interval $(t, t + \Delta t)$ is $\lambda\Delta t + O(\Delta t)$, while the probability of a single departure during the same interval $(t, t + \Delta t)$ is $\mu\Delta t + O(\Delta t)$. The probability of multiple arrivals or departures during the interval is negligible.

$$\begin{aligned} P_j(t+\Delta t) = {} & P_{j-1}(t)(\lambda\Delta t)(1-\mu\Delta t) \\ & + P_j(t)(1-\lambda\Delta t)(1-\mu\Delta t) \\ & + P_{j+1}(t)(1-\lambda\Delta t)(\mu\Delta t), \qquad j = 1, 2, \ldots \end{aligned}$$

Simplifying gives

$$P_j(t+\Delta t) = \lambda\Delta t P_{j-1}(t) + (1-(\lambda+\mu)\Delta t)P_j(t) + \mu\Delta t P_{j+1}(t), \qquad j = 1, 2, \ldots$$

Now rearranging terms gives

$$\frac{P_j(t+\Delta t) + P_j(t)}{\Delta t} = \lambda P_{j-1}(t) - (\lambda+\mu)P_j(t) + \mu P_{j+1}(t), \qquad j = 1, 2, \ldots$$

Taking the limit of both sides as $\Delta t \to 0$ gives

$$P_j'(t) = \lambda P_{j-1}(t) - (\lambda+\mu)P_j(t) + \mu P_{j+1}(t), \qquad j = 1, 2, \ldots$$

This equation holds for $j = 1, 2, \ldots$ The case $j = 0$ must be handled separately, since in this case S_{j-1} is not possible. Utilizing the same procedure gives

$$P_0'(t) = -\lambda P_0(t) + \mu P_1(t)$$

These equations can be summarized as a set of differential difference equations whose solution gives the distribution of the number of customers in the system.

$$\begin{aligned} & P_0'(t) = -\lambda P_0(t) + \mu P_1(t) \\ & P_j'(t) = \lambda P_{j-1}(t) - (\lambda+\mu)P_j(t) + \mu P_{j+1}(t), \qquad j \geq 1 \end{aligned}$$

This system of equations can be more readily solved once the system is in a steady state. **Steady state** has been reached if the probability P_j of finding j customers in the system approaches a limiting value. Then time is no longer of essence in computing the probabilities. Another way of expressing this is that the system has reached **statistical equilibrium**. Under the assumption of steady state, $dP_j(t)/dt = 0$ for $j = 0, 1, 2, \ldots$ Then the system of equations becomes a set of simple difference equations of the form

$$P_1 = \frac{\lambda}{\mu} P_0$$

$$P_{j+1} = \frac{\lambda + \mu}{\mu} P_j - \frac{\lambda}{\mu} P_{j-1}, \qquad j \geq 1$$

A number of approaches are useful in solving this set of difference equations. Two techniques are surveyed here. Both techniques work equally well for this simple system, but as the queueing system becomes more complex, one of the techniques may prove more advantageous over the other. An alternative derivation of these equations is also presented.

15.2.1 Solution by an Iterative Technique

Using the developed difference equations in an iterative manner gives

$$P_1 = \frac{\lambda}{\mu} P_0$$

$$P_2 = \frac{\lambda + \mu}{\mu} P_1 - \frac{\lambda}{\mu} P_0$$

$$= \left(\frac{\lambda + \mu}{\mu}\right)\left(\frac{\lambda}{\mu} P_0\right) - \frac{\lambda}{\mu} P_0 = \left(\frac{\lambda}{\mu}\right)^2 P_0$$

$$P_3 = \frac{\lambda + \mu}{\mu} P_2 - \frac{\lambda}{\mu} P_1$$

$$= \left(\frac{\lambda+\mu}{\mu}\right)\left(\left(\frac{\lambda}{\mu}\right)^2 P_0\right) - \left(\frac{\lambda}{\mu}\right)\left(\frac{\lambda}{\mu} P_0\right)$$

$$= \left(\frac{\lambda}{\mu}\right)^3 P_0$$

The general form of this iterative equation can be deduced using a straightforward induction argument.

$$\boxed{P_j = \left(\frac{\lambda}{\mu}\right)^j P_0}$$

To complete the solution of the steady-state equations, we need only to find P_0. Using this definition, we have

$$P_j = \rho^j P_0, \qquad j = 1, 2, \ldots$$

Now, as previously defined, the P_j, where $j = 0, 1, \ldots$, represent the probability that there are j customers in the system. By definition of probabilities,

$$\sum_{j=0}^{\infty} P_j = \sum_{j=0}^{\infty} \rho^j P_0 = 1$$

This equation can be rewritten to give a value of P_0, as required,

$$P_0 = \frac{1}{\sum_{j=0}^{\infty} \rho^j}$$

where $\sum_{j=0}^{\infty} \rho^j$ is a geometric series. This series converges if and only if $\lambda/\mu = \rho < 1$. When it converges, it converges to

$$\sum_{j=0}^{\infty} \rho^j = \frac{1}{1-\rho}$$

If we assume that $\rho < 1$, the necessary condition for a system to reach steady state will be $P_0 = 1 - \rho$, and the solution to the steady-state equations is given by

$$\boxed{P_j = \rho^j (1-\rho), \qquad j = 0, 1, 2, \ldots}$$

15.2.2 Solution Using Generating Functions

The **probability generating function**

$$P(z) = \sum_{j=0}^{\infty} P_j z^j$$

introduced in Chapter 3 can also be used to solve the steady-state equation. Using the previously defined utilization factor $\rho = \lambda/\mu$, we can write the steady-state equations as

$$P_1 = \rho P_0$$

$$P_{j+1} = (\rho+1)P_j - \rho P_{j-1}, \qquad j \geq 1$$

Multiplying both sides of the second equation by z^j and rewriting gives

$$z^{-1} P_{j+1} z^{j+1} = (\rho+1) P_j z^j - \rho z P_{j-1} z^{j-1}$$

Summing this equation from $j = 1$ to ∞ yields

$$z^{-1} \sum_{j=1}^{\infty} P_{j+1} z^{j+1} = (\rho+1) \sum_{j=1}^{\infty} P_j z^j - \rho z \sum_{j=1}^{\infty} P_{j-1} z^{j-1}$$

This equation can be rewritten

$$z^{-1}\left[\sum_{j=-1}^{\infty} P_{j+1}z^{j+1} - P_1 z - P_0\right] = (\rho+1)\left[\sum_{j=0}^{\infty} P_j z^j - P_0\right] - \rho z \sum_{j=1}^{\infty} P_{j-1}z^{j-1}$$

Now

$$\sum_{j=-1}^{\infty} P_{j+1}z^{j+1} = \sum_{j=0}^{\infty} P_j z^j = \sum_{j=1}^{\infty} P_{j-1}z^{j-1} = P(z)$$

so

$$z^{-1}[P(z) - P_1 z - P_0] = (\rho+1)[(P(z) - P_0] - \rho z P(z)$$

Since $P_1 = \rho P_0$, the equation can again be rewritten

$$z^{-1}[P(z) - (\rho z+1)P_0] = (\rho+1)[(P(z) - P_0] - \rho z P(z)$$

Solving this equation for $P(z)$ yields

$$\boxed{P(z) = \frac{P_0}{1 - z\rho}} \qquad [15.1]$$

Now $P(z) = \sum_{j=0}^{\infty} P_j z^j$, so

$$P(1) = \sum_{j=0}^{\infty} P_j 1^j = \sum_{j=0}^{\infty} P_j = 1$$

From equation 15.1, $P(1) = P_0/(1 - \rho)$, so $P_0 = 1 - \rho$, $\rho < 1$. Substituting this value into equation 15.1 yields the generating function

$$\boxed{P(z) = \frac{1-\rho}{1-z\rho}} \qquad [15.2]$$

To obtain probabilities from a generating function, one simply expands the generating function in a power series (sometimes more easily said then done) and utilizes the coefficients of the power of z. The probability $P(z)$ given in equation 15.2 can be easily expanded through long division. That is,

$$\frac{1-\rho}{1-z\rho} = (1-\rho)\frac{1}{1-z\rho}$$

$$= (1-\rho)(1 + z\rho + (z\rho)^2 + \ldots)$$

Then the infinite series expansion of the generating function is

$$P(z) = \sum_{j=0}^{\infty}(1-\rho)\rho^j z^j$$

The coefficient of the jth term gives the following relation:

$$\boxed{P_j = (1-\rho)\rho^j, \qquad j = 0, 1, 2, \ldots}$$

This is the result previously obtained using the iterative technique.

15.2.3 Derivation Using Stochastic Balance

The steady-state equations can also be easily derived using a method known as **stochastic balance**. This technique is based on the fact that in steady state, the expected rate of transitions out of state S_j is equal to the rate of transitions into state S_j This can be seen from the state diagram in Figure 15.2.

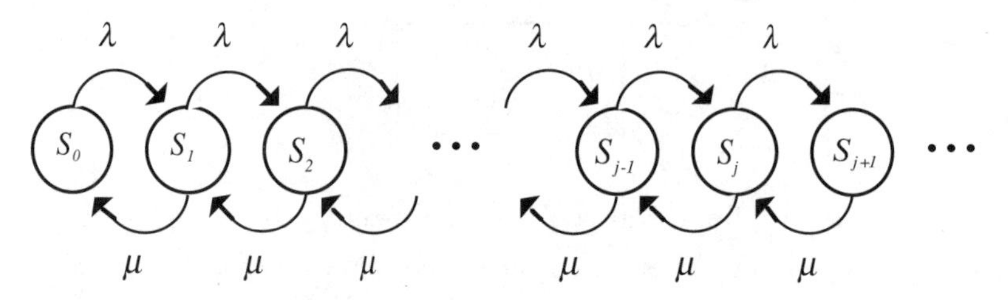

Figure 15.2. State transition diagram.

The system moves from state S_j to S_{j+1} only if an arrival but no departure occurs; from state S_j only if a departure but no arrival occurs; from state S_{j-1} to state S_j only if an arrival but no departure occurs; and from S_j to S_{j-1} only if a departure but no arrival occurs. The rate of transitions into state S_j is

$$\lambda P_{j-1} + \mu P_{j+1}$$

while the rate out of state S_j *is* $\mu P_j + \lambda P_j$ j. Equating the two gives

$$P_{j+1} = \frac{\lambda + \mu}{\mu} P_j - \frac{\lambda}{\mu} P_{j-1}$$

Using a similar approach, one can see that

$$P_1 = \frac{\lambda}{\mu} P_0$$

These are precisely the same equations obtained through the development of the differential difference equations. The foundations behind the two schemes are identical. This method is simply a shorthand version of the first technique. The equations must still be solved using either the iterative or the generating-function approach.

15.3 Summary Measures for the *M/M*/1/∞/FIFO System

Once the steady-state equations for the $M/M/1/\infty$/FIFO queueing system have been solved, the distribution of the number of customers in the system is known, at least in steady state. The distribution can then be used to calculate various summary measures such as the expected number in the system and expected waiting time, and these measures can then be used to characterize the behavior of the system.

15.3.1 Expected Number in the System

Let **X** denote the random variable that counts the number in the system. Then when the system is in a steady-state,

$$P(\mathbf{X} = x) = P_x = (1-\rho)\rho^x, \qquad x = 0,1,2,\ldots$$

The expected value of **X** can be calculated from

$$E[\mathbf{X}] = \sum_{x=0}^{\infty} xP_x = (1-\rho)\sum_{x=0}^{\infty} x\rho^x$$

Now

$$\sum_{x=0}^{\infty} x\rho^x = \rho\sum_{x=1}^{\infty} x\rho^{x-1} = \rho\frac{d}{d\rho}\left[\sum_{x=0}^{\infty}\rho^x\right]$$

But since $\rho < 1, \sum_{x=0}^{\infty}\rho^x = 1/(1-\rho)$. Then

$$\sum_{x=0}^{\infty} x\rho^x = \rho\frac{d}{d\rho}\left[\frac{1}{1-\rho}\right] = \frac{\rho}{(1-\rho)^2}$$

Then

$$E[\mathbf{X}] = (1-\rho)\left(\frac{\rho}{(1-\rho)^2}\right) = \frac{\rho}{1-\rho}$$

This average number in the system is sometimes denoted L.

$$\boxed{L = \frac{\rho}{1-\rho} = \frac{\lambda}{\mu-\lambda}}$$

15.3.2 Expected Number in the Queue

We sometimes want to calculate the average length of the queue rather than the average number in the system. Let **Q** denote the random variable that counts the customers waiting in the queue for service. Then

$$E(\mathbf{Q}) = 0P_0 + \sum_{j=1}^{\infty}(j-1)P_j$$

$$= \sum_{j=1}^{\infty} jP_j - \sum_{j=1}^{\infty} P_j$$

$$= L - \sum_{j=1}^{\infty} P_j$$

But $\sum_{j=0}^{\infty} P_j = 1$, so $\sum_{j=1}^{\infty} P_j = 1 - P_0$, and $E(\mathbf{Q}) = L - (1 - P_0)$. Since $P_0 = 1 - \rho$,

$$E(\mathbf{Q}) = L - \rho = \frac{\rho}{1-\rho} - \rho = \frac{\rho^2}{1-\rho}$$

This quantity is sometimes denoted L_q. So the average queue length is given by

$$L_q = \frac{\rho^2}{1-\rho} = \frac{\lambda^2}{\mu(\mu-\lambda)}$$

15.3.3 Expected Time in the System

Another measure important in summarizing the behavior of a queueing system is the average time that a customer spends in the system, both waiting for and receiving service. The distribution of waiting times can be readily obtained from an expectation argument. However, if all that is desired is the average time in the system, it is easier to compute it using a relationship known as **Little's formula** (15.3).

Let W be the expected time in the system, L the expected number in the system, and λ the arrival rate of customers to the service facility. Then

$$L = \lambda W \qquad [15.3]$$

The conditions under which this relationship holds are rather broad. For more details see Gross and Harris (15.1).

In Section 15.3.1, we established that $L = \rho/(1 - \rho) = \lambda/(\mu - \lambda)$. Using this relationship along with Little's formula gives the expected time in the system.

$$W = \frac{L}{\lambda} = \frac{1}{\lambda - \mu}$$

15.3.4 Expected Time in the Queue

The average time that a customer must wait for service is also important, since it is while waiting that a customer becomes discouraged and leaves the system. Once service begins, the customer is usually committed. The average waiting time in the system queue W_q is related to the average time in the system by

$$W = W_q + \frac{1}{\mu}$$

Although this relationship will not be proved here, it should be intuitive. Then since $W = 1/(\mu - \lambda)$,

$$W_q = \frac{1}{\mu - \lambda} - \frac{1}{\mu}$$

or

$$\boxed{W_q = \frac{\lambda}{\mu(\mu - \lambda)}}$$

Another version of Little's formula relates the average waiting time in the queue to the average length of the queue.

$$\boxed{L_q = \lambda W_q}$$

Example 15.1 Consider a queueing system composed of a message-switching center that serves five remote send-only terminals, as illustrated in Figure 15.3. Messages are received from each of the terminals according to the Poisson process with rates $\lambda_1 = 2$ per min, $\lambda_2 = 0.5$ per min, $\lambda_3 = \lambda_4 = 1$ per min, and $\lambda_5 = 3.5$ per min. The messages are then buffered before being processed and transmitted over a single line to another switching center. Assume that the time to process and transmit the messages can be considered exponentially distributed and with an average service time of four seconds. Assume also that the switching center has infinite available storage so that the system's capacity is not a factor. Assuming steady state, determine (1) the probability that there are fewer than five messages in the system, (2) the average number of messages in the system, (3) the average number of messages buffered, (4) the average time spent by each message in the system, and (5) the average time spent in the buffer by each message.

Since the arrival stream from each terminal is Poisson, the aggregate arrival stream will also be Poisson with a rate of $\lambda = \lambda_1 + \lambda_2 + \lambda_3 + \lambda_4 + \lambda_5 = 8$ per min. The service rate is $\mu = 60/4 = 15$ per min. Then the system can be considered an $M/M/1/\infty$/FIFO system with utilization factor $\rho = \lambda/\mu = 6/15 = 0.4000$. The probability that there are fewer than five messages in the system is

$$
\begin{aligned}
P(\mathbf{X}<5) &= P(\mathbf{X}=0)+P(\mathbf{X}=1)+P(\mathbf{X}=2)+P(\mathbf{X}=3)+P(\mathbf{X}=4) \\
&= P_0+P_1+P_2+P_3+P_4 \\
&= (1-\rho)+(1-\rho)\rho+(1-\rho)\rho^2+(1-\rho)\rho^3+(1-\rho)\rho^4 \\
&= (1-\rho)(1+\rho+\rho^2+\rho^3+\rho^4) \\
&= 0.4666(2.053) \\
&= 0.9567
\end{aligned}
$$

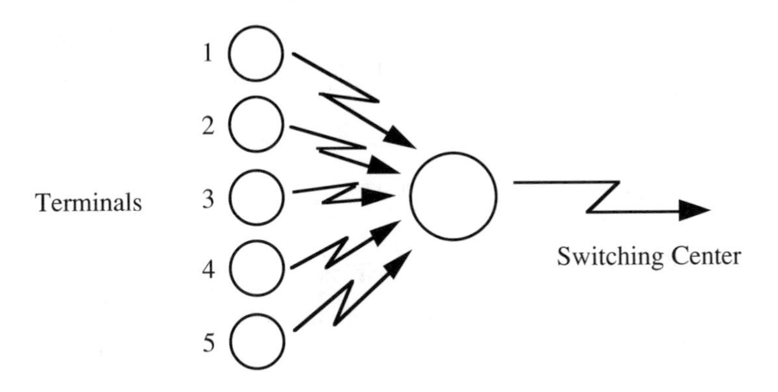

Figure 15.3. A message-switching diagram.

The expected number of customers in the system is

$$L = \frac{\rho}{1-\rho} = \frac{0.5333}{0.4666} = 1.1429$$

The average number of messages buffered in the system is

$$L_q = \frac{\rho^2}{1-\rho} = \frac{(0.5333)^2}{0.4666} = 0.6095$$

The average time spent by each message in the system is

$$W = \frac{1}{\mu-\lambda} = \frac{1}{15-8} = \frac{1}{7} \doteq 0.1429 \text{ min} \doteq 8.57 \text{ sec}$$

The average time spent by each message in the buffer is

$$W = \frac{\lambda}{\mu(\mu-\lambda)} = \frac{8}{15(7)} = 0.0762 \text{ min} \doteq 4.57 \text{ sec}$$

15.4 The *M*/*M*/1/*K*/FIFO System

Suppose that instead of allowing for an infinite system capacity as with the $M/M/1/\infty$/FIFO system, the system is restricted so that a maximum of K customers can be present at any given time. In most instances this is a more realistic model. Difference equations may be derived for this model just as for the $M/M/1/\infty$/FIFO system. In fact, as long as the number in the system is less than K, the equations are exactly the same as those previously developed. For $j = K$ the following difference equation must be added.

$$P_K(t+\Delta t) = P_K(t)(1-\mu\Delta t) + P_{K-1}(t)(\lambda\Delta t)(1-\mu\Delta t)$$

This equation must be included since transitions to state S_{K+1} are not possible. An arriving customer is turned away from the system when the system is full. If the system is assumed to be equilibrium, the steady-state equations are then

$$P_1 = \frac{\lambda}{\mu} P_0$$

$$P_{j+1} = \frac{\lambda+\mu}{\mu} P_j - \frac{\lambda}{\mu} P_{j-1}, \qquad 1 \le j \le K-1$$

$$P_K = \frac{\lambda}{\mu} P_{K-1}$$

The first two equations are identical to those of the $M/M/1/\infty$/FIFO system the third equation results from the additional difference equation.

Using the utilization factor ρ as well as the iterative procedure detailed in Section 15.2.1, one can show that

$$P_j = \rho^j P_0, \qquad 1 \le j \le K$$

In this case, $\sum_{j=0}^{K} P_j = 1$, so $\sum_{j=0}^{K} \rho^j P_0 = 1$ and $P_0 = 1 \Big/ \sum_{j=0}^{K} \rho^j$.

Now $\sum_{j=0}^{K} \rho^j$ is a finite geometric series whose sum is given by

$$\sum_{j=0}^{K} \rho^j = \begin{cases} \dfrac{1-\rho^{K+1}}{1-\rho}, & \rho \neq 1 \\[2ex] K+1, & \rho = 1 \end{cases}$$

Thus

$$P_0 = \begin{cases} \dfrac{1-\rho}{1-\rho^{K+1}}, & \rho \neq 1 \\[2ex] \dfrac{1}{K+1}, & \rho = 1 \end{cases}$$

from which the solution to the steady-state equations can be derived

$$P_n = \begin{cases} \dfrac{(1-\rho)\rho^n}{1-\rho^{K+1}}, & \rho \neq 1 \\[2ex] \dfrac{1}{K+1}, & \rho = 1 \qquad n = 0, 1, 2, \ldots, K \end{cases}$$

Thus in this case steady state will be reached regardless of the utilization factor. This should not be surprising, since constraining the maximum number of customers in the system prevents the queue from growing without bound, as happens in the $M/M/1/\infty$/FIFO system when $\rho \geq 1$.

The summary measures for the $M/M/1/K$/FIFO system can be developed in a manner analogous to those measures used for the $M/M/1/\infty$/FIFO system. Little's formula holds for the system, if the effective arrival rate $\lambda' = \lambda(1 - P_K)$ is used in place of λ. The following relationships can be easily derived by using the procedure used for the $M/M/1/\infty$/FIFO system.

Expected number in the system

$$L = \begin{cases} \dfrac{K}{2}, & \rho = 1 \\ \dfrac{\rho[1-(K+1)\rho^{K} + K\rho^{K+1}]}{(1-\rho^{K+1})(K-\rho)}, & \rho \neq 1 \end{cases}$$

Expected number in the queue

$$L_q = L - (1 - P_0)$$

Expected time in the system

$$W = \frac{L}{\lambda'} = \frac{L}{\lambda(1-P_K)}$$

Expected time in the queue

$$W_q = W - \frac{L}{\mu} = \frac{L_q}{\lambda'} = \frac{L_q}{\lambda(1-P_K)}$$

Example 15.2 Consider the system described in Example 15.1. Instead of assuming an infinite storage capacity at the switching center, assume that there is sufficient storage to keep a maximum of ten messages. With these assumptions the system is an $M/M/1/10/\text{FIFO}$ system.

The probability that there are fewer than five messages in the system is

$$P(\mathbf{X} < 5) = P_0 + P_1 + P_2 + P_3 + P_4$$

$$= \frac{(1-\rho)}{1-\rho^{11}}(1+\rho+\rho^2+\rho^3+\rho^4)$$

$$= \frac{0.4666}{0.9990}(2.0503)$$

$$= 0.9577$$

The expected number of customers in the system is

$$L = \frac{\rho[1-(K+1)\rho^2 + K\rho^{K+1}]}{(1-\rho^{K+1})(1-\rho)}$$

$$= \frac{0.5333[1-11(0.5333^{10})+10(0.5333^{11})]}{(1-(0.5333)^{11})(0.4666)}$$

$$= 1.1320$$

The average number of messages buffered in the system is

$$L_q = L-(1-P_0)$$

$$= 1.1320 - \left(1 - \frac{(1-\rho)}{(1-\rho^{11})}\right)$$

$$= 1.1320 - \left(1 - \frac{0.4666}{0.9990}\right)$$

$$= 1.1320 - 0.5329$$

$$= 0.5991$$

The average time spent by each message in the system is

$$W = \frac{L}{\lambda'} = \frac{L_q}{\lambda(1-P_K)}$$

$$= \frac{1.1320}{8(1.0-0.0009)} = 0.1416 \text{ min}$$

$$= 8.50 \text{ sec}$$

The average time spent by each message in the buffer is

$$W_q = W - \frac{1}{\mu} = 0.1416 - \frac{1}{15} = 0.0749 \text{ min}$$

$$= 4.496 \text{ sec}$$

As this example shows, when the system capacity is constrained, the summary measures are smaller than their respective counterparts of the $M/M/1/\infty$/FIFO system. This result should be expected, since in the $M/M/1/K$/FIFO system some customers are being turned away.

15.5 The $M/M/C/\infty$/FIFO System

Suppose that instead of having a single server as with the $M/M/1/\infty$/FIFO system, there are C servers, each with an independently and identically distributed exponential service time distribution with rate μ. The arrival process is still assumed to be Poisson. The system is illustrated in Figure 15.4.

First consider the mean service rate of the system. If there are more than C customers in the system, all the servers are busy; hence, the mean service rate is $C\mu$. If there are fewer than C customers in the system, say k, some of the servers are idle, so that the mean service rate is $k\mu$. Using an approach similar to that used for the analysis of the $M/M/1/\infty$/FIFO system, we can obtain the following relationship for P_j, the probability that j customers are in the system.

$$P_j = \begin{cases} \dfrac{\lambda^j}{j!\mu^j} P_0, & 1 \le j \le C \\[2ex] \dfrac{\lambda^j}{C^{j-C} C!\mu^j} P_0, & j > C \end{cases}$$

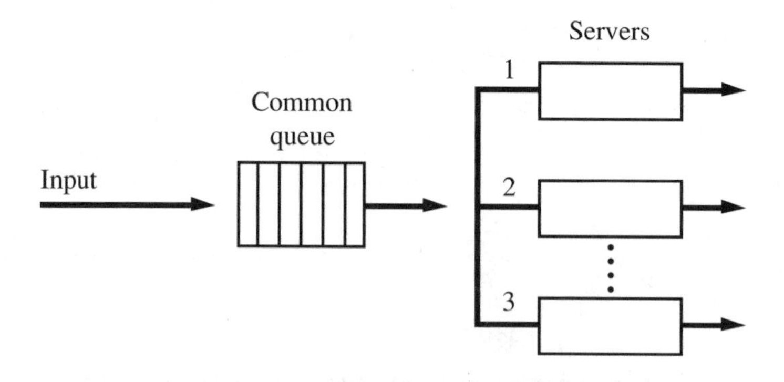

Figure 15.4. Multiserver queueing system.

The calculation of P_0 for this system is considerably more complicated than for the previous systems because of the more complex relation for P_j. Again using the condition $\sum_{j=0}^{\infty} P_j = 1$ yields

$$P_0\left[\sum_{j=0}^{C-1}\frac{\lambda^j}{j!\mu^j}+\sum_{j=C}^{\infty}\frac{\lambda^j}{C^{j-C}C!\mu^j}\right]=1$$

Define $r = \lambda/\mu$ and $\rho = \lambda/C\mu$; this relationship can then be rewritten

$$P_0\left[\sum_{j=0}^{C-1}\frac{r^j}{j!}+\sum_{j=C}^{\infty}\frac{r^j}{C^{j-C}C!}\right]=1$$

Thus

$$\sum_{j=C}^{\infty}\frac{r^j}{C^{j-C}C!}=\frac{r^C}{C!}\sum_{j=C}^{\infty}\left(\frac{r}{C}\right)^{j-C}$$

$$=\frac{r^C}{C!}\sum_{j=0}^{\infty}\left(\frac{r}{C}\right)^{j}$$

$$=\frac{r^C}{C!}\sum_{j=0}^{\infty}\rho^{j}$$

$$=\frac{r^C}{C!}\left(\frac{1}{1-\rho}\right) \qquad \text{if } \rho<1$$

Finally,

$$P_0=\left[\sum_{j=0}^{C-1}\frac{r^j}{j!}+\frac{Cr^C}{C!(C-r)}\right]^{-1}$$

or

$$P_0 = \left[\sum_{j=0}^{C-1} \frac{1}{j!} \left(\frac{\lambda}{\mu} \right)^j + \frac{1}{C!} \left(\frac{\lambda}{\mu} \right)^C \left(\frac{C\mu}{C\mu - \lambda} \right) \right]^{-1}$$

This equation appears quite formidable, and as more restrictions on the $M/M/1/\infty$/FIFO system are relaxed, the equations become worse to solve. Note that the requirement for the $M/M/C/\infty$/FIFO system to reach steady state is $\lambda/C\mu < 1$, rather than $\lambda/\mu < 1$ as with the $M/M/1/\infty$/FIFO system.

Obviously the development of the summary measures becomes somewhat cumbersome for the $M/M/C/\infty$/FIFO system. The derivation of these measures can be found in Gross and Harris (15.1).

Expected number in the system

$$L = \frac{\lambda}{\mu} + \left[\frac{(\lambda / \mu)^C \lambda\mu}{(C-1)!(C\mu - \lambda)^2} \right] P_0$$

Expected number in the queue

$$L_q = \left[\frac{(\lambda / \mu)^C \lambda\mu}{(C-1)!(C\mu - \lambda)^2} \right] P_0$$

Expected time in the system

$$W = \frac{1}{\mu} + \left[\frac{(\lambda / \mu)^C \mu}{(C-1)!(C\mu - \lambda)^2} \right] P_0$$

Expected time in the queue

$$W_q = \left[\frac{(\lambda / \mu)^C \mu}{(C-1)!(C\mu - \lambda)^2} \right] P_0$$

It should be apparent from these relationships that Little's formula holds for the *M/M/C/∞*/FIFO system.

Example 15.3 Consider the system described in Example 15.1. Suppose that there are four identical switch-line combinations at the switch-line center , each of which can service a message in an average of four seconds. The system can be analyzed as an *M/M/4/∞*/FIFO system. The quantities calculated in Examples 15.1 and 15.2 can be recalculated for this system as follows:

1.

$$P_0 = \left[\sum_{j=0}^{3} \frac{1}{j!}\left(\frac{\lambda}{\mu}\right)^j + \frac{1}{4!}\left(\frac{\lambda}{\mu}\right)^C \left(\frac{C\mu}{C\mu-\lambda}\right)\right]^{-1}$$

$$= \left[1 + \frac{\lambda}{\mu} + \frac{1}{2}\left(\frac{\lambda}{\mu}\right)^2 + \frac{1}{6}\left(\frac{\lambda}{\mu}\right)^3 + \frac{1}{24}\left(\frac{\lambda}{\mu}\right)^4 \left(\frac{4\mu}{4\mu-1}\right)\right]^{-1}$$

But $\lambda = 8$ and $\mu = 15$, so $\lambda/\mu = 8/15 = 0.5333$. Thus

$$P_0 = \left[1 + 0.5333 + \frac{1}{2}(0.5333)^2 + \frac{1}{6}(0.5333)^3 + \frac{1}{24}(0.5333)^4\left(\frac{60}{52}\right)\right]^{-1}$$

$$= 1/1.7047 = 0.5866$$

Finally

$$P(\mathbf{X} < 5) = P_0 + P_1 + P_2 + P_3 + P_4$$

$$= P_0\left(1 + \frac{\lambda}{\mu} + \frac{1}{2}\left(\frac{\lambda}{\mu}\right)^2 + \frac{1}{6}\left(\frac{\lambda}{\mu}\right)^3 + \frac{1}{24}\left(\frac{\lambda}{\mu}\right)^4\right)$$

$$= 0.5866\left(1 + 0.5333 + \frac{1}{2}(0.5333)^2 + \frac{1}{6}(0.5333)^3 + \frac{1}{24}(0.5333)^4\right)$$

$$= 0.5866(1.70415)$$
$$= 0.9997$$

2.

$$L = \frac{\lambda}{\mu} + \left[\frac{(\lambda / \mu)^4 \lambda\mu}{(3!)(4\mu - \lambda)} \right] P_0$$

$$= 0.5333 + \left[\frac{(0.5333)^4 (8)(15)}{(6)(52)} \right] (0.5866)$$

$$= 0.5333 + 0.0182 = 0.5515$$

3.

$$L_q = \left[\frac{(\lambda / \mu)^4 \lambda\mu}{(C-1)!(C\mu - \lambda)^2} \right] P_0$$

$$= 0.0182$$

4.

$$W = L / \lambda = 0.5515 / 8 = 0.0689 \text{ min} = 4.1363 \text{ sec}$$

5.

$$W_q = L_q / \lambda = 0.0182 / 8 = 0.0023 \text{ min} = 0.1365 \text{ sec}$$

Note that with this system the calculated summary measures have decreased dramatically from those calculated for the $M/M/1/\infty$/FIFO system.

15.6 Priority Queueing Systems

In all the previously surveyed queueing systems, the next customer selected for service in the system was the one at the head of the line. It is not uncommon for this selection to be based instead on a priority system, in which certain customers are given precedence over others. There are a number of reasons for applying priority disciplines. One is to reduce the average cost of the system. It may be more expensive to have certain customers wait in line rather than others. It would seem reasonable then to serve the high-cost customers first and thereby reduce the average total cost to the system. Another motivation might be to reduce the average number of customers in the system. The required service time for certain customers may be considerably shorter than for other customers in the system. By giving priority to customers who require the least service, it is possible to reduce the average number of customers in this system, an objective of particular interest for systems with a finite capacity.

Two general classes of priority queueing disciplines must be examined.

Nonpreemptive. Once the service of a given customer has started, it cannot be interrupted. If there are customers in the queue with different priorities, the next customer selected for service is the one with the highest priority. If there are customers in the queue with the same priority, some alternate discipline, normally first-in, first-out, is used to determine which customer is served next.

Preemptive. In this scheme, if an arriving customer has a higher priority than the customer currently being served, service is interrupted for the current customer, and the higher-priority customer gains control of the service facility. The interrupt customer rejoins the queue for service. The question then is, What happens when the interrupted customer is again selected for service? If the portion of service that the customer received is lost and service begins again, the discipline is known as a **preemptive repeat** discipline. If the service is resumed from the point of interruption, the discipline is known as a **preemptive resume** discipline.

Whether a given discipline is preemptive or nonpreemptive does not determine the priority of customers in the queue. Some techniques for assigning priorities to customers in the queue are the following.

Shortest service first. This scheme requires that the required service of each customer be known. Customers are then assigned priorities based on the required service, and the customer requiring the least amount of service is given the highest priority. This technique is generally used in nonpreemptive disciplines.

Willingness to pay. In some systems customers are allowed to buy a higher priority. Rates are set for various levels of priority, and a customer is charged according to the level of priority desired. This technique is normally used in nonpreemptive systems.

Round robin. Each customer in the queue is given some interval (quanta) of service before any customer receives a second interval. If the quanta is not sufficient to complete service on a given customer, service is interrupted and the customer rejoins the queue in a cyclic fashion. A number of techniques have been used to handle the customer who has received only part of the service. The customer can rejoin the original queue, for example, or join a second queue.

The development of the steady-state equations for priority queueing systems is a cumbersome job. For this reason, we refer the reader to Gross and Harris (15.1) for the details.

15.6.1 Nonpreemptive Priority System

Suppose that a total of r priority classes of customers are serviced by a single-channel service facility. Assume that the priority class numbers are assigned such that a lower number implies a higher priority. Suppose also that the arrival process of the kth class is Poisson with rate λ_k, while the service time required for the kth class customer is exponential with rate μ_k. Then define

$$\rho_k = \frac{\lambda_k}{\mu_k}, \qquad 1 \le k \le r$$

and the system will reach steady state if $\sum_{k=1}^{r} \rho_k < 1$. It is convenient to define $\sigma_k = \sum_{i=1}^{k} \rho_i$. Then the requirement for the system to reach steady state is $\sigma_r < 1$.

If the system has reached steady state, the expected waiting time (time in the queue) for a customer in the ith priority class is given by

$$W_q^{(i)} = \frac{\sum_{k=1}^{r} (\rho_k / \mu_k)}{(1-\sigma_{i-1})(1-\sigma_i)}$$

The expected number of customers of the ith class who are in the queue is

$$L_q^{(i)} = \lambda_i W_q^{(i)} = \frac{\lambda_i \sum_{k=1}^{r} (\rho_k / \mu_k)}{(1-\sigma_{i-1})(1-\sigma_i)}$$

Then the total expected number in the queue is given by

$$L_q = \sum_{i=1}^{r} L_q^{(i)} = \sum_{i=1}^{r} \frac{\lambda_i \sum_{k=1}^{r} (\rho_k / \mu_k)}{(1 - \sigma_{i-1})(1 - \sigma_i)}$$

15.6.2 Preemptive Priority System

The major difference between the preemptive and the nonpreemptive systems is that the **preemptive priority system** allows higher-priority customers to preempt lower-priority customers. The next consideration is whether the preemptive-repeat or the preemptive-resume technique is being used. If the service distribution is assumed to be exponential, the techniques used are immaterial because the distribution is memoryless.

With the same definitions as in previous sections, the expected number of customers in the ith priority class waiting in the queue is given by

$$L_q^{(i)} = \frac{\rho_i}{\left(1 - \sum_{n=1}^{i-1} \rho_n\right)\left(1 - \sum_{n=1}^{i} \rho_n\right)}$$

With this discipline a customer in the ith priority class is unaffected by customers in the $i + 1, i + 2, ..., r$ priority classes, so these priority classes have no effect on the expected waiting time. The expected waiting time for a customer in the *i*th priority class is identical to that of a customer in a nonpreemptive model with *i* priority classes. Thus

$$W_q^{(i)} = \frac{\sum_{k=1}^{i} (\rho_k / \mu_k)}{(1 - \sigma_{i-1})(1 - \sigma_i)}$$

Example 15.4 Consider the system described in Example 15.1. Assume that messages arriving to the switching center are assigned priorities on the basis of their originating terminal, with messages received from terminal 1 assigned the highest priority, those from terminal 2 the second highest, and so on. Also assume that $\mu_1 = \mu_2 = \mu_3 = \mu_4 = \mu_5 = \mu$. Using the preemptive model, calculate the average number of terminal-3 messages in the buffer awaiting retransmission and the average time that each of these messages remain in the buffer.

$$\rho_1 = 2/15 = 0.1333$$
$$\rho_2 = 0.5/15 = 0.0333$$
$$\rho_3 = 1/15 = 0.0667$$
$$\rho_4 = 1/15 = 0.0667$$
$$\rho_5 = 1.5/15 = 0.1000$$

From these we get

$$\sigma_2 = \rho_1 + \rho_2 = 0.1666, \qquad \sigma_3 = \rho_1 + \rho_2 + \rho_3 = 0.2333$$

Then

$$W_q^{(3)} = \frac{\sum_{k=1}^{5} (\rho_k / \mu_k)}{(1-\sigma_2)(1-\sigma_3)}$$

$$= \frac{1/15(0.1333 + 0.0333 + 0.0667 + 0.0667 + 0.10000)}{(1-0.1666)(1-0.2333)}$$

$$= \frac{0.0267}{(0.8334)(0.7667)}$$

$$= 0.0418 \text{ min } = 2.5 \text{ secs}$$

and

$$L_q^{(3)} = \lambda W_q^{(3)} = (0.0418)(1) = 0.0418$$

15.7 Summary

This chapter has surveyed some of the more common queueing systems, deriving summary measures for some and merely listing those for others. We have of course ignored many systems. For example, we did not consider models in which arrivals or services occur in batches. We also ignored series and parallel queueing systems. As queueing systems become more complex, the mathematics involved becomes nearly intractable, and this complexity calls for simulation. If a system cannot be analyzed analytically using queueing models, it is normally simulated. Even if it can be analyzed analytically, it is sometimes more convenient to simulate it and use the analytic model to validate the simulation.

15.8 Exercises

15.1 The manager of a single-channel service facility wishes to study the operation of the facility. To this end the manager collects data for one hour. The data collected are as follows:

Customer	*Arrival time*	*Service required*
1	8:01	6
2	8:06	4
3	8:07	3
4	8:11	5
5	8:12	3
6	8:14	2
7	8:15	1
8	8:16	7
9	8:17	4
10	8:21	2

11	8:24	6
12	8:27	3
13	8:28	3
14	8:29	2
15	8:30	1
16	8:32	1
17	8:36	11
18	8:41	1
19	8:42	4
20	8:44	3
21	8:45	2
22	8:50	3
23	8:52	2
24	8:54	2
25	8:58	1

Construct an empirical distribution for the interarrival times and service times by connecting the points with straight-line segments, as described in Section 5.1.

15.2 The manager of the system described in Exercise 15.1 suspects that the arrival distribution is Poisson with an average arrival rate of 0.5 customers per minute, and that the service time distribution is exponential with an average of three minutes per customer. Do the data collected support these conjectures?

15.3 Suppose that the service facility described in Exercise 15.1 closes at 9:00 but continues to service customers until all customers in line at the time of closing are served. Assuming a FIFO queueing discipline, determine the system state (number of customers in the system) at each point in time that this state changes. Determine the average queue length and the system idle time. How many customers were waiting at 9:00?

15.4 Repeat Exercise 15.3 under the assumption that the shortest-job-first (SIF) queueing discipline is used. In this discipline the customer in the queue requiring the least service is selected when the server finishes serving the present customer.

15.5 Suppose that the service facility described in Exercise 15.1 is composed of two identical servers working from a common queue. Repeat Exercise 15.3 under this assumption.

15.6 Automobiles arrive at a single-bay service station according to a Poisson process at a rate of ten per hour. The time required to service the cars is exponentially distributed with an average of five minutes per car. Assuming infinite system capacity and steady state, determine the following:

a. The probability that the service station is idle
b. The average number of cars waiting for service
c. The average time each spends waiting for service

15.7 If the service station has only enough ramp space for ten cars, how will the answers to Exercise 15.6 change?

15.8 If there are four bays with identically distributed service times, how will the answers to Exercise 15.6 change?

15.9 Use the method of stochastic balance to derive the steady-state difference equations for the $M/M/1/K$/FIFO system.

15.10 Failures of a given machine occur according to a Poisson process at a rate of five machines per hour. It costs $10 for each hour of downtime per machine. The manager of the facility must decide between two repair services. One service charge $5 per hour and can repair an average of six machines per hour. The other charges $6 per hour but can repair an average of eight machines per hour. Assuming that the service times are exponentially distributed, which service should be used?

15.11 A given service facility uses a nonpreemptive priority queueing discipline. There are three priority classes, with arrivals in each class following a Poisson process at rates $\lambda_1 = 2$ per hour, $\lambda_2 = 3$ per hour and $\lambda_3 = 1$ per hour. These service requirements for each class are exponentially distributed, with rates of four per hour, eight per hour, and ten per hour. Determine the expected waiting time and average queue length for each priority class.

15.12 To simulate the operation of an $M/M/1/\infty/\text{FIFO}$ system, one would need a source of Poisson random numbers with which to simulate the arrival process and a source of exponential random numbers with which to simulate service completions. In addition, a clock would be necessary to control the simulator. Write a FORTRAN program to simulate this system using the random number generators developed in Chapter 7. Collect statistics on the average number in the system and compare with the theoretical results.

15.9 References

15.1 Gross, D., and Harris, C., *Fundamentals of Queueing Theory*. New York: John Wiley and Sons, 1974.

15.2 Kendall, D.G. , "Stochastic Processes Occurring in the Theory of Queues and Their Analysis by the Method of Imbedded Markov Chains." *Ann. Math. Statistics*, 24 (1953):338-354.

15.3 Little, J. D. C., "A Proof for the Queueing Formula $L = \lambda W$." *Oper. Res.*, 16, (1961): 651-665.

APPENDIX TABLES

I. Normal Distribution Function

This table gives values of:

a) $f(x)$ = the probability density of a standardized random variable

$$= \frac{1}{\sqrt{2\pi}} e^{-\frac{1}{2}x^2}$$

For negative values of x, one uses the fact that $f(-x) = f(x)$.

b) $F(x)$ = the cumulative distribution function of a standardized normal random variable

$$= \int_{-\infty}^{x} \frac{1}{\sqrt{2\pi}} e^{-\frac{1}{2}t^2}\, dt$$

For negative values of x, one uses the relationship $F(-x) = 1 - F(x)$. Values of x corresponding to a few special values of $F(x)$ are given in a separate table following the main table.

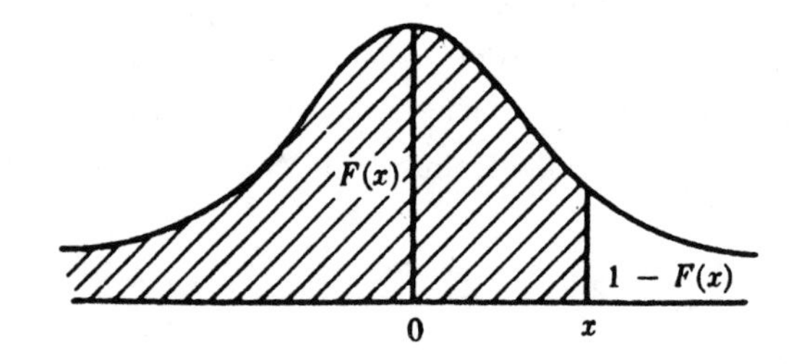

c) $f'(x)$ = the first derivative of $f(x)$ with respect to x

$$= -\frac{x}{\sqrt{2\pi}} e^{-\frac{1}{2}x^2} = -xf(x)$$

d) $f''(x)$ = the second derivative of $f(x)$ with respect to x

$$= \frac{(x^2 - 1)}{\sqrt{2\pi}} e^{-\frac{1}{2}x^2} = (x^2 - 1)f(x)$$

e) $f'''(x)$ = the third derivative of $f(x)$ with respect to x

$$= \frac{3x - x^3}{\sqrt{2\pi}} e^{-\frac{1}{2}x^2} = (3x - x^3)f(x)$$

f) $f^{\text{iv}}(x)$ = the fourth derivative of $f(x)$ with respect to x

$$= \frac{x^4 - 6x^2 + 3}{\sqrt{2\pi}} e^{-\frac{1}{2}x^2} = (x^4 - 6x^2 + 3)f(x)$$

It should be noted that other probability integrals can be evaluated by the use of these tables. For example,

$$\int_0^x f(t)dt = \tfrac{1}{2}\,\mathrm{erf}\left(\frac{x}{\sqrt{2}}\right),$$

where $\mathrm{erf}\left(\frac{x}{\sqrt{2}}\right)$ represents the error function associated with the normal curve.

To evaluate erf (2.3) one proceeds as follows: Since $\frac{x}{\sqrt{2}} = 2.3$, one finds $x = (2.3)(\sqrt{2}) = 3.25$. In the entry opposite $x = 3.25$, the value 0.9994 is given. Subtracting 0.5000 from the tabular value, one finds the value 0.4994. Thus erf $(2.3) = 2(0.4994) = 0.9988$.

NORMAL DISTRIBUTION AND RELATED FUNCTIONS

x	$F(x)$	$1 - F(x)$	$f(x)$	$f'(x)$	$f''(x)$	$f'''(x)$	$f^{iv}(x)$
.00	.5000	.5000	.3989	−.0000	−.3989	.0000	1.1968
.01	.5040	.4960	.3989	−.0040	−.3989	.0120	1.1965
.02	.5080	.4920	.3989	−.0080	−.3987	.0239	1.1956
.03	.5120	.4880	.3988	−.0120	−.3984	.0359	1.1941
.04	.5160	.4840	.3986	−.0159	−.3980	.0478	1.1920
.05	.5199	.4801	.3984	−.0199	−.3975	.0597	1.1894
.06	.5239	.4761	.3982	−.0239	−.3968	.0716	1.1861
.07	.5279	.4721	.3980	−.0279	−.3960	.0834	1.1822
.08	.5319	.4681	.3977	−.0318	−.3951	.0952	1.1778
.09	.5359	.4641	.3973	−.0358	−.3941	.1070	1.1727
.10	.5398	.4602	.3970	−.0397	−.3930	.1187	1.1671
.11	.5438	.4562	.3965	−.0436	−.3917	.1303	1.1609
.12	.5478	.4522	.3961	−.0475	−.3904	.1419	1.1541
.13	.5517	.4483	.3956	−.0514	−.3889	.1534	1.1468
.14	.5557	.4443	.3951	−.0553	−.3873	.1648	1.1389
.15	.5596	.4404	.3945	−.0592	−.3856	.1762	1.1304
.16	.5636	.4364	.3939	−.0630	−.3838	.1874	1.1214
.17	.5675	.4325	.3932	−.0668	−.3819	.1986	1.1118
.18	.5714	.4286	.3925	−.0707	−.3798	.2097	1.1017
.19	.5753	.4247	.3918	−.0744	−.3777	.2206	1.0911
.20	.5793	.4207	.3910	−.0782	−.3754	.2315	1.0799
.21	.5832	.4168	.3902	−.0820	−.3730	.2422	1.0682
.22	.5871	.4129	.3894	−.0857	−.3706	.2529	1.0560
.23	.5910	.4090	.3885	−.0894	−.3680	.2634	1.0434
.24	.5948	.4052	.3876	−.0930	−.3653	.2737	1.0302
.25	.5987	.4013	.3867	−.0967	−.3625	.2840	1.0165
.26	.6026	.3974	.3857	−.1003	−.3596	.2941	1.0024
.27	.6064	.3936	.3847	−.1039	−.3566	.3040	0.9878
.28	.6103	.3897	.3836	−.1074	−.3535	.3138	0.9727
.29	.6141	.3859	.3825	−.1109	−.3504	.3235	0.9572
.30	.6179	.3821	.3814	−.1144	−.3471	.3330	0.9413
.31	.6217	.3783	.3802	−.1179	−.3437	.3423	0.9250
.32	.6255	.3745	.3790	−.1213	−.3402	.3515	0.9082
.33	.6293	.3707	.3778	−.1247	−.3367	.3605	0.8910
.34	.6331	.3669	.3765	−.1280	−.3330	.3693	0.8735
.35	.6368	.3632	.3752	−.1313	−.3293	.3779	0.8556
.36	.6406	.3594	.3739	−.1346	−.3255	.3864	0.8373
.37	.6443	.3557	.3725	−.1378	−.3216	.3947	0.8186
.38	.6480	.3520	.3712	−.1410	−.3176	.4028	0.7996
.39	.6517	.3483	.3697	−.1442	−.3135	.4107	0.7803
.40	.6554	.3446	.3683	−.1473	−.3094	.4184	0.7607
.41	.6591	.3409	.3668	−.1504	−.3051	.4259	0.7408
.42	.6628	.3372	.3653	−.1534	−.3008	.4332	0.7206
.43	.6664	.3336	.3637	−.1564	−.2965	.4403	0.7001
.44	.6700	.3300	.3621	−.1593	−.2920	.4472	0.6793
.45	.6736	.3264	.3605	−.1622	−.2875	.4539	0.6583
.46	.6772	.3228	.3589	−.1651	−.2830	.4603	0.6371
.47	.6808	.3192	.3572	−.1679	−.2783	.4666	0.6156
.48	.6844	.3156	.3555	−.1707	−.2736	.4727	0.5940
.49	.6879	.3121	.3538	−.1734	−.2689	.4785	0.5721
.50	.6915	.3085	.3521	−.1760	−.2641	.4841	0.5501

NORMAL DISTRIBUTION AND RELATED FUNCTIONS

x	$F(x)$	$1 - F(x)$	$f(x)$	$f'(x)$	$f''(x)$	$f'''(x)$	$f^{iv}(x)$
.50	.6915	.3085	.3521	−.1760	−.2641	.4841	.5501
.51	.6950	.3050	.3503	−.1787	−.2592	.4895	.5279
.52	.6985	.3015	.3485	−.1812	−.2543	.4947	.5056
.53	.7019	.2981	.3467	−.1837	−.2493	.4996	.4831
.54	.7054	.2946	.3448	−.1862	−.2443	.5043	.4605
.55	.7088	.2912	.3429	−.1886	−.2392	.5088	.4378
.56	.7123	.2877	.3410	−.1920	−.2341	.5131	.4150
.57	.7157	.2843	.3391	−.1933	−.2289	.5171	.3921
.58	.7190	.2810	.3372	−.1956	−.2238	.5209	.3691
.59	.7224	.2776	.3352	−.1978	−.2185	.5245	.3461
.60	.7257	.2743	.3332	−.1999	−.2133	.5278	.3231
.61	.7291	.2709	.3312	−.2020	−.2080	.5309	.3000
.62	.7324	.2676	.3292	−.2041	−.2027	.5338	.2770
.63	.7357	.2643	.3271	−.2061	−.1973	.5365	.2539
.64	.7389	.2611	.3251	−.2080	−.1919	.5389	.2309
.65	.7422	.2578	.3230	−.2099	−.1865	.5411	.2078
.66	.7454	.2546	.3209	−.2118	−.1811	.5431	.1849
.67	.7486	.2514	.3187	−.2136	−.1757	.5448	.1620
.68	.7517	.2483	.3166	−.2153	−.1702	.5463	.1391
.69	.7549	.2451	.3144	−.2170	−.1647	.5476	.1164
.70	.7580	.2420	.3123	−.2186	−.1593	.5486	.0937
.71	.7611	.2389	.3101	−.2201	−.1538	.5495	.0712
.72	.7642	.2358	.3079	−.2217	−.1483	.5501	.0487
.73	.7673	.2327	.3056	−.2231	−.1428	.5504	.0265
.74	.7704	.2296	.3034	−.2245	−.1373	.5506	.0043
.75	.7734	.2266	.3011	−.2259	−.1318	.5505	−.0176
.76	.7764	.2236	.2989	−.2271	−.1262	.5502	−.0394
.77	.7794	.2206	.2966	−.2284	−.1207	.5497	−.0611
.78	.7823	.2177	.2943	−.2296	−.1153	.5490	−.0825
.79	.7852	.2148	.2920	−.2307	−.1098	.5481	−.1037
.80	.7881	.2119	.2897	−.2318	−.1043	.5469	−.1247
.81	.7910	.2090	.2874	−.2328	−.0988	.5456	−.1455
.82	.7939	.2061	.2850	−.2337	−.0934	.5440	−.1660
.83	.7967	.2033	.2827	−.2346	−.0880	.5423	−.1862
.84	.7995	.2005	.2803	−.2355	−.0825	.5403	−.2063
.85	.8023	.1977	.2780	−.2363	−.0771	.5381	−.2260
.86	.8051	.1949	.2756	−.2370	−.0718	.5358	−.2455
.87	.8078	.1922	.2732	−.2377	−.0664	.5332	−.2646
.88	.8106	.1894	.2709	−.2384	−.0611	.5305	−.2835
.89	.8133	.1867	.2685	−.2389	−.0558	.5276	−.3021
.90	.8159	.1841	.2661	−.2395	−.0506	.5245	−.3203
.91	.8186	.1814	.2637	−.2400	−.0453	.5212	−.3383
.92	.8212	.1788	.2613	−.2404	−.0401	.5177	−.3559
.93	.8238	.1762	.2589	−.2408	−.0350	.5140	−.3731
.94	.8264	.1736	.2565	−.2411	−.0299	.5102	−.3901
.95	.8289	.1711	.2541	−.2414	−.0248	.5062	−.4066
.96	.8315	.1685	.2516	−.2416	−.0197	.5021	−.4228
.97	.8340	.1660	.2492	−.2417	−.0147	.4978	−.4387
.98	.8365	.1635	.2468	−.2419	−.0098	.4933	−.4541
.99	.8389	.1611	.2444	−.2420	−.0049	.4887	−.4692
1.00	.8413	.1587	.2420	−.2420	.0000	.4839	−.4839

NORMAL DISTRIBUTION AND RELATED FUNCTIONS

x	$F(x)$	$1 - F(x)$	$f(x)$	$f'(x)$	$f''(x)$	$f'''(x)$	$f^{iv}(x)$
1.00	.8413	.1587	.2420	−.2420	.0000	.4839	−.4839
1.01	.8438	.1562	.2396	−.2420	.0048	.4790	−.4983
1.02	.8461	.1539	.2371	−.2419	.0096	.4740	−.5122
1.03	.8485	.1515	.2347	−.2418	.0143	.4688	−.5257
1.04	.8508	.1492	.2323	−.2416	.0190	.4635	−.5389
1.05	.8531	.1469	.2299	−.2414	.0236	.4580	−.5516
1.06	.8554	.1446	.2275	−.2411	.0281	.4524	−.5639
1.07	.8577	.1423	.2251	−.2408	.0326	.4467	−.5758
1.08	.8599	.1401	.2227	−.2405	.0371	.4409	−.5873
1.09	.8621	.1379	.2203	−.2401	.0414	.4350	−.5984
1.10	.8643	.1357	.2179	−.2396	.0458	.4290	−.6091
1.11	.8665	.1335	.2155	−.2392	.0500	.4228	−.6193
1.12	.8686	.1314	.2131	−.2386	.0542	.4166	−.6292
1.13	.8708	.1292	.2107	−.2381	.0583	.4102	−.6386
1.14	.8729	.1271	.2083	−.2375	.0624	.4038	−.6476
1.15	.8749	.1251	.2059	−.2368	.0664	.3973	−.6561
1.16	.8770	.1230	.2036	−.2361	.0704	.3907	−.6643
1.17	.8790	.1210	.2012	−.2354	.0742	.3840	−.6720
1.18	.8810	.1190	.1989	−.2347	.0780	.3772	−.6792
1.19	.8830	.1170	.1965	−.2339	.0818	.3704	−.6861
1.20	.8849	.1151	.1942	−.2330	.0854	.3635	−.6926
1.21	.8869	.1131	.1919	−.2322	.0890	.3566	−.6986
1.22	.8888	.1112	.1895	−.2312	.0926	.3496	−.7042
1.23	.8907	.1093	.1872	−.2303	.0960	.3425	−.7094
1.24	.8925	.1075	.1849	−.2293	.0994	.3354	−.7141
1.25	.8944	.1056	.1826	−.2283	.1027	.3282	−.7185
1.26	.8962	.1038	.1804	−.2273	.1060	.3210	−.7224
1.27	.8980	.1020	.1781	−.2262	.1092	.3138	−.7259
1.28	.8997	.1003	.1758	−.2251	.1123	.3065	−.7291
1.29	.9015	.0985	.1736	−.2240	.1153	.2992	−.7318
1.30	.9032	.0968	.1714	−.2228	.1182	.2918	−.7341
1.31	.9049	.0951	.1691	−.2216	.1211	.2845	−.7361
1.32	.9066	.0934	.1669	−.2204	.1239	.2771	−.7376
1.33	.9082	.0918	.1647	−.2191	.1267	.2697	−.7388
1.34	.9099	.0901	.1626	−.2178	.1293	.2624	−.7395
1.35	.9115	.0885	.1604	−.2165	.1319	.2550	−.7399
1.36	.9131	.0869	.1582	−.2152	.1344	.2476	−.7400
1.37	.9147	.0853	.1561	−.2138	.1369	.2402	−.7396
1.38	.9162	.0838	.1539	−.2125	.1392	.2328	−.7389
1.39	.9177	.0823	.1518	−.2110	.1415	.2254	−.7378
1.40	.9192	.0808	.1497	−.2096	.1437	.2180	−.7364
1.41	.9207	.0793	.1476	−.2082	.1459	.2107	−.7347
1.42	.9222	.0778	.1456	−.2067	.1480	.2033	−.7326
1.43	.9236	.0764	.1435	−.2052	.1500	.1960	−.7301
1.44	.9251	.0749	.1415	−.2037	.1519	.1887	−.7274
1.45	.9265	.0735	.1394	−.2022	.1537	.1815	−.7243
1.46	.9279	.0721	.1374	−.2006	.1555	.1742	−.7209
1.47	.9292	.0708	.1354	−.1991	.1572	.1670	−.7172
1.48	.9306	.0694	.1334	−.1975	.1588	.1599	−.7132
1.49	.9319	.0681	.1315	−.1959	.1604	.1528	−.7089
1.50	.9332	.0668	.1295	−.1943	.1619	.1457	−.7043

NORMAL DISTRIBUTION AND RELATED FUNCTIONS

x	$F(x)$	$1 - F(x)$	$f(x)$	$f'(x)$	$f''(x)$	$f'''(x)$	$f^{iv}(x)$
1.50	.9332	.0668	.1295	−.1943	.1619	.1457	−.7043
1.51	.9345	.0655	.1276	−.1927	.1633	.1387	−.6994
1.52	.9357	.0643	.1257	−.1910	.1647	.1317	−.6942
1.53	.9370	.0630	.1238	−.1894	.1660	.1248	−.6888
1.54	.9382	.0618	.1219	−.1877	.1672	.1180	−.6831
1.55	.9394	.0606	.1200	−.1860	.1683	.1111	−.6772
1.56	.9406	.0594	.1182	−.1843	.1694	.1044	−.6710
1.57	.9418	.0582	.1163	−.1826	.1704	.0977	−.6646
1.58	.9429	.0571	.1145	−.1809	.1714	.0911	−.6580
1.59	.9441	.0559	.1127	−.1792	.1722	.0846	−.6511
1.60	.9452	.0548	.1109	−.1775	.1730	.0781	−.6441
1.61	.9463	.0537	.1092	−.1757	.1738	.0717	−.6368
1.62	.9474	.0526	.1074	−.1740	.1745	.0654	−.6293
1.63	.9484	.0516	.1057	−.1723	.1751	.0591	−.6216
1.64	.9495	.0505	.1040	−.1705	.1757	.0529	−.6138
1.65	.9505	.0495	.1023	−.1687	.1762	.0468	−.6057
1.66	.9515	.0485	.1006	−.1670	.1766	.0408	−.5975
1.67	.9525	.0475	.0989	−.1652	.1770	.0349	−.5891
1.68	.9535	.0465	.0973	−.1634	.1773	.0290	−.5806
1.69	.9545	.0455	.0957	−.1617	.1776	.0233	−.5720
1.70	.9554	.0446	.0940	−.1599	.1778	.0176	−.5632
1.71	.9564	.0436	.0925	−.1581	.1779	.0120	−.5542
1.72	.9573	.0427	.0909	−.1563	.1780	.0065	−.5452
1.73	.9582	.0418	.0893	−.1546	.1780	.0011	−.5360
1.74	.9591	.0409	.0878	−.1528	.1780	−.0042	−.5267
1.75	.9599	.0401	.0863	−.1510	.1780	−.0094	−.5173
1.76	.9608	.0392	.0848	−.1492	.1778	−.0146	−.5079
1.77	.9616	.0384	.0833	−.1474	.1777	−.0196	−.4983
1.78	.9625	.0375	.0818	−.1457	.1774	−.0245	−.4887
1.79	.9633	.0367	.0804	−.1439	.1772	−.0294	−.4789
1.80	.9641	.0359	.0790	−.1421	.1769	−.0341	−.4692
1.81	.9649	.0351	.0775	−.1403	.1765	−.0388	−.4593
1.82	.9656	.0344	.0761	−.1386	.1761	−.0433	−.4494
1.83	.9664	.0336	.0748	−.1368	.1756	−.0477	−.4395
1.84	.9671	.0329	.0734	−.1351	.1751	−.0521	−.4295
1.85	.9678	.0322	.0721	−.1333	.1746	−.0563	−.4195
1.86	.9686	.0314	.0707	−.1316	.1740	−.0605	−.4095
1.87	.9693	.0307	.0694	−.1298	.1734	−.0645	−.3995
1.88	.9699	.0301	.0681	−.1281	.1727	−.0685	−.3894
1.89	.9706	.0294	.0669	−.1264	.1720	−.0723	−.3793
1.90	.9713	.0287	.0656	−.1247	.1713	−.0761	−.3693
1.91	.9719	.0281	.0644	−.1230	.1705	−.0797	−.3592
1.92	.9726	.0274	.0632	−.1213	.1697	−.0832	−.3492
1.93	.9732	.0268	.0620	−.1196	.1688	−.0867	−.3392
1.94	.9738	.0262	.0608	−.1179	.1679	−.0900	−.3292
1.95	.9744	.0256	.0596	−.1162	.1670	−.0933	−.3192
1.96	.9750	.0250	.0584	−.1145	.1661	−.0964	−.3093
1.97	.9756	.0244	.0573	−.1129	.1651	−.0994	−.2994
1.98	.9761	.0239	.0562	−.1112	.1641	−.1024	−.2895
1.99	.9767	.0233	.0551	−.1096	.1630	−.1052	−.2797
2.00	.9772	.0228	.0540	−.1080	.1620	−.1080	−.2700

NORMAL DISTRIBUTION AND RELATED FUNCTIONS

x	$F(x)$	$1 - F(x)$	$f(x)$	$f'(x)$	$f''(x)$	$f'''(x)$	$f^{iv}(x)$
2.00	.9773	.0227	.0540	−.1080	.1620	−.1080	−.2700
2.01	.9778	.0222	.0529	−.1064	.1609	−.1106	−.2603
2.02	.9783	.0217	.0519	−.1048	.1598	−.1132	−.2506
2.03	.9788	.0212	.0508	−.1032	.1586	−.1157	−.2411
2.04	.9793	.0207	.0498	−.1016	.1575	−.1180	−.2316
2.05	.9798	.0202	.0488	−.1000	.1563	−.1203	−.2222
2.06	.9803	.0197	.0478	−.0985	.1550	−.1225	−.2129
2.07	.9808	.0192	.0468	−.0969	.1538	−.1245	−.2036
2.08	.9812	.0188	.0459	−.0954	.1526	−.1265	−.1945
2.09	.9817	.0183	.0449	−.0939	.1513	−.1284	−.1854
2.10	.9821	.0179	.0440	−.0924	.1500	−.1302	−.1765
2.11	.9826	.0174	.0431	−.0909	.1487	−.1320	−.1676
2.12	.9830	.0170	.0422	−.0894	.1474	−.1336	−.1588
2.13	.9834	.0166	.0413	−.0879	.1460	−.1351	−.1502
2.14	.9838	.0162	.0404	−.0865	.1446	−.1366	−.1416
2.15	.9842	.0158	.0396	−.0850	.1433	−.1380	−.1332
2.16	.9846	.0154	.0387	−.0836	.1419	−.1393	−.1249
2.17	.9850	.0150	.0379	−.0822	.1405	−.1405	−.1167
2.18	.9854	.0146	.0371	−.0808	.1391	−.1416	−.1086
2.19	.9857	.0143	.0363	−.0794	.1377	−.1426	−.1006
2.20	.9861	.0139	.0355	−.0780	.1362	−.1436	−.0927
2.21	.9864	.0136	.0347	−.0767	.1348	−.1445	−.0850
2.22	.9868	.0132	.0339	−.0754	.1333	−.1453	−.0774
2.23	.9871	.0129	.0332	−.0740	.1319	−.1460	−.0700
2.24	.9875	.0125	.0325	−.0727	.1304	−.1467	−.0626
2.25	.9878	.0122	.0317	−.0714	.1289	−.1473	−.0554
2.26	.9881	.0119	.0310	−.0701	.1275	−.1478	−.0484
2.27	.9884	.0116	.0303	−.0689	.1260	−.1483	−.0414
2.28	.9887	.0113	.0297	−.0676	.1245	−.1486	−.0346
2.29	.9890	.0110	.0290	−.0664	.1230	−.1490	−.0279
2.30	.9893	.0107	.0283	−.0652	.1215	−.1492	−.0214
2.31	.9896	.0104	.0277	−.0639	.1200	−.1494	−.0150
2.32	.9898	.0102	.0270	−.0628	.1185	−.1495	−.0088
2.33	.9901	.0099	.0264	−.0616	.1170	−.1496	−.0027
2.34	.9904	.0096	.0258	−.0604	.1155	−.1496	.0033
2.35	.9906	.0094	.0252	−.0593	.1141	−.1495	.0092
2.36	.9909	.0091	.0246	−.0581	.1126	−.1494	.0149
2.37	.9911	.0089	.0241	−.0570	.1111	−.1492	.0204
2.38	.9913	.0087	.0235	−.0559	.1096	−.1490	.0258
2.39	.9916	.0084	.0229	−.0548	.1081	−.1487	.0311
2.40	.9918	.0082	.0224	−.0538	.1066	−.1483	.0362
2.41	.9920	.0080	.0219	−.0527	.1051	−.1480	.0412
2.42	.9922	.0078	.0213	−.0516	.1036	−.1475	.0461
2.43	.9925	.0075	.0208	−.0506	.1022	−.1470	.0508
2.44	.9927	.0073	.0203	−.0496	.1007	−.1465	.0554
2.45	.9929	.0071	.0198	−.0486	.0992	−.1459	.0598
2.46	.9931	.0069	.0194	−.0476	.0978	−.1453	.0641
2.47	.9932	.0068	.0189	−.0467	.0963	−.1446	.0683
2.48	.9934	.0066	.0184	−.0457	.0949	−.1439	.0723
2.49	.9936	.0064	.0180	−.0448	.0935	−.1432	.0762
2.50	.9938	.0062	.0175	−.0438	.0920	−.1424	.0800

NORMAL DISTRIBUTION AND RELATED FUNCTIONS

x	$F(x)$	$1 - F(x)$	$f(x)$	$f'(x)$	$f''(x)$	$f'''(x)$	$f^{iv}(x)$
2.50	.9938	.0062	.0175	−.0438	.0920	−.1424	.0800
2.51	.9940	.0060	.0171	−.0429	.0906	−.1416	.0836
2.52	.9941	.0059	.0167	−.0420	.0892	−.1408	.0871
2.53	.9943	.0057	.0163	−.0411	.0878	−.1399	.0905
2.54	.9945	.0055	.0158	−.0403	.0864	−.1389	.0937
2.55	.9946	.0054	.0155	−.0394	.0850	−.1380	.0968
2.56	.9948	.0052	.0151	−.0386	.0836	−.1370	.0998
2.57	.9949	.0051	.0147	−.0377	.0823	−.1360	.1027
2.58	.9951	.0049	.0143	−.0369	.0809	−.1350	.1054
2.59	.9952	.0048	.0139	−.0361	.0796	−.1339	.1080
2.60	.9953	.0047	.0136	−.0353	.0782	−.1328	.1105
2.61	.9955	.0045	.0132	−.0345	.0769	−.1317	.1129
2.62	.9956	.0044	.0129	−.0338	.0756	−.1305	.1152
2.63	.9957	.0043	.0126	−.0330	.0743	−.1294	.1173
2.64	.9959	.0041	.0122	−.0323	.0730	−.1282	.1194
2.65	.9960	.0040	.0119	−.0316	.0717	−.1270	.1213
2.66	.9961	.0039	.0116	−.0309	.0705	−.1258	.1231
2.67	.9962	.0038	.0113	−.0302	.0692	−.1245	.1248
2.68	.9963	.0037	.0110	−.0295	.0680	−.1233	.1264
2.69	.9964	.0036	.0107	−.0288	.0668	−.1220	.1279
2.70	.9965	.0035	.0104	−.0281	.0656	−.1207	.1293
2.71	.9966	.0034	.0101	−.0275	.0644	−.1194	.1306
2.72	.9967	.0033	.0099	−.0269	.0632	−.1181	.1317
2.73	.9968	.0032	.0096	−.0262	.0620	−.1168	.1328
2.74	.9969	.0031	.0093	−.0256	.0608	−.1154	.1338
2.75	.9970	.0030	.0091	−.0250	.0597	−.1141	.1347
2.76	.9971	.0029	.0088	−.0244	.0585	−.1127	.1356
2.77	.9972	.0028	.0086	−.0238	.0574	−.1114	.1363
2.78	.9973	.0027	.0084	−.0233	.0563	−.1100	.1369
2.79	.9974	.0026	.0081	−.0227	.0552	−.1087	.1375
2.80	.9974	.0026	.0079	−.0222	.0541	−.1073	.1379
2.81	.9975	.0025	.0077	−.0216	.0531	−.1059	.1383
2.82	.9976	.0024	.0075	−.0211	.0520	−.1045	.1386
2.83	.9977	.0023	.0073	−.0206	.0510	−.1031	.1389
2.84	.9977	.0023	.0071	−.0201	.0500	−.1017	.1390
2.85	.9978	.0022	.0069	−.0196	.0490	−.1003	.1391
2.86	.9979	.0021	.0067	−.0191	.0480	−.0990	.1391
2.87	.9979	.0021	.0065	−.0186	.0470	−.0976	.1391
2.88	.9980	.0020	.0063	−.0182	.0460	−.0962	.1389
2.89	.9981	.0019	.0061	−.0177	.0451	−.0948	.1388
2.90	.9981	.0019	.0060	−.0173	.0441	−.0934	.1385
2.91	.9982	.0018	.0058	−.0168	.0432	−.0920	.1382
2.92	.9982	.0018	.0056	−.0164	.0423	−.0906	.1378
2.93	.9983	.0017	.0055	−.0160	.0414	−.0893	.1374
2.94	.9984	.0016	.0053	−.0156	.0405	−.0879	.1369
2.95	.9984	.0016	.0051	−.0152	.0396	−.0865	.1364
2.96	.9985	.0015	.0050	−.0148	.0388	−.0852	.1358
2.97	.9985	.0015	.0048	−.0144	.0379	−.0838	.1352
2.98	.9986	.0014	.0047	−.0140	.0371	−.0825	.1345
2.99	.9986	.0014	.0046	−.0137	.0363	−.0811	.1337
3.00	.9987	.0013	.0044	−.0133	.0355	−.0798	.1330

NORMAL DISTRIBUTION AND RELATED FUNCTIONS

x	$F(x)$	$1 - F(x)$	$f(x)$	$f'(x)$	$f''(x)$	$f'''(x)$	$f^{iv}(x)$
3.00	.9987	.0013	.0044	−.0133	.0355	−.0798	.1330
3.01	.9987	.0013	.0043	−.0130	.0347	−.0785	.1321
3.02	.9987	.0013	.0042	−.0126	.0339	−.0771	.1313
3.03	.9988	.0012	.0040	−.0123	.0331	−.0758	.1304
3.04	.9988	.0012	.0039	−.0119	.0324	−.0745	.1294
3.05	.9989	.0011	.0038	−.0116	.0316	−.0732	.1285
3.06	.9989	.0011	.0037	−.0113	.0309	−.0720	.1275
3.07	.9989	.0011	.0036	−.0110	.0302	−.0707	.1264
3.08	.9990	.0010	.0035	−.0107	.0295	−.0694	.1254
3.09	.9990	.0010	.0034	−.0104	.0288	−.0682	.1243
3.10	.9990	.0010	.0033	−.0101	.0281	−.0669	.1231
3.11	.9991	.0009	.0032	−.0099	.0275	−.0657	.1220
3.12	.9991	.0009	.0031	−.0096	.0268	−.0645	.1208
3.13	.9991	.0009	.0030	−.0093	.0262	−.0633	.1196
3.14	.9992	.0008	.0029	−.0091	.0256	−.0621	.1184
3.15	.9992	.0008	.0028	−.0088	.0249	−.0609	.1171
3.16	.9992	.0008	.0027	−.0086	.0243	−.0598	.1159
3.17	.9992	.0008	.0026	−.0083	.0237	−.0586	.1146
3.18	.9993	.0007	.0025	−.0081	.0232	−.0575	.1133
3.19	.9993	.0007	.0025	−.0079	.0226	−.0564	.1120
3.20	.9993	.0007	.0024	−.0076	.0220	−.0552	.1107
3.21	.9993	.0007	.0023	−.0074	.0215	−.0541	.1093
3.22	.9994	.0006	.0022	−.0072	.0210	−.0531	.1080
3.23	.9994	.0006	.0022	−.0070	.0204	−.0520	.1066
3.24	.9994	.0006	.0021	−.0068	.0199	−.0509	.1053
3.25	.9994	.0006	.0020	−.0066	.0194	−.0499	.1039
3.26	.9994	.0006	.0020	−.0064	.0189	−.0488	.1025
3.27	.9995	.0005	.0019	−.0062	.0184	−.0478	.1011
3.28	.9995	.0005	.0018	−.0060	.0180	−.0468	.0997
3.29	.9995	.0005	.0018	−.0059	.0175	−.0458	.0983
3.30	.9995	.0005	.0017	−.0057	.0170	−.0449	.0969
3.31	.9995	.0005	.0017	−.0055	.0166	−.0439	.0955
3.32	.9995	.0005	.0016	−.0054	.0162	−.0429	.0941
3.33	.9996	.0004	.0016	−.0052	.0157	−.0420	.0927
3.34	.9996	.0004	.0015	−.0050	.0153	−.0411	.0913
3.35	.9996	.0004	.0015	−.0049	.0149	−.0402	.0899
3.36	.9996	.0004	.0014	−.0047	.0145	−.0393	.0885
3.37	.9996	.0004	.0014	−.0046	.0141	−.0384	.0871
3.38	.9996	.0004	.0013	−.0045	.0138	−.0376	.0857
3.39	.9997	.0003	.0013	−.0043	.0134	−.0367	.0843
3.40	.9997	.0003	.0012	−.0042	.0130	−.0359	.0829
3.41	.9997	.0003	.0012	−.0041	.0127	−.0350	.0815
3.42	.9997	.0003	.0012	−.0039	.0123	−.0342	.0801
3.43	.9997	.0003	.0011	−.0038	.0120	−.0334	.0788
3.44	.9997	.0003	.0011	−.0037	.0116	−.0327	.0774
3.45	.9997	.0003	.0010	−.0036	.0113	−.0319	.0761
3.46	.9997	.0003	.0010	−.0035	.0110	−.0311	.0747
3.47	.9997	.0003	.0010	−.0034	.0107	−.0304	.0734
3.48	.9997	.0003	.0009	−.0033	.0104	−.0297	.0721
3.49	.9998	.0002	.0009	−.0032	.0101	−.0290	.0707
3.50	.9998	.0002	.0009	−.0031	.0098	−.0283	.0694

NORMAL DISTRIBUTION AND RELATED FUNCTIONS

x	$F(x)$	$1 - F(x)$	$f(x)$	$f'(x)$	$f''(x)$	$f'''(x)$	$f^{iv}(x)$
3.50	.9998	.0002	.0009	−.0031	.0098	−.0283	.0694
3.51	.9998	.0002	.0008	−.0030	.0095	−.0276	.0681
3.52	.9998	.0002	.0008	−.0029	.0093	−.0269	.0669
3.53	.9998	.0002	.0008	−.0028	.0090	−.0262	.0656
3.54	.9998	.0002	.0008	−.0027	.0087	−.0256	.0643
3.55	.9998	.0002	.0007	−.0026	.0085	−.0249	.0631
3.56	.9998	.0002	.0007	−.0025	.0082	−.0243	.0618
3.57	.9998	.0002	.0007	−.0024	.0080	−.0237	.0606
3.58	.9998	.0002	.0007	−.0024	.0078	−.0231	.0594
3.59	.9998	.0002	.0006	−.0023	.0075	−.0225	.0582
3.60	.9998	.0002	.0006	−.0022	.0073	−.0219	.0570
3.61	.9998	.0002	.0006	−.0021	.0071	−.0214	.0559
3.62	.9999	.0001	.0006	−.0021	.0069	−.0208	.0547
3.63	.9999	.0001	.0005	−.0020	.0067	−.0203	.0536
3.64	.9999	.0001	.0005	−.0019	.0065	−.0198	.0524
3.65	.9999	.0001	.0005	−.0019	.0063	−.0192	.0513
3.66	.9999	.0001	.0005	−.0018	.0061	−.0187	.0502
3.67	.9999	.0001	.0005	−.0017	.0059	−.0182	.0492
3.68	.9999	.0001	.0005	−.0017	.0057	−.0177	.0481
3.69	.9999	.0001	.0004	−.0016	.0056	−.0173	.0470
3.70	.9999	.0001	.0004	−.0016	.0054	−.0168	.0460
3.71	.9999	.0001	.0004	−.0015	.0052	−.0164	.0450
3.72	.9999	.0001	.0004	−.0015	.0051	−.0159	.0440
3.73	.9999	.0001	.0004	−.0014	.0049	−.0155	.0430
3.74	.9999	.0001	.0004	−.0014	.0048	−.0150	.0420
3.75	.9999	.0001	.0004	−.0013	.0046	−.0146	.0410
3.76	.9999	.0001	.0003	−.0013	.0045	−.0142	.0401
3.77	.9999	.0001	.0003	−.0012	.0043	−.0138	.0392
3.78	.9999	.0001	.0003	−.0012	.0042	−.0134	.0382
3.79	.9999	.0001	.0003	−.0012	.0041	−.0131	.0373
3.80	.9999	.0001	.0003	−.0011	.0039	−.0127	.0365
3.81	.9999	.0001	.0003	−.0011	.0038	−.0123	.0356
3.82	.9999	.0001	.0003	−.0010	.0037	−.0120	.0347
3.83	.9999	.0001	.0003	−.0010	.0036	−.0116	.0339
3.84	.9999	.0001	.0003	−.0010	.0034	−.0113	.0331
3.85	.9999	.0001	.0002	−.0009	.0033	−.0110	.0323
3.86	.9999	.0001	.0002	−.0009	.0032	−.0107	.0315
3.87	.9999	.0001	.0002	−.0009	.0031	−.0104	.0307
3.88	.9999	.0001	.0002	−.0008	.0030	−.0100	.0299
3.89	1.0000	.0000	.0002	−.0008	.0029	−.0098	.0292
3.90	1.0000	.0000	.0002	−.0008	.0028	−.0095	.0284
3.91	1.0000	.0000	.0002	−.0008	.0027	−.0092	.0277
3.92	1.0000	.0000	.0002	−.0007	.0026	−.0089	.0270
3.93	1.0000	.0000	.0002	−.0007	.0026	−.0086	.0263
3.94	1.0000	.0000	.0002	−.0007	.0025	−.0084	.0256
3.95	1.0000	.0000	.0002	−.0006	.0024	−.0081	.0250
3.96	1.0000	.0000	.0002	−.0006	.0023	−.0079	.0243
3.97	1.0000	.0000	.0002	−.0006	.0022	−.0076	.0237
3.98	1.0000	.0000	.0001	−.0006	.0022	−.0074	.0230
3.99	1.0000	.0000	.0001	−.0006	.0021	−.0072	.0224
4.00	1.0000	.0000	.0001	−.0005	.0020	−.0070	.0218

x	1.282	1.645	1.960	2.326	2.576	3.090
$F(x)$	.90	.95	.975	.99	.995	.999
$2[1 - F(x)]$	.20	.10	.05	.02	.01	.002

II. Student's *t*- distribution Function

This table gives values of t such that

$$F(t) = \int_{-\infty}^{t} \frac{\Gamma\left(\frac{n+1}{2}\right)}{\sqrt{n\pi}\,\Gamma\left(\frac{n}{2}\right)} \left(1 + \frac{x^2}{n}\right)^{-\frac{n+1}{2}} dx$$

for n, the number of degrees of freedom, equal to 1, 2, . . ., 30, 40, 60, 120, ∞; and for $F(t)$ = 0.60, 0.75, 0.90, 0.95, 0.975, 0.99, 0.995, and 0.9995. The t-distribution is symmetrical, so that $F(-t) = 1 - F(t)$

n \ F	.60	.75	.90	.95	.975	.99	.995	.9995
1	.325	1.000	3.078	6.314	12.706	31.821	63.657	636.619
2	.289	.816	1.886	2.920	4.303	6.965	9.925	31.598
3	.277	.765	1.638	2.353	3.182	4.541	5.841	12.924
4	.271	.741	1.533	2.132	2.776	3.747	4.604	8.610
5	.267	.727	1.476	2.015	2.571	3.365	4.032	6.869
6	.265	.718	1.440	1.943	2.447	3.143	3.707	5.959
7	.263	.711	1.415	1.895	2.365	2.998	3.499	5.408
8	.262	.706	1.397	1.860	2.306	2.896	3.355	5.041
9	.261	.703	1.383	1.833	2.262	2.821	3.250	4.781
10	.260	.700	1.372	1.812	2.228	2.764	3.169	4.587
11	.260	.697	1.363	1.796	2.201	2.718	3.106	4.437
12	.259	.695	1.356	1.782	2.179	2.681	3.055	4.318
13	.259	.694	1.350	1.771	2.160	2.650	3.012	4.221
14	.258	.692	1.345	1.761	2.145	2.624	2.977	4.140
15	.258	.691	1.341	1.753	2.131	2.602	2.947	4.073
16	.258	.690	1.337	1.746	2.120	2.583	2.921	4.015
17	.257	.689	1.333	1.740	2.110	2.567	2.898	3.965
18	.257	.688	1.330	1.734	2.101	2.552	2.878	3.922
19	.257	.688	1.328	1.729	2.093	2.539	2.861	3.883
20	.257	.687	1.325	1.725	2.086	2.528	2.845	3.850
21	.257	.686	1.323	1.721	2.080	2.518	2.831	3.819
22	.256	.686	1.321	1.717	2.074	2.508	2.819	3.792
23	.256	.685	1.319	1.714	2.069	2.500	2.807	3.767
24	.256	.685	1.318	1.711	2.064	2.492	2.797	3.745
25	.256	.684	1.316	1.708	2.060	2.485	2.787	3.725
26	.256	.684	1.315	1.706	2.056	2.479	2.779	3.707
27	.256	.684	1.314	1.703	2.052	2.473	2.771	3.690
28	.256	.683	1.313	1.701	2.048	2.467	2.763	3.674
29	.256	.683	1.311	1.699	2.045	2.462	2.756	3.659
30	.256	.683	1.310	1.697	2.042	2.457	2.750	3.646
40	.255	.681	1.303	1.684	2.021	2.423	2.704	3.551
60	.254	.679	1.296	1.671	2.000	2.390	2.660	3.460
120	.254	.677	1.289	1.658	1.980	2.358	2.617	3.373
∞	.253	.674	1.282	1.645	1.960	2.326	2.576	3.291

* This table is abridged from the "Statistical Tables" of R. A. Fisher and Frank Yates published by Oliver & Boyd. Ltd., Edinburgh and London, 1938. It is here published with the kind permission of the authors and their publishers.

III. Chi–square Distribution Function

This table gives values of χ^2 such that

$$F(\chi^2) = \int_0^{\chi^2} \frac{1}{2^{\frac{n}{2}}\,\Gamma\left(\frac{n}{2}\right)} x^{\frac{n-2}{2}} e^{-\frac{x}{2}}\,dx$$

for n, the number of degrees of freedom, equal to 1, 2, . . . , 30. For $n > 30$, a normal approximation is quite accurate. The expression $\sqrt{2\chi^2} - \sqrt{2n-1}$ is approximately normally distributed as the standard normal distribution. Thus χ^2_α, the α-point of the distribution, may be computed by the formula

$$\chi^2_\alpha = \tfrac{1}{2}[x_\alpha + \sqrt{2n-1}]^2,$$

where x_α is the α-point of the cumulative normal distribution. For even values of n, $F(\chi^2)$ can be written as

$$1 - F(\chi^2) = \sum_{x=0}^{x'-1} \frac{e^{-\lambda}\lambda^x}{x!}$$

with $\lambda = \frac{1}{2}\chi^2$ and $x' = \frac{1}{2}n$. Thus the cumulative Chi-Square distribution is related to the cumulative Poisson distribution.

PERCENTAGE POINTS, CHI-SQUARE DISTRIBUTION

$$F(\chi^2) = \int_0^{\chi^2} \frac{1}{2^{\frac{n}{2}}\Gamma\left(\frac{n}{2}\right)} x^{\frac{n-2}{2}} e^{-\frac{x}{2}}\, dx$$

n \ F	.005	.010	.025	.050	.100	.250	.500	.750	.900	.950	.975	.990	.995
1	.0000393	.000157	.000982	.00393	.0158	.102	.455	1.32	2.71	3.84	5.02	6.63	7.88
2	.0100	.0201	.0506	.103	.211	.575	1.39	2.77	4.61	5.99	7.38	9.21	10.6
3	.0717	.115	.216	.352	.584	1.21	2.37	4.11	6.25	7.81	9.35	11.3	12.8
4	.207	.297	.484	.711	1.06	1.92	3.36	5.39	7.78	9.49	11.1	13.3	14.9
5	.412	.554	.831	1.15	1.61	2.67	4.35	6.63	9.24	11.1	12.8	15.1	16.7
6	.676	.872	1.24	1.64	2.20	3.45	5.35	7.84	10.6	12.6	14.4	16.8	18.5
7	.989	1.24	1.69	2.17	2.83	4.25	6.35	9.04	12.0	14.1	16.0	18.5	20.3
8	1.34	1.65	2.18	2.73	3.49	5.07	7.34	10.2	13.4	15.5	17.5	20.1	22.0
9	1.73	2.09	2.70	3.33	4.17	5.90	8.34	11.4	14.7	16.9	19.0	21.7	23.6
10	2.16	2.56	3.25	3.94	4.87	6.74	9.34	12.5	16.0	18.3	20.5	23.2	25.2
11	2.60	3.05	3.82	4.57	5.58	7.58	10.3	13.7	17.3	19.7	21.9	24.7	26.8
12	3.07	3.57	4.40	5.23	6.30	8.44	11.3	14.8	18.5	21.0	23.3	26.2	28.3
13	3.57	4.11	5.01	5.89	7.04	9.30	12.3	16.0	19.8	22.4	24.7	27.7	29.8
14	4.07	4.66	5.63	6.57	7.79	10.2	13.3	17.1	21.1	23.7	26.1	29.1	31.3
15	4.60	5.23	6.26	7.26	8.55	11.0	.4.3	18.2	22.3	25.0	27.5	30.6	32.8
16	5.14	5.81	6.91	7.96	9.31	11.9	15.3	19.4	23.5	26.3	28.8	32.0	34.3
17	5.70	6.41	7.56	8.67	10.1	12.8	16.3	20.5	24.8	27.6	30.2	33.4	35.7
18	6.26	7.01	8.23	9.39	10.9	13.7	17.3	21.6	26.0	28.9	31.5	34.8	37.2
19	6.84	7.63	8.91	10.1	11.7	14.6	18.3	22.7	27.2	30.1	32.9	36.2	38.6
20	7.43	8.26	9.59	10.9	12.4	15.5	19.3	23.8	28.4	31.4	34.2	37.6	40.0
21	8.03	8.90	10.3	11.6	13.2	16.3	20.3	24.9	29.6	32.7	35.5	38.9	41.4
22	8.64	9.54	11.0	12.3	14.0	17.2	21.3	26.0	30.8	33.9	36.8	40.3	42.8
23	9.26	10.2	11.7	13.1	14.8	18.1	22.3	27.1	32.0	35.2	38.1	41.6	44.2
24	9.89	10.9	12.4	13.8	15.7	19.0	23.3	28.2	33.2	36.4	39.4	43.0	45.6
25	10.5	11.5	13.1	14.6	16.5	19.9	24.3	29.3	34.4	37.7	40.6	44.3	46.9
26	11.2	12.2	13.8	15.4	17.3	20.8	25.3	30.4	35.6	38.9	41.9	45.6	48.3
27	11.8	12.9	14.6	16.2	18.1	21.7	26.3	31.5	36.7	40.1	43.2	47.0	49.6
28	12.5	13.6	15.3	16.9	18.9	22.7	27.3	32.6	37.9	41.3	44.5	48.3	51.0
29	13.1	14.3	16.0	17.7	19.8	23.6	28.3	33.7	39.1	42.6	45.7	49.6	52.3
30	13.8	15.0	16.8	18.5	20.6	24.5	29.3	34.8	40.3	43.8	47.0	50.9	53.7

IV. *F*- distribution Function

This table gives values of F such that

$$F(F) = \int_0^F \frac{\Gamma\left(\frac{m+n}{2}\right)}{\Gamma\left(\frac{m}{2}\right)\Gamma\left(\frac{n}{2}\right)} m^{\frac{m}{2}} n^{\frac{n}{2}} x^{\frac{m-2}{2}} (n + mx)^{-\frac{m+n}{2}} dx$$

for selected values of m, the number of degrees of freedom of the numerator of F; and for selected values of n, the number of degrees of freedom of the denominator of F. The table also provides values corresponding to $F(F) = .10, .05, .025, .01, .005, .001$ since $F_{1-\alpha}$ for m and n degrees of freedom is the reciprocal of F_α for n and m degrees of freedom. Thus

$$F_{.05}(4, 7) = \frac{1}{F_{.95}(7, 4)} = \frac{1}{6.09} = .164 \ .$$

PERCENTAGE POINTS, F-DISTRIBUTION

$$F(F) = \int_0^F \frac{\Gamma\left(\frac{m+n}{2}\right)}{\Gamma\left(\frac{m}{2}\right)\Gamma\left(\frac{n}{2}\right)} m^{\frac{m}{2}} n^{\frac{n}{2}} x^{\frac{m}{2}-1} (n+mx)^{-\frac{m+n}{2}} dx = .95$$

n \ m	1	2	3	4	5	6	7	8	9	10	12	15	20	24	30	40	60	120	∞
1	161.4	199.5	215.7	224.6	230.2	234.0	236.8	238.9	240.5	241.9	243.9	245.9	248.0	249.1	250.1	251.1	252.2	253.3	254.3
2	18.51	19.00	19.16	19.25	19.30	19.33	19.35	19.37	19.38	19.40	19.41	19.43	19.45	19.45	19.46	19.47	19.48	19.49	19.50
3	10.13	9.55	9.28	9.12	9.01	8.94	8.89	8.85	8.81	8.79	8.74	8.70	8.66	8.64	8.62	8.59	8.57	8.55	8.53
4	7.71	6.94	6.59	6.39	6.26	6.16	6.09	6.04	6.00	5.96	5.91	5.86	5.80	5.77	5.75	5.72	5.69	5.66	5.63
5	6.61	5.79	5.41	5.19	5.05	4.95	4.88	4.82	4.77	4.74	4.68	4.62	4.56	4.53	4.50	4.46	4.43	4.40	4.36
6	5.99	5.14	4.76	4.53	4.39	4.28	4.21	4.15	4.10	4.06	4.00	3.94	3.87	3.84	3.81	3.77	3.74	3.70	3.67
7	5.59	4.74	4.35	4.12	3.97	3.87	3.79	3.73	3.68	3.64	3.57	3.51	3.44	3.41	3.38	3.34	3.30	3.27	3.23
8	5.32	4.46	4.07	3.84	3.69	3.58	3.50	3.44	3.39	3.35	3.28	3.22	3.15	3.12	3.08	3.04	3.01	2.97	2.93
9	5.12	4.26	3.86	3.63	3.48	3.37	3.29	3.23	3.18	3.14	3.07	3.01	2.94	2.90	2.86	2.83	2.79	2.75	2.71
10	4.96	4.10	3.71	3.48	3.33	3.22	3.14	3.07	3.02	2.98	2.91	2.85	2.77	2.74	2.70	2.66	2.62	2.58	2.54
11	4.84	3.98	3.59	3.36	3.20	3.09	3.01	2.95	2.90	2.85	2.79	2.72	2.65	2.61	2.57	2.53	2.49	2.45	2.40
12	4.75	3.89	3.49	3.26	3.11	3.00	2.91	2.85	2.80	2.75	2.69	2.62	2.54	2.51	2.47	2.43	2.38	2.34	2.30
13	4.67	3.81	3.41	3.18	3.03	2.92	2.83	2.77	2.71	2.67	2.60	2.53	2.46	2.42	2.38	2.34	2.30	2.25	2.21
14	4.60	3.74	3.34	3.11	2.96	2.85	2.76	2.70	2.65	2.60	2.53	2.46	2.39	2.35	2.31	2.27	2.22	2.18	2.13
15	4.54	3.68	3.29	3.06	2.90	2.79	2.71	2.64	2.59	2.54	2.48	2.40	2.33	2.29	2.25	2.20	2.16	2.11	2.07
16	4.49	3.63	3.24	3.01	2.85	2.74	2.66	2.59	2.54	2.49	2.42	2.35	2.28	2.24	2.19	2.15	2.11	2.06	2.01
17	4.45	3.59	3.20	2.96	2.81	2.70	2.61	2.55	2.49	2.45	2.38	2.31	2.23	2.19	2.15	2.10	2.06	2.01	1.96
18	4.41	3.55	3.16	2.93	2.77	2.66	2.58	2.51	2.46	2.41	2.34	2.27	2.19	2.15	2.11	2.06	2.02	1.97	1.92
19	4.38	3.52	3.13	2.90	2.74	2.63	2.54	2.48	2.42	2.38	2.31	2.23	2.16	2.11	2.07	2.03	1.98	1.93	1.88
20	4.35	3.49	3.10	2.87	2.71	2.60	2.51	2.45	2.39	2.35	2.28	2.20	2.12	2.08	2.04	1.99	1.95	1.90	1.84
21	4.32	3.47	3.07	2.84	2.68	2.57	2.49	2.42	2.37	2.32	2.25	2.18	2.10	2.05	2.01	1.96	1.92	1.87	1.81
22	4.30	3.44	3.05	2.82	2.66	2.55	2.46	2.40	2.34	2.30	2.23	2.15	2.07	2.03	1.98	1.94	1.89	1.84	1.78
23	4.28	3.42	3.03	2.80	2.64	2.53	2.44	2.37	2.32	2.27	2.20	2.13	2.05	2.01	1.96	1.91	1.86	1.81	1.76
24	4.26	3.40	3.01	2.78	2.62	2.51	2.42	2.36	2.30	2.25	2.18	2.11	2.03	1.98	1.94	1.89	1.84	1.79	1.73
25	4.24	3.39	2.99	2.76	2.60	2.49	2.40	2.34	2.28	2.24	2.16	2.09	2.01	1.96	1.92	1.87	1.82	1.77	1.71
26	4.23	3.37	2.98	2.74	2.59	2.47	2.39	2.32	2.27	2.22	2.15	2.07	1.99	1.95	1.90	1.85	1.80	1.75	1.69
27	4.21	3.35	2.96	2.73	2.57	2.46	2.37	2.31	2.25	2.20	2.13	2.06	1.97	1.93	1.88	1.84	1.79	1.73	1.67
28	4.20	3.34	2.95	2.71	2.56	2.45	2.36	2.29	2.24	2.19	2.12	2.04	1.96	1.91	1.87	1.82	1.77	1.71	1.65
29	4.18	3.33	2.93	2.70	2.55	2.43	2.35	2.28	2.22	2.18	2.10	2.03	1.94	1.90	1.85	1.81	1.75	1.70	1.64
30	4.17	3.32	2.92	2.69	2.53	2.42	2.33	2.27	2.21	2.16	2.09	2.01	1.93	1.89	1.84	1.79	1.74	1.68	1.62
40	4.08	3.23	2.84	2.61	2.45	2.34	2.25	2.18	2.12	2.08	2.00	1.92	1.84	1.79	1.74	1.69	1.64	1.58	1.51
60	4.00	3.15	2.76	2.53	2.37	2.25	2.17	2.10	2.04	1.99	1.92	1.84	1.75	1.70	1.65	1.59	1.53	1.47	1.39
120	3.92	3.07	2.68	2.45	2.29	2.17	2.09	2.02	1.96	1.91	1.83	1.75	1.66	1.61	1.55	1.50	1.43	1.35	1.25
∞	3.84	3.00	2.60	2.37	2.21	2.10	2.01	1.94	1.88	1.83	1.75	1.67	1.57	1.52	1.46	1.39	1.32	1.22	1.00

$F = \frac{s_1^2}{s_2^2} = \frac{S_1}{m} \Big/ \frac{S_2}{n}$, where $s_1^2 = S_1/m$ and $s_2^2 = S_2/n$ are independent mean squares estimating a common variance σ^2 and based on m and n degrees of freedom, respectively.

PERCENTAGE POINTS, *F*-DISTRIBUTION

$$F(F) = \int_0^F \frac{\Gamma\left(\frac{m+n}{2}\right)}{\Gamma\left(\frac{m}{2}\right)\Gamma\left(\frac{n}{2}\right)} m^{\frac{m}{2}} n^{\frac{n}{2}} x^{\frac{m}{2}-1} (n+mx)^{-\frac{m+n}{2}} dx = .95$$

n \ m	1	2	3	4	5	6	7	8	9	10	12	15	20	24	30	40	60	120	∞
1	161.4	199.5	215.7	224.6	230.2	234.0	236.8	238.9	240.5	241.9	243.9	245.9	248.0	249.1	250.1	251.1	252.2	253.3	254.3
2	18.51	19.00	19.16	19.25	19.30	19.33	19.35	19.37	19.38	19.40	19.41	19.43	19.45	19.45	19.46	19.47	19.48	19.49	19.50
3	10.13	9.55	9.28	9.12	9.01	8.94	8.89	8.85	8.81	8.79	8.74	8.70	8.66	8.64	8.62	8.59	8.57	8.55	8.53
4	7.71	6.94	6.59	6.39	6.26	6.16	6.09	6.04	6.00	5.96	5.91	5.86	5.80	5.77	5.75	5.72	5.69	5.66	5.63
5	6.61	5.79	5.41	5.19	5.05	4.95	4.88	4.82	4.77	4.74	4.68	4.62	4.56	4.53	4.50	4.46	4.43	4.40	4.36
6	5.99	5.14	4.76	4.53	4.39	4.28	4.21	4.15	4.10	4.06	4.00	3.94	3.87	3.84	3.81	3.77	3.74	3.70	3.67
7	5.59	4.74	4.35	4.12	3.97	3.87	3.79	3.73	3.68	3.64	3.57	3.51	3.44	3.41	3.38	3.34	3.30	3.27	3.23
8	5.32	4.46	4.07	3.84	3.69	3.58	3.50	3.44	3.39	3.35	3.28	3.22	3.15	3.12	3.08	3.04	3.01	2.97	2.93
9	5.12	4.26	3.86	3.63	3.48	3.37	3.29	3.23	3.18	3.14	3.07	3.01	2.94	2.90	2.86	2.83	2.79	2.75	2.71
10	4.96	4.10	3.71	3.48	3.33	3.22	3.14	3.07	3.02	2.98	2.91	2.85	2.77	2.74	2.70	2.66	2.62	2.58	2.54
11	4.84	3.98	3.59	3.36	3.20	3.09	3.01	2.95	2.90	2.85	2.79	2.72	2.65	2.61	2.57	2.53	2.49	2.45	2.40
12	4.75	3.89	3.49	3.26	3.11	3.00	2.91	2.85	2.80	2.75	2.69	2.62	2.54	2.51	2.47	2.43	2.38	2.34	2.30
13	4.67	3.81	3.41	3.18	3.03	2.92	2.83	2.77	2.71	2.67	2.60	2.53	2.46	2.42	2.38	2.34	2.30	2.25	2.21
14	4.60	3.74	3.34	3.11	2.96	2.85	2.76	2.70	2.65	2.60	2.53	2.46	2.39	2.35	2.31	2.27	2.22	2.18	2.13
15	4.54	3.68	3.29	3.06	2.90	2.79	2.71	2.64	2.59	2.54	2.48	2.40	2.33	2.29	2.25	2.20	2.16	2.11	2.07
16	4.49	3.63	3.24	3.01	2.85	2.74	2.66	2.59	2.54	2.49	2.42	2.35	2.28	2.24	2.19	2.15	2.11	2.06	2.01
17	4.45	3.59	3.20	2.96	2.81	2.70	2.61	2.55	2.49	2.45	2.38	2.31	2.23	2.19	2.15	2.10	2.06	2.01	1.96
18	4.41	3.55	3.16	2.93	2.77	2.66	2.58	2.51	2.46	2.41	2.34	2.27	2.19	2.15	2.11	2.06	2.02	1.97	1.92
19	4.38	3.52	3.13	2.90	2.74	2.63	2.54	2.48	2.42	2.38	2.31	2.23	2.16	2.11	2.07	2.03	1.98	1.93	1.88
20	4.35	3.49	3.10	2.87	2.71	2.60	2.51	2.45	2.39	2.35	2.28	2.20	2.12	2.08	2.04	1.99	1.95	1.90	1.84
21	4.32	3.47	3.07	2.84	2.68	2.57	2.49	2.42	2.37	2.32	2.25	2.18	2.10	2.05	2.01	1.96	1.92	1.87	1.81
22	4.30	3.44	3.05	2.82	2.66	2.55	2.46	2.40	2.34	2.30	2.23	2.15	2.07	2.03	1.98	1.94	1.89	1.84	1.78
23	4.28	3.42	3.03	2.80	2.64	2.53	2.44	2.37	2.32	2.27	2.20	2.13	2.05	2.01	1.96	1.91	1.86	1.81	1.76
24	4.26	3.40	3.01	2.78	2.62	2.51	2.42	2.36	2.30	2.25	2.18	2.11	2.03	1.98	1.94	1.89	1.84	1.79	1.73
25	4.24	3.39	2.99	2.76	2.60	2.49	2.40	2.34	2.28	2.24	2.16	2.09	2.01	1.96	1.92	1.87	1.82	1.77	1.71
26	4.23	3.37	2.98	2.74	2.59	2.47	2.39	2.32	2.27	2.22	2.15	2.07	1.99	1.95	1.90	1.85	1.80	1.75	1.69
27	4.21	3.35	2.96	2.73	2.57	2.46	2.37	2.31	2.25	2.20	2.13	2.06	1.97	1.93	1.88	1.84	1.79	1.73	1.67
28	4.20	3.34	2.95	2.71	2.56	2.45	2.36	2.29	2.24	2.19	2.12	2.04	1.96	1.91	1.87	1.82	1.77	1.71	1.65
29	4.18	3.33	2.93	2.70	2.55	2.43	2.35	2.28	2.22	2.18	2.10	2.03	1.94	1.90	1.85	1.81	1.75	1.70	1.64
30	4.17	3.32	2.92	2.69	2.53	2.42	2.33	2.27	2.21	2.16	2.09	2.01	1.93	1.89	1.84	1.79	1.74	1.68	1.62
40	4.08	3.23	2.84	2.61	2.45	2.34	2.25	2.18	2.12	2.08	2.00	1.92	1.84	1.79	1.74	1.69	1.64	1.58	1.51
60	4.00	3.15	2.76	2.53	2.37	2.25	2.17	2.10	2.04	1.99	1.92	1.84	1.75	1.70	1.65	1.59	1.53	1.47	1.39
120	3.92	3.07	2.68	2.45	2.29	2.17	2.09	2.02	1.96	1.91	1.83	1.75	1.66	1.61	1.55	1.50	1.43	1.35	1.25
∞	3.84	3.00	2.60	2.37	2.21	2.10	2.01	1.94	1.88	1.83	1.75	1.67	1.57	1.52	1.46	1.39	1.32	1.22	1.00

$F = \frac{s_1^2}{s_2^2} = \frac{S_1}{m} \Big/ \frac{S_2}{n}$, where $s_1^2 = S_1/m$ and $s_2^2 = S_2/n$ are independent mean squares estimating a common variance σ^2 and based on m and n degrees of freedom, respectively.

PERCENTAGE POINTS, F-DISTRIBUTION

$$F(F) = \int_0^F \frac{\Gamma\left(\frac{m+n}{2}\right)}{\Gamma\left(\frac{m}{2}\right)\Gamma\left(\frac{n}{2}\right)} m^{\frac{m}{2}} n^{\frac{n}{2}} x^{\frac{m}{2}-1} (n+mx)^{-\frac{m+n}{2}} dx = .975$$

n \ m	1	2	3	4	5	6	7	8	9	10	12	15	20	24	30	40	60	120	∞
1	647.8	799.5	864.2	899.6	921.8	937.1	948.2	956.7	963.3	968.6	976.7	984.9	993.1	997.2	1001	1006	1010	1014	1018
2	38.51	39.00	39.17	39.25	39.30	39.33	39.36	39.37	39.39	39.40	39.41	39.43	39.45	39.46	39.46	39.47	39.48	39.49	39.50
3	17.44	16.04	15.44	15.10	14.88	14.73	14.62	14.54	14.47	14.42	14.34	14.25	14.17	14.12	14.08	14.04	13.99	13.95	13.90
4	12.22	10.65	9.98	9.60	9.36	9.20	9.07	8.98	8.90	8.84	8.75	8.66	8.56	8.51	8.46	8.41	8.36	8.31	8.26
5	10.01	8.43	7.76	7.39	7.15	6.98	6.85	6.76	6.68	6.62	6.52	6.43	6.33	6.28	6.23	6.18	6.12	6.07	6.02
6	8.81	7.26	6.60	6.23	5.99	5.82	5.70	5.60	5.52	5.46	5.37	5.27	5.17	5.12	5.07	5.01	4.96	4.90	4.85
7	8.07	6.54	5.89	5.52	5.29	5.12	4.99	4.90	4.82	4.76	4.67	4.57	4.47	4.42	4.36	4.31	4.25	4.20	4.14
8	7.57	6.06	5.42	5.05	4.82	4.65	4.53	4.43	4.36	4.30	4.20	4.10	4.00	3.95	3.89	3.84	3.78	3.73	3.67
9	7.21	5.71	5.08	4.72	4.48	4.32	4.20	4.10	4.03	3.96	3.87	3.77	3.67	3.61	3.56	3.51	3.45	3.39	3.33
10	6.94	5.46	4.83	4.47	4.24	4.07	3.95	3.85	3.78	3.72	3.62	3.52	3.42	3.37	3.31	3.26	3.20	3.14	3.08
11	6.72	5.26	4.63	4.28	4.04	3.88	3.76	3.66	3.59	3.53	3.43	3.33	3.23	3.17	3.12	3.06	3.00	2.94	2.88
12	6.55	5.10	4.47	4.12	3.89	3.73	3.61	3.51	3.44	3.37	3.28	3.18	3.07	3.02	2.96	2.91	2.85	2.79	2.72
13	6.41	4.97	4.35	4.00	3.77	3.60	3.48	3.39	3.31	3.25	3.15	3.05	2.95	2.89	2.84	2.78	2.72	2.66	2.60
14	6.30	4.86	4.24	3.89	3.66	3.50	3.38	3.29	3.21	3.15	3.05	2.95	2.84	2.79	2.73	2.67	2.61	2.55	2.49
15	6.20	4.77	4.15	3.80	3.58	3.41	3.29	3.20	3.12	3.06	2.96	2.86	2.76	2.70	2.64	2.59	2.52	2.46	2.40
16	6.12	4.69	4.08	3.73	3.50	3.34	3.22	3.12	3.05	2.99	2.89	2.79	2.68	2.63	2.57	2.51	2.45	2.38	2.32
17	6.04	4.62	4.01	3.66	3.44	3.28	3.16	3.06	2.98	2.92	2.82	2.72	2.62	2.56	2.50	2.44	2.38	2.32	2.25
18	5.98	4.56	3.95	3.61	3.38	3.22	3.10	3.01	2.93	2.87	2.77	2.67	2.56	2.50	2.44	2.38	2.32	2.26	2.19
19	5.92	4.51	3.90	3.56	3.33	3.17	3.05	2.96	2.88	2.82	2.72	2.62	2.51	2.45	2.39	2.33	2.27	2.20	2.13
20	5.87	4.46	3.86	3.51	3.29	3.13	3.01	2.91	2.84	2.77	2.68	2.57	2.46	2.41	2.35	2.29	2.22	2.16	2.09
21	5.83	4.42	3.82	3.48	3.25	3.09	2.97	2.87	2.80	2.73	2.64	2.53	2.42	2.37	2.31	2.25	2.18	2.11	2.04
22	5.79	4.38	3.78	3.44	3.22	3.05	2.93	2.84	2.76	2.70	2.60	2.50	2.39	2.33	2.27	2.21	2.14	2.08	2.00
23	5.75	4.35	3.75	3.41	3.18	3.02	2.90	2.81	2.73	2.67	2.57	2.47	2.36	2.30	2.24	2.18	2.11	2.04	1.97
24	5.72	4.32	3.72	3.38	3.15	2.99	2.87	2.78	2.70	2.64	2.54	2.44	2.33	2.27	2.21	2.15	2.08	2.01	1.94
25	5.69	4.29	3.69	3.35	3.13	2.97	2.85	2.75	2.68	2.61	2.51	2.41	2.30	2.24	2.18	2.12	2.05	1.98	1.91
26	5.66	4.27	3.67	3.33	3.10	2.94	2.82	2.73	2.65	2.59	2.49	2.39	2.28	2.22	2.16	2.09	2.03	1.95	1.88
27	5.63	4.24	3.65	3.31	3.08	2.92	2.80	2.71	2.63	2.57	2.47	2.36	2.25	2.19	2.13	2.07	2.00	1.93	1.85
28	5.61	4.22	3.63	3.29	3.06	2.90	2.78	2.69	2.61	2.55	2.45	2.34	2.23	2.17	2.11	2.05	1.98	1.91	1.83
29	5.59	4.20	3.61	3.27	3.04	2.88	2.76	2.67	2.59	2.53	2.43	2.32	2.21	2.15	2.09	2.03	1.96	1.89	1.81
30	5.57	4.18	3.59	3.25	3.03	2.87	2.75	2.65	2.57	2.51	2.41	2.31	2.20	2.14	2.07	2.01	1.94	1.87	1.79
40	5.42	4.05	3.46	3.13	2.90	2.74	2.62	2.53	2.45	2.39	2.29	2.18	2.07	2.01	1.94	1.88	1.80	1.72	1.64
60	5.29	3.93	3.34	3.01	2.79	2.63	2.51	2.41	2.33	2.27	2.17	2.06	1.94	1.88	1.82	1.74	1.67	1.58	1.48
120	5.15	3.80	3.23	2.89	2.67	2.52	2.39	2.30	2.22	2.16	2.05	1.94	1.82	1.76	1.69	1.61	1.53	1.43	1.31
∞	5.02	3.69	3.12	2.79	2.57	2.41	2.29	2.19	2.11	2.05	1.94	1.83	1.71	1.64	1.57	1.48	1.39	1.27	1.00

$F = \frac{s_1^2}{s_2^2} = \frac{S_1}{m} \Big/ \frac{S_2}{n}$, where $s_1^2 = S_1/m$ and $s_2^2 = S_2/n$ are independent mean squares estimating a common variance σ^2 and based on m and n degrees of freedom, respectively.

PERCENTAGE POINTS, F-DISTRIBUTION

$$F(F) = \int_0^F \frac{\Gamma\left(\frac{m+n}{2}\right)}{\Gamma\left(\frac{m}{2}\right)\Gamma\left(\frac{n}{2}\right)} m^{\frac{m}{2}} n^{\frac{n}{2}} x^{\frac{m}{2}-1} (n+mx)^{-\frac{m+n}{2}} dx = .99$$

n \ m	1	2	3	4	5	6	7	8	9	10	12	15	20	24	30	40	60	120	∞
1	4052	4999.5	5403	5625	5764	5859	5928	5982	6022	6056	6106	6157	6209	6235	6261	6287	6313	6339	6366
2	98.50	99.00	99.17	99.25	99.30	99.33	99.36	99.37	99.39	99.40	99.42	99.43	99.45	99.46	99.47	99.47	99.48	99.49	99.50
3	34.12	30.82	29.46	28.71	28.24	27.91	27.67	27.49	27.35	27.23	27.05	26.87	26.69	26.60	26.50	26.41	26.32	26.22	26.13
4	21.20	18.00	16.69	15.98	15.52	15.21	14.98	14.80	14.66	14.55	14.37	14.20	14.02	13.93	13.84	13.75	13.65	13.56	13.46
5	16.26	13.27	12.06	11.39	10.97	10.67	10.46	10.29	10.16	10.05	9.89	9.72	9.55	9.47	9.38	9.29	9.20	9.11	9.02
6	13.75	10.92	9.78	9.15	8.75	8.47	8.26	8.10	7.98	7.87	7.72	7.56	7.40	7.31	7.23	7.14	7.06	6.97	6.88
7	12.25	9.55	8.45	7.85	7.46	7.19	6.99	6.84	6.72	6.62	6.47	6.31	6.16	6.07	5.99	5.91	5.82	5.74	5.65
8	11.26	8.65	7.59	7.01	6.63	6.37	6.18	6.03	5.91	5.81	5.67	5.52	5.36	5.28	5.20	5.12	5.03	4.95	4.86
9	10.56	8.02	6.99	6.42	6.06	5.80	5.61	5.47	5.35	5.26	5.11	4.96	4.81	4.73	4.65	4.57	4.48	4.40	4.31
10	10.04	7.56	6.55	5.99	5.64	5.39	5.20	5.06	4.94	4.85	4.71	4.56	4.41	4.33	4.25	4.17	4.08	4.00	3.91
11	9.65	7.21	6.22	5.67	5.32	5.07	4.89	4.74	4.63	4.54	4.40	4.25	4.10	4.02	3.94	3.86	3.78	3.69	3.60
12	9.33	6.93	5.95	5.41	5.06	4.82	4.64	4.50	4.39	4.30	4.16	4.01	3.86	3.78	3.70	3.62	3.54	3.45	3.36
13	9.07	6.70	5.74	5.21	4.86	4.62	4.44	4.30	4.19	4.10	3.96	3.82	3.66	3.59	3.51	3.43	3.34	3.25	3.17
14	8.86	6.51	5.56	5.04	4.69	4.46	4.28	4.14	4.03	3.94	3.80	3.66	3.51	3.43	3.35	3.27	3.18	3.09	3.00
15	8.68	6.36	5.42	4.89	4.56	4.32	4.14	4.00	3.89	3.80	3.67	3.52	3.37	3.29	3.21	3.13	3.05	2.96	2.87
16	8.53	6.23	5.29	4.77	4.44	4.20	4.03	3.89	3.78	3.69	3.55	3.41	3.26	3.18	3.10	3.02	2.93	2.84	2.75
17	8.40	6.11	5.18	4.67	4.34	4.10	3.93	3.79	3.68	3.59	3.46	3.31	3.16	3.08	3.00	2.92	2.83	2.75	2.65
18	8.29	6.01	5.09	4.58	4.25	4.01	3.84	3.71	3.60	3.51	3.37	3.23	3.08	3.00	2.92	2.84	2.75	2.66	2.57
19	8.18	5.93	5.01	4.50	4.17	3.94	3.77	3.63	3.52	3.43	3.30	3.15	3.00	2.92	2.84	2.76	2.67	2.58	2.49
20	8.10	5.85	4.94	4.43	4.10	3.87	3.70	3.56	3.46	3.37	3.23	3.09	2.94	2.86	2.78	2.69	2.61	2.52	2.42
21	8.02	5.78	4.87	4.37	4.04	3.81	3.64	3.51	3.40	3.31	3.17	3.03	2.88	2.80	2.72	2.64	2.55	2.46	2.36
22	7.95	5.72	4.82	4.31	3.99	3.76	3.59	3.45	3.35	3.26	3.12	2.98	2.83	2.75	2.67	2.58	2.50	2.40	2.31
23	7.88	5.66	4.76	4.26	3.94	3.71	3.54	3.41	3.30	3.21	3.07	2.93	2.78	2.70	2.62	2.54	2.45	2.35	2.26
24	7.82	5.61	4.72	4.22	3.90	3.67	3.50	3.36	3.26	3.17	3.03	2.89	2.74	2.66	2.58	2.49	2.40	2.31	2.21
25	7.77	5.57	4.68	4.18	3.85	3.63	3.46	3.32	3.22	3.13	2.99	2.85	2.70	2.62	2.54	2.45	2.36	2.27	2.17
26	7.72	5.53	4.64	4.14	3.82	3.59	3.42	3.29	3.18	3.09	2.96	2.81	2.66	2.58	2.50	2.42	2.33	2.23	2.13
27	7.68	5.49	4.60	4.11	3.78	3.56	3.39	3.26	3.15	3.06	2.93	2.78	2.63	2.55	2.47	2.38	2.29	2.20	2.10
28	7.64	5.45	4.57	4.07	3.75	3.53	3.36	3.23	3.12	3.03	2.90	2.75	2.60	2.52	2.44	2.35	2.26	2.17	2.06
29	7.60	5.42	4.54	4.04	3.73	3.50	3.33	3.20	3.09	3.00	2.87	2.73	2.57	2.49	2.41	2.33	2.23	2.14	2.03
30	7.56	5.39	4.51	4.02	3.70	3.47	3.30	3.17	3.07	2.98	2.84	2.70	2.55	2.47	2.39	2.30	2.21	2.11	2.01
40	7.31	5.18	4.31	3.83	3.51	3.29	3.12	2.99	2.89	2.80	2.66	2.52	2.37	2.29	2.20	2.11	2.02	1.92	1.80
60	7.08	4.98	4.13	3.65	3.34	3.12	2.95	2.82	2.72	2.63	2.50	2.35	2.20	2.12	2.03	1.94	1.84	1.73	1.60
120	6.85	4.79	3.95	3.48	3.17	2.96	2.79	2.66	2.56	2.47	2.34	2.19	2.03	1.95	1.86	1.76	1.66	1.53	1.38
∞	6.63	4.61	3.78	3.32	3.02	2.80	2.64	2.51	2.41	2.32	2.18	2.04	1.88	1.79	1.70	1.59	1.47	1.32	1.00

$F = \frac{s_1^2}{s_2^2} = \frac{S_1}{m} \Big/ \frac{S_2}{n}$, where $s_1^2 = S_1/m$ and $s_2^2 = S_2/n$ are independent mean squares estimating a common variance σ^2 and based on m and n degrees of freedom, respectively.

PERCENTAGE POINTS, F-DISTRIBUTION

$$F(F) = \int_0^F \frac{\Gamma\left(\frac{m+n}{2}\right)}{\Gamma\left(\frac{m}{2}\right)\Gamma\left(\frac{n}{2}\right)} m^{\frac{m}{2}} n^{\frac{n}{2}} x^{\frac{m}{2}-1} (n+mx)^{-\frac{m+n}{2}} dx = .995$$

n \ m	1	2	3	4	5	6	7	8	9	10	12	15	20	24	30	40	60	120	∞
1	16211	20000	21615	22500	23056	23437	23715	23925	24091	24224	24426	24630	24836	24940	25044	25148	25253	25359	25465
2	198.5	199.0	199.2	199.2	199.3	199.3	199.4	199.4	199.4	199.4	199.4	199.4	199.4	199.5	199.5	199.5	199.5	199.5	199.5
3	55.55	49.80	47.47	46.19	45.39	44.84	44.43	44.13	43.88	43.69	43.39	43.08	42.78	42.62	42.47	42.31	42.15	41.99	41.83
4	31.33	26.28	24.26	23.15	22.46	21.97	21.62	21.35	21.14	20.97	20.70	20.44	20.17	20.03	19.89	19.75	19.61	19.47	19.32
5	22.78	18.31	16.53	15.56	14.94	14.51	14.20	13.96	13.77	13.62	13.38	13.15	12.90	12.78	12.66	12.53	12.40	12.27	12.14
6	18.63	14.54	12.92	12.03	11.46	11.07	10.79	10.57	10.39	10.25	10.03	9.81	9.59	9.47	9.36	9.24	9.12	9.00	8.88
7	16.24	12.40	10.88	10.05	9.52	9.16	8.89	8.68	8.51	8.38	8.18	7.97	7.75	7.65	7.53	7.42	7.31	7.19	7.08
8	14.69	11.04	9.60	8.81	8.30	7.95	7.69	7.50	7.34	7.21	7.01	6.81	6.61	6.50	6.40	6.29	6.18	6.06	5.95
9	13.61	10.11	8.72	7.96	7.47	7.13	6.88	6.69	6.54	6.42	6.23	6.03	5.83	5.73	5.62	5.52	5.41	5.30	5.19
10	12.83	9.43	8.08	7.34	6.87	6.54	6.30	6.12	5.97	5.85	5.66	5.47	5.27	5.17	5.07	4.97	4.86	4.75	4.64
11	12.23	8.91	7.60	6.88	6.42	6.10	5.86	5.68	5.54	5.42	5.24	5.05	4.86	4.76	4.65	4.55	4.44	4.34	4.23
12	11.75	8.51	7.23	6.52	6.07	5.76	5.52	5.35	5.20	5.09	4.91	4.72	4.53	4.43	4.33	4.23	4.12	4.01	3.90
13	11.37	8.19	6.93	6.23	5.79	5.48	5.25	5.08	4.94	4.82	4.64	4.46	4.27	4.17	4.07	3.97	3.87	3.76	3.65
14	11.06	7.92	6.68	6.00	5.56	5.26	5.03	4.86	4.72	4.60	4.43	4.25	4.06	3.96	3.86	3.76	3.66	3.55	3.44
15	10.80	7.70	6.48	5.80	5.37	5.07	4.85	4.67	4.54	4.42	4.25	4.07	3.88	3.79	3.69	3.58	3.48	3.37	3.26
16	10.58	7.51	6.30	5.64	5.21	4.91	4.69	4.52	4.38	4.27	4.10	3.92	3.73	3.64	3.54	3.44	3.33	3.22	3.11
17	10.38	7.35	6.16	5.50	5.07	4.78	4.56	4.39	4.25	4.14	3.97	3.79	3.61	3.51	3.41	3.31	3.21	3.10	2.98
18	10.22	7.21	6.03	5.37	4.96	4.66	4.44	4.28	4.14	4.03	3.86	3.68	3.50	3.40	3.30	3.20	3.10	2.99	2.87
19	10.07	7.09	5.92	5.27	4.85	4.56	4.34	4.18	4.04	3.93	3.76	3.59	3.40	3.31	3.21	3.11	3.00	2.89	2.78
20	9.94	6.99	5.82	5.17	4.76	4.47	4.26	4.09	3.96	3.85	3.68	3.50	3.32	3.22	3.12	3.02	2.92	2.81	2.69
21	9.83	6.89	5.73	5.09	4.68	4.39	4.18	4.01	3.88	3.77	3.60	3.43	3.24	3.15	3.05	2.95	2.84	2.73	2.61
22	9.73	6.81	5.65	5.02	4.61	4.32	4.11	3.94	3.81	3.70	3.54	3.36	3.18	3.08	2.98	2.88	2.77	2.66	2.55
23	9.63	6.73	5.58	4.95	4.54	4.26	4.05	3.88	3.75	3.64	3.47	3.30	3.12	3.02	2.92	2.82	2.71	2.60	2.48
24	9.55	6.66	5.52	4.89	4.49	4.20	3.99	3.83	3.69	3.59	3.42	3.25	3.06	2.97	2.87	2.77	2.66	2.55	2.43
25	9.48	6.60	5.46	4.84	4.43	4.15	3.94	3.78	3.64	3.54	3.37	3.20	3.01	2.92	2.82	2.72	2.61	2.50	2.38
26	9.41	6.54	5.41	4.79	4.38	4.10	3.89	3.73	3.60	3.49	3.33	3.15	2.97	2.87	2.77	2.67	2.56	2.45	2.33
27	9.34	6.49	5.36	4.74	4.34	4.06	3.85	3.69	3.56	3.45	3.28	3.11	2.93	2.83	2.73	2.63	2.52	2.41	2.25
28	9.28	6.44	5.32	4.70	4.30	4.02	3.81	3.65	3.52	3.41	3.25	3.07	2.89	2.79	2.69	2.59	2.48	2.37	2.29
29	9.23	6.40	5.28	4.66	4.26	3.98	3.77	3.61	3.48	3.38	3.21	3.04	2.86	2.76	2.66	2.56	2.45	2.33	2.24
30	9.18	6.35	5.24	4.62	4.23	3.95	3.74	3.58	3.45	3.34	3.18	3.01	2.82	2.73	2.63	2.52	2.42	2.30	2.18
40	8.83	6.07	4.98	4.37	3.99	3.71	3.51	3.35	3.22	3.12	2.95	2.78	2.60	2.50	2.40	2.30	2.18	2.06	1.93
60	8.49	5.79	4.73	4.14	3.76	3.49	3.29	3.13	3.01	2.90	2.74	2.57	2.39	2.29	2.19	2.08	1.96	1.83	1.69
120	8.18	5.54	4.50	3.92	3.55	3.28	3.09	2.93	2.81	2.71	2.54	2.37	2.19	2.09	1.98	1.87	1.75	1.61	1.43
∞	7.88	5.30	4.28	3.72	3.35	3.09	2.90	2.74	2.62	2.52	2.36	2.19	2.00	1.90	1.79	1.67	1.53	1.36	1.00

$F = \frac{s_1^2}{s_2^2} = \frac{S_1}{m} \Big/ \frac{S_2}{n}$, where $s_1^2 = S_1/m$ and $s_2^2 = S_2/n$ are independent mean squares estimating a common variance σ^2 and based on m and n degrees of freedom, respectively.

PERCENTAGE POINTS, F-DISTRIBUTION

$$F(F) = \int_0^F \frac{\Gamma\left(\frac{m+n}{2}\right)}{\Gamma\left(\frac{m}{2}\right)\Gamma\left(\frac{n}{2}\right)} m^{\frac{m}{2}} n^{\frac{n}{2}} x^{\frac{m}{2}-1} (n+mx)^{-\frac{m+n}{2}} \, dx = .999$$

n \ m	1	2	3	4	5	6	7	8	9	10	12	15	20	24	30	40	60	120	∞
1	4053*	5000*	5404*	5625*	5764*	5859*	5929*	5981*	6023*	6056*	6107*	6158*	6209*	6235*	6261*	6287*	6313*	6340*	6366*
2	998.5	999.0	999.2	999.2	999.3	999.3	999.4	999.4	999.4	999.4	999.4	999.4	999.4	999.5	999.5	999.5	999.5	999.5	999.5
3	167.0	148.5	141.1	137.1	134.6	132.8	131.6	130.6	129.9	129.2	128.3	127.4	126.4	125.9	125.4	125.0	124.5	124.0	123.5
4	74.14	61.25	56.18	53.44	51.71	50.53	49.66	49.00	48.47	48.05	47.41	46.76	46.10	45.77	45.43	45.09	44.75	44.40	44.05
5	47.18	37.12	33.20	31.09	29.75	28.84	28.16	27.64	27.24	26.92	26.42	25.91	25.39	25.14	24.87	24.60	24.33	24.06	23.79
6	35.51	27.00	23.70	21.92	20.81	20.03	19.46	19.03	18.69	18.41	17.99	17.56	17.12	16.89	16.67	16.44	16.21	15.99	15.75
7	29.25	21.69	18.77	17.19	16.21	15.52	15.02	14.63	14.33	14.08	13.71	13.32	12.93	12.73	12.53	12.33	12.12	11.91	11.70
8	25.42	18.49	15.83	14.39	13.49	12.86	12.40	12.04	11.77	11.54	11.19	10.84	10.48	10.30	10.11	9.92	9.73	9.53	9.33
9	22.86	16.39	13.90	12.56	11.71	11.13	10.70	10.37	10.11	9.89	9.57	9.24	8.90	8.72	8.55	8.37	8.19	8.00	7.81
10	21.04	14.91	12.55	11.28	10.48	9.92	9.52	9.20	8.96	8.75	8.45	8.13	7.80	7.64	7.47	7.30	7.12	6.94	6.76
11	19.69	13.81	11.56	10.35	9.58	9.05	8.66	8.35	8.12	7.92	7.63	7.32	7.01	6.85	6.68	6.52	6.35	6.17	6.00
12	18.64	12.97	10.80	9.63	8.89	8.38	8.00	7.71	7.48	7.29	7.00	6.71	6.40	6.25	6.09	5.93	5.76	5.59	5.42
13	17.81	12.31	10.21	9.07	8.35	7.86	7.49	7.21	6.98	6.80	6.52	6.23	5.93	5.78	5.63	5.47	5.30	5.14	4.97
14	17.14	11.78	9.73	8.62	7.92	7.43	7.08	6.80	6.58	6.40	6.13	5.85	5.56	5.41	5.25	5.10	4.94	4.77	4.60
15	16.59	11.34	9.34	8.25	7.57	7.09	6.74	6.47	6.26	6.08	5.81	5.54	5.25	5.10	4.95	4.80	4.64	4.47	4.31
16	16.12	10.97	9.00	7.94	7.27	6.81	6.46	6.19	5.98	5.81	5.55	5.27	4.99	4.85	4.70	4.54	4.39	4.23	4.06
17	15.72	10.66	8.73	7.68	7.02	6.56	6.22	5.96	5.75	5.58	5.32	5.05	4.78	4.63	4.48	4.33	4.18	4.02	3.85
18	15.38	10.39	8.49	7.46	6.81	6.35	6.02	5.76	5.56	5.39	5.13	4.87	4.59	4.45	4.30	4.15	4.00	3.84	3.67
19	15.08	10.16	8.28	7.26	6.62	6.18	5.85	5.59	5.39	5.22	4.97	4.70	4.43	4.29	4.14	3.99	3.84	3.68	3.51
20	14.82	9.95	8.10	7.10	6.46	6.02	5.69	5.44	5.24	5.08	4.82	4.56	4.29	4.15	4.00	3.86	3.70	3.54	3.38
21	14.59	9.77	7.94	6.95	6.32	5.88	5.56	5.31	5.11	4.95	4.70	4.44	4.17	4.03	3.88	3.74	3.58	3.42	3.26
22	14.38	9.61	7.80	6.81	6.19	5.76	5.44	5.19	4.99	4.83	4.58	4.33	4.06	3.92	3.78	3.63	3.48	3.32	3.15
23	14.19	9.47	7.67	6.69	6.08	5.65	5.33	5.09	4.89	4.73	4.48	4.23	3.96	3.82	3.68	3.53	3.38	3.22	3.05
24	14.03	9.34	7.55	6.59	5.98	5.55	5.23	4.99	4.80	4.64	4.39	4.14	3.87	3.74	3.59	3.45	3.29	3.14	2.97
25	13.88	9.22	7.45	6.49	5.88	5.46	5.15	4.91	4.71	4.56	4.31	4.06	3.79	3.66	3.52	3.37	3.22	3.06	2.89
26	13.74	9.12	7.36	6.41	5.80	5.38	5.07	4.83	4.64	4.48	4.24	3.99	3.72	3.59	3.44	3.30	3.15	2.99	2.82
27	13.61	9.02	7.27	6.33	5.73	5.31	5.00	4.76	4.57	4.41	4.17	3.92	3.66	3.52	3.38	3.23	3.08	2.92	2.75
28	13.50	8.93	7.19	6.25	5.66	5.24	4.93	4.69	4.50	4.35	4.11	3.86	3.60	3.46	3.32	3.18	3.02	2.86	2.69
29	13.39	8.85	7.12	6.19	5.59	5.18	4.87	4.64	4.45	4.29	4.05	3.80	3.54	3.41	3.27	3.12	2.97	2.81	2.64
30	13.29	8.77	7.05	6.12	5.53	5.12	4.82	4.58	4.39	4.24	4.00	3.75	3.49	3.36	3.22	3.07	2.92	2.76	2.59
40	12.61	8.25	6.60	5.70	5.13	4.73	4.44	4.21	4.02	3.87	3.64	3.40	3.15	3.01	2.87	2.73	2.57	2.41	2.23
60	11.97	7.76	6.17	5.31	4.76	4.37	4.09	3.87	3.69	3.54	3.31	3.08	2.83	2.69	2.55	2.41	2.25	2.08	1.89
120	11.38	7.32	5.79	4.95	4.42	4.04	3.77	3.55	3.38	3.24	3.02	2.78	2.53	2.40	2.26	2.11	1.95	1.76	1.54
∞	10.83	6.91	5.42	4.62	4.10	3.74	3.47	3.27	3.10	2.96	2.74	2.51	2.27	2.13	1.99	1.84	1.66	1.45	1.00

* Multiply these entries by 100.

V. Poisson Distribution Function

$$F(x:\lambda) = \sum_{k=0}^{x} e^{-\lambda} \frac{\lambda^k}{k!}$$

λ \ x	0	1	2	3	4	5	6	7	8	9
0.02	0.980	1.000								
0.04	0.961	0.999	1.000							
0.06	0.942	0.998	1.000							
0.08	0.923	0.997	1.000							
0.10	0.905	0.995	1.000							
0.15	0.861	0.990	0.999	1.000						
0.20	0.819	0.982	0.999	1.000						
0.25	0.779	0.974	0.998	1.000						
0.30	0.741	0.963	0.996	1.000						
0.35	0.705	0.951	0.994	1.000						
0.40	0.670	0.938	0.992	0.999	1.000					
0.45	0.638	0.925	0.989	0.999	1.000					
0.50	0.607	0.910	0.986	0.998	1.000					
0.55	0.577	0.894	0.982	0.998	1.000					
0.60	0.549	0.878	0.997	0.977	1.000					
0.65	0.522	0.861	0.972	0.996	0.999	1.000				
0.70	0.497	0.844	0.966	0.994	0.999	1.000				
0.75	0.472	0.827	0.959	0.993	0.999	1.000				
0.80	0.449	0.809	0.953	0.991	0.999	1.000				
0.85	0.427	0.791	0.945	0.989	0.998	1.000				
0.90	0.407	0.772	0.937	0.987	0.998	1.000				
0.95	0.387	0.754	0.929	0.984	0.997	1.000				
1.00	0.368	0.736	0.920	0.981	0.996	0.999	1.000			
1.1	0.333	0.699	0.900	0.974	0.995	0.999	1.000			
1.2	0.301	0.663	0.879	0.966	0.992	0.998	1.000			
1.3	0.273	0.627	0.857	0.957	0.989	0.998	1.000			
1.4	0.247	0.592	0.833	0.946	0.986	0.997	0.999	1.000		
1.5	0.223	0.558	0.809	0.934	0.981	0.996	0.999	1.000		
1.6	0.202	0.525	0.783	0.921	0.976	0.994	0.999	1.000		
1.7	0.183	0.493	0.757	0.907	0.970	0.992	0.998	1.000		
1.8	0.165	0.463	0.731	0.891	0.964	0.990	0.997	0.999	1.000	
1.9	0.150	0.434	0.704	0.875	0.956	0.987	0.997	0.999	1.000	
2.0	0.135	0.406	0.677	0.857	0.947	0.983	0.995	0.999	1.000	
2.2	0.111	0.355	0.623	0.819	0.928	0.975	0.993	0.998	1.000	
2.4	0.091	0.308	0.570	0.779	0.904	0.964	0.988	0.997	0.999	1.000
2.6	0.074	0.267	0.518	0.736	0.877	0.951	0.983	0.995	0.999	1.000
2.8	0.061	0.231	0.469	0.692	0.848	0.935	0.976	0.992	0.998	0.999
3.0	0.050	0.199	0.423	0.647	0.815	0.916	0.966	0.988	0.996	0.999
3.2	0.041	0.171	0.380	0.603	0.781	0.895	0.955	0.983	0.994	0.998
3.4	0.033	0.147	0.340	0.558	0.744	0.871	0.942	0.977	0.992	0.997
3.6	0.027	0.126	0.303	0.515	0.706	0.844	0.927	0.969	0.988	0.996
3.8	0.022	0.107	0.269	0.473	0.668	0.816	0.909	0.960	0.984	0.994
4.0	0.018	0.092	0.238	0.433	0.629	0.785	0.889	0.949	0.979	0.992

λ \ x	0	1	2	3	4	5	6	7	8	9
4.2	0.015	0.078	0.210	0.395	0.590	0.753	0.867	0.936	0.972	0.989
4.4	0.012	0.066	0.185	0.359	0.551	0.720	0.844	0.921	0.964	0.985
4.6	0.010	0.056	0.163	0.326	0.513	0.686	0.818	0.905	0.955	0.980
4.8	0.008	0.048	0.143	0.294	0.476	0.651	0.791	0.887	0.944	0.975
5.0	0.007	0.040	0.125	0.265	0.440	0.616	0.762	0.867	0.932	0.968
5.2	0.006	0.034	0.109	0.238	0.406	0.581	0.732	0.845	0.918	0.960
5.4	0.005	0.029	0.095	0.213	0.373	0.546	0.702	0.822	0.903	0.951
5.6	0.004	0.024	0.082	0.191	0.342	0.512	0.670	0.797	0.886	0.941
5.8	0.003	0.021	0.072	0.170	0.313	0.478	0.638	0.771	0.867	0.929
6.0	0.002	0.017	0.062	0.151	0.285	0.446	0.606	0.744	0.847	0.916

	10	11	12	13	14	15	16
2.8	1.000						
3.0	1.000						
3.2	1.000						
3.4	0.999	1.000					
3.6	0.999	1.000					
3.8	0.998	0.999	1.000				
4.0	0.997	0.999	1.000				
4.2	0.996	0.999	1.000				
4.4	0.994	0.998	0.999	1.000			
4.6	0.992	0.997	0.999	1.000			
4.8	0.990	0.996	0.999	1.000			
5.0	0.986	0.995	0.998	0.999	1.000		
5.2	0.982	0.993	0.997	0.999	1.000		
5.4	0.977	0.990	0.996	0.999	1.000		
5.6	0.972	0.988	0.995	0.998	0.999	1.000	
5.8	0.965	0.984	0.993	0.997	0.999	1.000	
6.0	0.957	0.980	0.991	0.996	0.999	0.999	1.000

λ \ x	0	1	2	3	4	5	6	7	8	9
6.2	0.002	0.015	0.054	0.134	0.259	0.414	0.574	0.716	0.826	0.902
6.4	0.002	0.012	0.046	0.119	0.235	0.384	0.542	0.687	0.803	0.886
6.6	0.001	0.010	0.040	0.105	0.213	0.355	0.511	0.658	0.780	0.869
6.8	0.001	0.009	0.034	0.093	0.192	0.327	0.480	0.628	0.755	0.850
7.0	0.001	0.007	0.030	0.082	0.173	0.301	0.450	0.599	0.729	0.830
7.2	0.001	0.006	0.025	0.072	0.156	0.276	0.420	0.569	0.703	0.810
7.4	0.001	0.005	0.022	0.063	0.140	0.253	0.392	0.539	0.676	0.788
7.6	0.001	0.004	0.019	0.055	0.125	0.231	0.365	0.510	0.648	0.765
7.8	0.000	0.004	0.016	0.048	0.112	0.210	0.338	0.481	0.620	0.741
8.0	0.000	0.003	0.014	0.042	0.100	0.191	0.313	0.453	0.593	0.717

λ \ x	0	1	2	3	4	5	6	7	8	9
8.5	0.000	0.002	0.009	0.030	0.074	0.150	0.256	0.386	0.523	0.653
9.0	0.000	0.001	0.006	0.021	0.055	0.116	0.207	0.324	0.456	0.587
9.5	0.000	0.001	0.004	0.015	0.040	0.089	0.165	0.269	0.392	0.522
10.0	0.000	0.000	0.003	0.010	0.029	0.067	0.130	0.220	0.333	0.458

	10	11	12	13	14	15	16	17	18	19
6.2	0.949	0.975	0.989	0.995	0.998	0.999	1.000			
6.4	0.939	0.969	0.986	0.994	0.997	0.999	1.000			
6.6	0.927	0.963	0.982	0.992	0.997	0.999	0.999	1.000		
6.8	0.915	0.955	0.978	0.990	0.996	0.998	0.999	1.000		
7.0	0.901	0.947	0.973	0.987	0.994	0.998	0.999	1.000		
7.2	0.887	0.937	0.967	0.984	0.993	0.997	0.999	0.999	1.000	
7.4	0.871	0.926	0.961	0.980	0.991	0.996	0.998	0.999	1.000	
7.6	0.854	0.915	0.954	0.976	0.989	0.995	0.998	0.999	1.000	
7.8	0.835	0.902	0.945	0.971	0.986	0.993	0.997	0.999	1.000	
8.0	0.816	0.888	0.936	0.966	0.983	0.992	0.996	0.998	0.999	1.000
8.5	0.763	0.849	0.909	0.949	0.973	0.986	0.993	0.997	0.999	1.999
9.0	0.706	0.803	0.876	0.926	0.959	0.978	0.989	0.995	0.998	0.999
9.5	0.645	0.752	0.836	0.898	0.940	0.967	0.982	0.991	0.996	0.998
10.0	0.583	0.697	0.792	0.864	0.917	0.951	0.973	0.986	0.993	0.997

	20	21	22
8.5	1.000		
9.0	1.000		
9.5	0.999	1.000	
10.0	0.998	0.999	1.000

λ \ x	0	1	2	3	4	5	6	7	8	9
10.5	0.000	0.000	0.002	0.007	0.021	0.050	0.102	0.179	0.279	0.397
11.0	0.000	0.000	0.001	0.005	0.015	0.038	0.079	0.143	0.232	0.341
11.5	0.000	0.000	0.001	0.003	0.011	0.028	0.060	0.114	0.191	0.289
12.0	0.000	0.000	0.001	0.002	0.008	0.020	0.046	0.090	0.155	0.242
12.5	0.000	0.000	0.000	0.002	0.005	0.015	0.035	0.070	0.125	0.201
13.0	0.000	0.000	0.000	0.001	0.004	0.011	0.026	0.054	0.100	0.166
13.5	0.000	0.000	0.000	0.001	0.003	0.008	0.019	0.041	0.079	0.135
14.0	0.000	0.000	0.000	0.000	0.002	0.006	0.014	0.032	0.062	0.109
14.5	0.000	0.000	0.000	0.000	0.001	0.004	0.010	0.024	0.048	0.088
15.0	0.000	0.000	0.000	0.000	0.001	0.003	0.008	0.018	0.037	0.070

	10	11	12	13	14	15	16	17	18	19
10.5	0.521	0.639	0.742	0.825	0.888	0.932	0.960	0.978	0.988	0.994
11.0	0.460	0.579	0.689	0.781	0.854	0.907	0.944	0.968	0.982	0.991
11.5	0.402	0.520	0.633	0.733	0.815	0.878	0.924	0.954	0.974	0.986
12.0	0.347	0.462	0.576	0.682	0.772	0.844	0.899	0.937	0.963	0.979
12.5	0.297	0.406	0.519	0.628	0.725	0.806	0.869	0.916	0.948	0.969
13.0	0.252	0.353	0.463	0.573	0.675	0.764	0.835	0.890	0.930	0.957
13.5	0.211	0.304	0.409	0.518	0.623	0.718	0.798	0.861	0.908	0.942
14.0	0.176	0.260	0.358	0.464	0.570	0.669	0.756	0.827	0.883	0.923
14.5	0.145	0.220	0.311	0.413	0.518	0.619	0.711	0.790	0.853	0.901
15.0	0.118	0.185	0.268	0.363	0.466	0.568	0.664	0.749	0.819	0.875

	20	21	22	23	24	25	26	27	28	29
10.5	0.997	0.999	1.999	1.000						
11.0	0.995	0.998	0.999	1.000						
11.5	0.992	0.996	0.998	0.999	1.000					
12.0	0.988	0.994	0.997	0.999	0.999	1.000				
12.5	0.983	0.991	0.995	0.998	0.999	0.999	1.000			
13.0	0.975	0.986	0.992	0.996	0.998	0.999	1.000			
13.5	0.965	0.980	0.989	0.994	0.997	0.998	0.999	1.000		
14.0	0.952	0.971	0.983	0.991	0.995	0.997	0.999	0.999	1.000	
14.5	0.936	0.960	0.976	0.986	0.992	0.996	0.998	0.999	0.999	1.000
15.0	0.917	0.947	0.967	0.981	0.989	0.994	0.997	0.998	0.999	1.000

λ \ x	4	5	6	7	8	9	10	11	12	13
16	0.000	0.001	0.004	0.010	0.022	0.043	0.077	0.127	0.193	0.275
17	0.000	0.001	0.002	0.005	0.013	0.026	0.049	0.085	0.135	0.201
18	0.000	0.000	0.001	0.003	0.007	0.015	0.030	0.055	0.092	0.143
19	0.000	0.000	0.001	0.002	0.004	0.009	0.018	0.035	0.061	0.098
20	0.000	0.000	0.000	0.001	0.002	0.005	0.011	0.021	0.039	0.066
21	0.000	0.000	0.000	0.000	0.001	0.003	0.006	0.013	0.025	0.043
22	0.000	0.000	0.000	0.000	0.001	0.002	0.004	0.008	0.015	0.028
23	0.000	0.000	0.000	0.000	0.000	0.001	0.002	0.004	0.009	0.017
24	0.000	0.000	0.000	0.000	0.000	0.000	0.001	0.003	0.005	0.011
25	0.000	0.000	0.000	0.000	0.000	0.000	0.001	0.001	0.003	0.006

	14	15	16	17	18	19	20	21	22	23
16	0.368	0.467	0.566	0.659	0.742	0.812	0.868	0.911	0.942	0.963
17	0.281	0.371	0.468	0.564	0.655	0.736	0.805	0.861	0.905	0.937
18	0.208	0.287	0.375	0.469	0.562	0.651	0.731	0.799	0.855	0.899
19	0.150	0.215	0.292	0.378	0.469	0.561	0.647	0.725	0.793	0.849
20	0.105	0.157	0.221	0.297	0.381	0.470	0.559	0.644	0.721	0.787
21	0.072	0.111	0.163	0.227	0.302	0.384	0.471	0.558	0.640	0.716

	14	15	16	17	18	19	20	21	22	23
22	0.048	0.077	0.117	0.169	0.232	0.306	0.387	0.472	0.556	0.637
23	0.031	0.052	0.082	0.123	0.175	0.238	0.310	0.389	0.472	0.555
24	0.020	0.034	0.056	0.087	0.128	0.180	0.243	0.314	0.392	0.473
25	0.012	0.022	0.038	0.060	0.092	0.134	0.185	0.247	0.318	0.394

	24	25	26	27	28	29	30	31	32	33
16	0.978	0.987	0.993	0.996	0.998	0.999	0.999	1.000		
17	0.959	0.975	0.985	0.991	0.995	0.997	0.999	0.999	1.000	
18	0.932	0.955	0.972	0.983	0.990	0.994	0.997	0.998	0.999	1.000
19	0.893	0.927	0.951	0.969	0.980	0.988	0.993	0.996	0.998	0.999
20	0.843	0.888	0.922	0.948	0.966	0.978	0.987	0.992	0.995	0.997
21	0.782	0.838	0.883	0.917	0.944	0.963	0.976	0.985	0.991	0.994
22	0.712	0.777	0.832	0.877	0.913	0.940	0.959	0.973	0.983	0.989
23	0.635	0.708	0.772	0.827	0.873	0.908	0.936	0.956	0.971	0.981
24	0.554	0.632	0.704	0.768	0.823	0.868	0.904	0.932	0.953	0.969
25	0.473	0.553	0.629	0.700	0.763	0.818	0.863	0.900	0.929	0.950

	34	35	36	37	38	39	40	41	42	43
19	0.909	1.000								
20	0.999	0.999	1.000							
21	0.997	0.998	0.999	1.999	1.000					
22	0.994	0.996	0.998	0.999	0.999	1.000				
23	0.998	0.993	0.996	0.997	0.999	0.999	1.000			
24	0.979	0.987	0.992	0.995	0.997	0.998	0.999	0.999		
25	0.966	0.978	0.985	0.991	0.994	0.997	0.998	0.999	1.000	

VI. Critical Values for the Kolmogorov–Smirnov Test

A sample of size n is drawn from a population with cumulative distribution function $F(x)$. Define the empirical distribution function $F_n(x)$ to be the step function

$$F_n(x) = \frac{k}{n} \quad \text{for} \quad x_{(i)} \leq x \leq x_{(i+1)} ,$$

where k is the number of observations not greater than x. $x_{(1)}, \ldots, x_{(n)}$ denote the sample values arranged in ascending order. Under the null hypothesis that the sample has been drawn from the specified distribution, $F_n(x)$ should be fairly close to $F(x)$. Define

$$D = \max |F_n(x) - F(x)| .$$

For a two-tailed test this table gives critical values of the sampling distribution of D under the null hypothesis. Reject the hypothetical distribution if D exceeds the tabulated value. If n is over 35, determine the critical values of D by the divisions indicated in the table.

A one-tailed test is provided by the statistic

$$D^+ = \max [F_n(x) - F(x)] .$$

CRITICAL VALUES FOR THE KOLMOGOROV-SMIRNOV TEST OF GOODNESS OF FIT

One-sided test Two-sided test	p = 0.90 p = 0.80	0.95 0.90	0.975 0.95	0.99 0.98	0.995 0.99
n = 1	0.900	0.950	0.975	0.990	0.995
2	0.684	0.776	0.842	0.900	0.929
3	0.565	0.636	0.708	0.785	0.829
4	0.493	0.565	0.624	0.689	0.734
5	0.447	0.509	0.563	0.627	0.669
6	0.410	0.468	0.519	0.577	0.617
7	0.381	0.436	0.483	0.538	0.576
8	0.358	0.410	0.454	0.507	0.542
9	0.339	0.387	0.430	0.480	0.513
10	0.323	0.369	0.409	0.457	0.489
11	0.308	0.352	0.391	0.437	0.468
12	0.296	0.338	0.375	0.419	0.449
13	0.285	0.325	0.361	0.404	0.432
14	0.275	0.314	0.349	0.390	0.418
15	0.266	0.304	0.338	0.377	0.404
16	0.258	0.295	0.327	0.366	0.392
17	0.250	0.286	0.318	0.355	0.381
18	0.244	0.279	0.309	0.346	0.371
19	0.237	0.271	0.301	0.337	0.361
20	0.232	0.265	0.294	0.329	0.352
21	0.226	0.259	0.287	0.321	0.344
22	0.221	0.253	0.281	0.314	0.337
23	0.216	0.247	0.275	0.307	0.330
24	0.212	0.242	0.269	0.301	0.323
25	0.208	0.238	0.264	0.295	0.317
26	0.204	0.233	0.259	0.290	0.311
27	0.200	0.229	0.254	0.284	0.305
28	0.197	0.225	0.250	0.279	0.300
29	0.193	0.221	0.246	0.275	0.295
30	0.190	0.218	0.242	0.270	0.290
31	0.187	0.214	0.238	0.266	0.285
32	0.184	0.211	0.234	0.262	0.281
33	0.182	0.208	0.231	0.258	0.277
34	0.179	0.205	0.227	0.254	0.273
35	0.177	0.202	0.224	0.251	0.269
36	0.174	0.199	0.221	0.247	0.265
37	0.172	0.196	0.218	0.244	0.262
38	0.170	0.194	0.215	0.241	0.258
39	0.168	0.191	0.213	0.238	0.255
40	0.165	0.189	0.210	0.235	0.252
Approximation for $n > 40$:	$\frac{1.07}{\sqrt{n}}$	$\frac{1.22}{\sqrt{n}}$	$\frac{1.36}{\sqrt{n}}$	$\frac{1.52}{\sqrt{n}}$	$\frac{1.63}{\sqrt{n}}$

Index